Closed Loop Control and Management

Serge Zacher

Closed Loop Control and Management

Introduction to Feedback Control Theory with Data Stream Managers

 Springer

Serge Zacher
Stuttgart, Baden-Württemberg, Germany

ISBN 978-3-031-13482-1   ISBN 978-3-031-13483-8   (eBook)
https://doi.org/10.1007/978-3-031-13483-8

This Springer imprint is published by the registered company Springer Nature Switzerland AG
The registered company address is: Gewerbestrasse 11, 6330 Cham, Switzerland

Preface

This book is very important to me because it summarizes concepts, methods, and applications that have been developed and published in various places over the years. I hope that the book will also be important and useful for students, for my colleagues' professors all over the world and their students, as well as for graduates who are already working in the industry. I am certainly sure that it will, because the conceptions are new, the methods are practicable, and the applications allow to enormously simplify the engineering of closed loop control.

The questions I asked myself about 30 years ago were

- Is it possible to replace the established functional block diagram of a closed loop with the elements of modern automation technology (bus, client/server, etc.) without losing the mathematical basics of control theory?
- How can a closed loop be not only controlled but also managed like the technical devices are managed through their whole life cycle?
- Is it possible to reduce the number of variables and the number of controllers which are needed for multivariable control systems without converting their transfer functions into the corresponding state space models?

The answers to these and more questions are given in the book, and I hope, in an understandable and exhaustive form. I was surprised myself by some of the results, e.g., that a controller could be adjusted based on a single point in the Bode diagram. Even the Bode diagram can be determined using a single step response instead of as usual a series of frequency responses. The set point behavior of a closed loop could be controlled with one non-linear controller even if the parameters of the plant are far shifted from its initial constant values. The disturbance behavior is not controlled with only one conventional controller but supervised with a Data Stream Manager called "Terminator" that destroyed at once as well all external as internal disturbances.

Altogether, the book describes almost 20 Data Stream Managers (DSM), most of which are designed as visually human-like silhouettes as in the case of software-agents. The DSM symbolized with these silhouettes are control elements, which include the open loop control (e.g., switches and logical operations) and/or closed loop control (i.e., sensors, actuators, and controllers). Some of the DSMs are developed on the principles of neuro-networks and fuzzy logic, and another on the new approaches like bus-approach or Antisystem Approach (ASA).

As mentioned above, almost all the topics covered in this book have already been published in many books, articles, and papers in German. When translating formulas and indices, I took the liberty of sticking to their original designations and hope for your understanding. Some transfer functions are left the same as in the original publications, e.g., a controller is G_R, oriented on the word "Regler" (controller). The "S" by the transfer function G_s indicates "Strecke" (plant). The settling time is T_{aus} and not T_{set}, as is expected in English literature. Turning over such an enormous number of formulas and figures is practically impossible for me alone.

The book is aimed at students at technical universities and colleges as well as at technology engineers of automation. More documents on the topics of the book can be found on my author's website https://www.zacher-automation.de or directly asked via mail: info@szacher.de.

▶   All numerical data, test results, curves, etc. available in book are not real but fictitious data. The further development or use of the publication without reference to the author is not permitted. The scripts and Simulink models created with MATLAB® are subject to the copyright reserved by the author and the publisher. The author and the publisher accept no liability for use in practical cases or for any damage that may result from incomplete or incorrect information during implementation.

Finally, I would like to always thank the employees at Springer Vieweg Verlag for their friendly atmosphere and constructive cooperation, namely the chief editor for electrical engineering/IT/computer science, Mr. Reinhard Dapper, and the editorial assistant, Ms. Andrea Brossler. My thanks go to the entire team at Springer-Nature Verlag's German Books Production, especially the project coordinator, Ms. Anupriya Harikrishnan.

Stuttgart, Germany   Serge Zacher
May 2022

Contents

Closed Loop Control and Management

1

> **"Well begun is half done."** Quote by *Artistotle*.
>
> *Source*: https://www.brainyquote.com/quotes/aristotle_109750
> accessed Jan 13, 2022

1.1 Introduction

1.1.1 The aim of this book

The aim of this book is to design the closed loop control (CLC) not only with the well-known methods of classical control theory but with the engineering tools, which are commonly used today in the automation systems, such as programmable logic control (PLC), bus-connection, and information exchange. Therefore, a bridge must be built between the control theory and the management of the control systems. This bridge is the application of the live cycles of the CLC to its design and control.

It is generally known that the whole live cycle of the CLC as well as of the other industrial products is divided into two stages:

- Engineering (Design of CLC):
  - a) conception,
  - b) planning, supply of components, assembly, commission,
  - c) test, measurements, experiments,
  - d) identification of plants,
  - e) tuning of controllers,
  - f) simulation of control

- Implementation (Application of designed CLC)
  - a) control
  - b) supervision of control
  - c) maintenance

The engineering and the implementation are carried out books in two domains:

- Real world = the world of real devices and physical signals.
- Virtual world = the world of mathematical descriptions and simulation.

The exchange of physical (electrical) signals in the real world and the exchange of information in the virtual world as well as the exchange of information between both worlds are called in the book unified *Data Streams*.

For engineering tasks (d) till (f) and for some implementation tasks (a) and (b) mentioned above are in the book introduced so-called *Data Stream Manager* (DSM). There are control elements, which include the open loop control (e.g., switches, logical operations) and closed loop control (i.e., sensors, actuators, controllers). The DSM exchange signals of the real world, which relate to the real bus, and information of the virtual world with the virtual bus.

The DSM, described in this book, is listed below.

- Chapter 7:
  - AFIC (Adaptive Filter for Identification and Control),
  - Ident,
  - Tuner.
- Chapter 8:
  - Axon,
  - FFF (Feed Forward Fuzzy),
  - Override,
  - Overset,
  - SFC (Surf Feedback Control),
  - SPFC (Simplified Predictive Functional Control),
- Chapter 9:
  - Terminator,
  - Plant Guard,
  - LTV (Linear Time Varying) Supervisor
  - CLIMB & HOLD.
- Chapter 10:
  - Router.
- Chapter 11:
  - ASA (Antisystem-Approach),
  - Bypass,

- Predictor,
- Turbo.

1.1.2   Content of this book

The book is conditionally divided into the following three parts:

Part 1: Classic conceptions and methods of closed loop control (CLC)
- Overview of control and management of CLC and industrial automatic systems (Chap. 1).
- Basics of dynamics and feedback control (Chap. 2).
- Classical controller's tuning rules (Chap. 3).

Part 2: New conceptions and methods of control of signals and management of data streams
- Bus-Approach (Chap. 4),
- ASA: Antisystem-Approach (Chap. 5),
- BAD: Bode-Aided Design (Chap. 6).

Part 2: Implementation of the Data Stream Management (DSM) for CLC
- Identification (Chap. 7),
- Setpoint behavior (Chap. 8),
- Disturbance behavior (Chap. 9)
- Multivariable control (Chap. 10)
- Model-based control (Chap. 11).

1.2   From Single Controller to Automation System

1.2.1   Historical Overview

The development of automation control is briefly considered in this section in order to properly appreciate the events of today's technical progress and to assess it for the goals of this book. Only a few prominent results are listed below with approximate dates over the years of inventions or developments.

First worldwide feedback controller
The essence of a closed loop control consists in feedback, which serves the *controlled variable* (the variable to be controlled) to the controller input independent of *disturbing influences* so that the controlled variable always maintains a predetermined value, which

is called a *set point*. In technical systems, the variables to be controlled are of a physical nature, e.g., pressure, temperature, speed, flow, liquid level, current, voltage, etc.

The first worldwide feedback controller was invented around 3000 years ago and ever since till now has been the basic principle of control. Let us consider only one of these controllers, namely the level control of the oil lamp (Fig. 1.1), which function we describe below according to today's terms of the closed loop control theory.

The oil lamp has all elements of today's feedback control, which are shown in Fig. 1.2. The actual level (process variable PV) is x, and the desired level (set point SP) is w.

The oil jug is the plant, who's level x should be controlled. The air pressure is an actor, and the aperture between oil jug and bowl is a sensor. If the oil level sinks low, then the following condition is fulfilled

$$e = (w - x) > 0$$

and air will enter the container through this aperture. The above condition has the function of a two-point controller so that the oil will flow out of the oil jug until the actual level x (process value) and target level w (set point) become the same so that the condition

$$e = 0$$

will be achieved.

Fig. 1.1 Worldwide first feedback level controller of an oil lamp

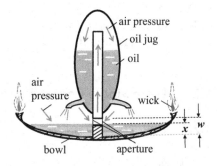

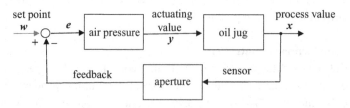

Fig. 1.2 Block diagram of the oil lamp level control

First worldwide industrial feedback controllers

The need for automation was found in the sixteenth century. However, systematic industrial automation begins after the development of windmill control (*Lee*, 1747), water level control of a steam boiler (*Polsunow*, 1765), and speed control of a steam engine (*James Watt*, 1788), as shown in Fig. 1.3. In the middle of this figure is represented the functional structure of the closed loop control with the established terms and graphic symbols of today's control theory, which is common for all shown examples from the sixteenth century till now.

Era of mechanical automation and electric drives

Between 1828 and 1888 the mathematical theory of feedback control was developed by *Maxwell, Wyschnegradski, Routh*, and *Linke*. Later *Ljapunov* (1892) and *Hurwitz* (1895) developed the methods of stability analysis. This was the era in which control theory was shaped by mechanics.

Since 1890, the control systems have been equipped with electric drives. The control theory was tailored to electrical machines and relays that came from electrical engineering. The theoretical investigations were based on the analytical solution of the differential equations using Laplace transformation or graphic circle diagrams, which also came from electrical engineering. As a result, follows 1922 the patent of *Minorski* for an PID-

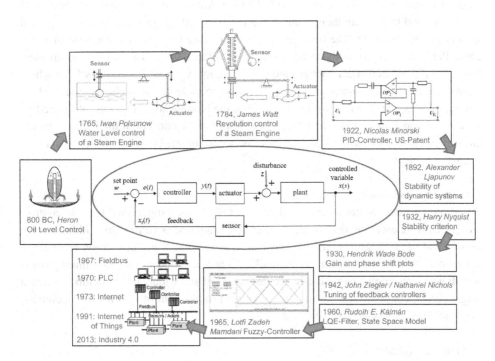

Fig. 1.3 Progress by the development of closed loop control

controller (proportional-integral-derivative controller) and the tuning method of Ziegler-Nichols (1942). Both are commonly used until today.

Mathematical methods of control
The progress in communications technology was of great importance for the further development of the control theory. For the first time used *Kuepfmueller* in 1928 the functional block diagram to describe the open control loop. *Nyquist, Mikhailow, Bode,* and many other scientists introduced frequency methods into automation technology in 1930–1940. In addition to the frequency methods, the methods of the poles and zeros of the transfer function were developed and their influence on the stability and control quality of a control loop was examined.

In the years that followed (1945–1960), on the one hand, the frequency methods were further developed (*Solodownikow, Neimark, Bromberg*, 1945 to 1953), and on the other hand, began the era of digital control technology. The feedback electronic operational amplifiers were replaced by microprocessors and microcontrollers as DDC (direct digital control).

Computational intelligence
Fuzzy logic was first introduced by *Zadeh* in 1965, further developed by *Kosko*, and implemented by *Mamdani* and *Sugeno* for control purposes. The design and use of fuzzy controllers have been described by *Frank, Kahlert, Kindl*, and *Tilli*. The fuzzy controllers are robust, and they retain their stable behavior even if the parameters of the controlled system are not constant. The time and the costs required for the development of fuzzy controllers are lower than those of classic controllers. But as far as fuzzy controllers have no strict mathematical backgrounds, they are mostly used for systems from which simple and robust behavior is expected, e.g., in motor vehicles, households, and medical devices. The first artificial neuron as a model of the biological neuron was introduced by *McCulloch* and *Pitts* in 1943. In 1949, *Hebb* proposed the concept of the learning mechanism, which states that the learning process occurs in the brain by changing synapse strengths (weights).

Information technology
Based on the fundamental invention of *Morley* (1959), the programmable logic controllers (PLC) were used and later configured and programmed as controllers with microprocessors and personal computers. The enormous development in information technology between 1970 and 1980 has made it possible to program and to integrate microprocessors directly into controllers. Finally, the development and widely use of fieldbus grows a single central control system to the decentralized process control system (PCS), consisting of several interconnected control stations.

- Field Control Station (FCS)
- Operator Station

Today, the focus of the entire automation technology is changing significantly in the direction of software engineering, data processing, and networking. Instead of the simple, single-loop control circuits of the past, meshed, intelligent controls with multiple variables are predominately used today.

Summary

The technology developed rapidly, and the generations changed, but the representation of a control circuit in the form of a functional block diagram in the middle of Fig. 1.3 remained unaffected. Today, industrial networks contain hundreds of controllers in one system, but the classical graphic representation based on the functional block diagram is possible with at most two coupled control loops.

The representation of a control loop with signals, which has been established for decades and which historically originates from electrical engineering, is useful for understanding its dynamic behavior, but no longer fits the current state of the control engineering. Today's automation control is based on the processing of messages, data, and information of an entire automation system, while the classic functional block diagram only describes the signals of an isolated closed loop.

The industrial automation is inconceivable without a HMI (man-machine interface), whereas the functional block diagram is purely signal flow diagrams.

The aim of this book is to eliminate the discrepancy between an idealized isolated functional block diagram and interconnected real control systems.

1.2.2 Automation Levels

The best way to describe functions and hierarchical structure of a complex industrial automation system is its representation as a pyramid consisting of the different levels. There are commonly known five levels of an industrial automation system, which are shown in Fig. 1.4.

- Production Level
- Field Level
- Control Level
- Supervisory Level
- Management Level

On each level are summarized the similar control functions.

Production Level

The production level, which is also called the *process level*, is the original goal of the automation system. Here are devices and machines of a physical technical process. The

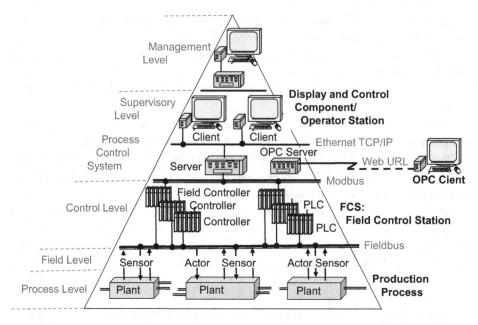

Fig. 1.4 Automation "Pyramid" of an industrial company

main part of this level is devices, which should be controlled, e.g., chemical reactors, furnaces, heaters, etc. They are called *plants*.

Field Level

The field level can be found all the components that are part of the machinery and that are required for the control of plants. There are sensors of pressure, flow, temperature, position, etc., which are needed to obtain the information about the process and actuators to influence the process (valves, electric motors, switches, etc.). In some books the production level and the field level are not separated and represented as one level.

Control Level

At this level are placed control devices, which receive the information from the previous level, and which automatically generate the decisions on how to perform the control tasks: controllers, PLCs (programmable logic controllers), and microcontrollers. The necessary part of the control level is a fieldbus, such as PROFIBUS, PROFINET, Modbus, etc. Some of the control level devices are interconnected and build so-called Field Control Stations (FCS), which are part of the whole decentralized process control system (PCS).

Supervisory Level

The PCS, mentioned above, extends in two levels, namely:

- Field Control Station (FCS) in the control level
- Operator Station at the supervisory level

This supervisory level includes the display components of the Process Control System (PCS), which are interconnected by a system bus like Ethernet TCP/IP, also called SCADA (short for Supervisory Control And Data Acquisition). SCADA usually adds a graphical user interface called an HMI (Human Machine Interface) to remotely function control.

The connection between display components is called *client/server* connection. The server is a passive participant and is used to connect to the process. A server makes its services available to several clients. The clients are active participants who take over the control functions. The OPC client is also an active participant but from another PCS of another industrial company.

Management Level

The top of the pyramid is what is called the management level. This level includes the monitoring of machines as well as the planning and administration of the entire industrial company and actions for the safety of people and machines. In the last 10–15 years the management level has evolved so much that the automation "pyramid" of Fig. 1.4 has lost its original geometric shape, as shown in Fig. 1.5

The functions of the management level are not considered in this book. Let us only briefly explain the meaning of the following terms shown in Fig. 1.5.

- *ASSET Management*: An asset is a control device, e.g., a controller, or a component of a control system, such as a valve, with or without its own diagnostic functions. Thus, ASSET management is a system for the administration of assets and has the following functions:
  - Diagnosis and maintenance
  - Determination of the operating status of the controller
  - Determination of the status of the sensors and actuators
  - Quality Management.
- *Manufacturing Enterprise Solutions* (MES): MES monitors the entire manufacturing process in a plant or factory from the raw materials to the finished product and is also called a production control system. It enables management, steering, control, and monitoring of production in real time. This includes all processes that have a real-time

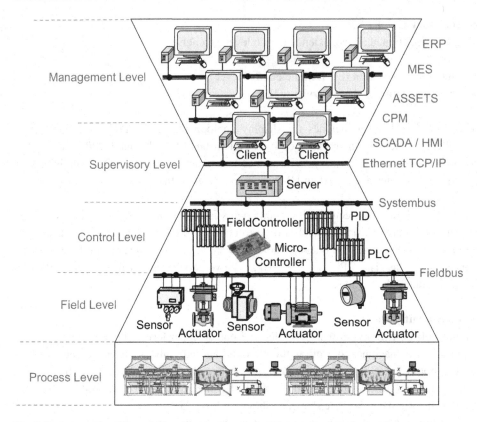

Fig. 1.5 Automation "Pyramid" under consideration of the management's development

impact on the production process, as well as the classic data acquisition functions
such as:

- Production data acquisition
- Machine data acquisition
- Personal data collection.

- *Enterprise Resource Planning* (ERP): With the ERP, the planning and control of business processes should be optimized. The ERP allows to monitor all levels of the business from manufacturing to sales, to purchasing, to finance and payroll. An ERP system
is a complex solution for planning the following resources of an entire company:
  - capital
  - operating resources
  - staff.

- *Condition and Performance Monitoring* (CPM) is the predictive maintenance of
machines in an automation system. The condition of machines and systems will be
constantly recorded and evaluated by sensors. Repairs and replacements are only initi-

ated when an imminent failure is to be feared. The CPM has significant advantages over other well-known maintenance approaches:

- *reactive maintenance*, in which the systems are operated until a failure or malfunction leads to a system standstill.
- *preventive maintenance*, in which the wear-prone components are replaced at regular intervals as a precaution and regardless of the condition of care and damage.

1.2.3   Automation Subsystems and Functions

The aim of the book, as it was written above, is to provide a possibly simple representation of a closed loop control as part of an overall automation system that should be understandable in today's technical terms.

Let us spread out for this purpose the "double pyramid" of an automation system of Fig. 1.5 into subsystems and functions as it is shown in Fig. 1.6.

There are clearly seen three subsystems:

- *Control*, which includes three levels (Process, Field, Control):
  - Open Loop Control
  - Closed Loop Control
  - Bus-Interconnection
- *Monitoring* of data transfer, which consists of one supervisory level:
  - SCADA / HMI

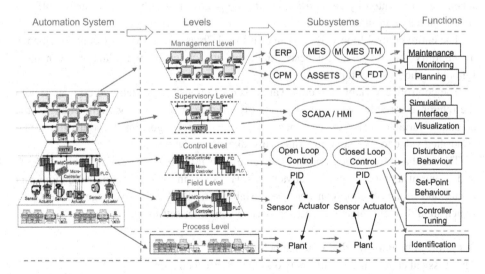

Fig. 1.6 Overview of control levels and management functions of an industrial automation system

Simulation

Visualization

Human-Operator

- *Management* consists of only one level, but with many interconnected functions (ASSETS, MES, ERP, etc.), which penetrate down to the lowest levels.

In this chapter we will discuss only the first and third subsystems, namely *Control* and *Management*.

The second subsystem, *monitoring of data transfer*, is not treated in this book. The *Simulation* and *Visualization* will be simply used without an explanation of their algorithms and programming basics.

1.3   Control

1.3.1   Closed Loop Control (CLC)

The basics of feedback control are given in books [1] till [22]. These classical methods are well known and widely used. Nevertheless, the most important results are given as a collection of formulas in Chap. 2. The general features of the CLC are given below.

Functional block diagram of CLC

The elements of the CLC are symbolized by small rectangular boxes, called functional blocks, as shown in Fig. 1.7a. The input/output signals are represented by action lines whose arrowheads indicate the direction of action. The points, at which several signals meet, are represented by an *addition point*. The points, at which signal branches off, are represented by a *branching point*.

The entire control loop is represented in Fig. 1.7b as a series of blocks and elements. To simplify the mathematical description, the blocks sensor and actuator are usually ignored in the classical theory (Fig. 1.7c).

The controlled variable x (actual value) is continuously recorded and compared with the reference variable (set point) w. If the disturbance or reference variable changes, the actuating variable y also changes according to a control algorithm to adjust the actual controlled value to the set point value. This takes place automatically.

The feedback from plant output to the controller input is the essential feature of CLC and is one of its most important advantages. However, in a closed loop, oscillations can occur, and a control loop can become unstable. This is the disadvantage of the CLC, which should be eliminated by the appropriate controller setting.

Set point behavior and disturbance behavior

The classical control theory differs two kinds of CLC behavior:

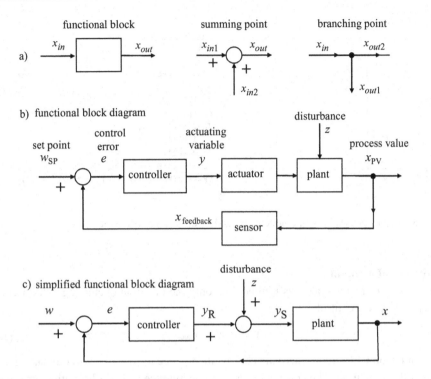

Fig. 1.7 Elements and structures of CLC

- set point behavior, when only the set point value w changes, while the disturbance z kept on constant,
- disturbance behavior, when in opposite the disturbance z acts but set point value w does not change.

Both kinds of behavior are controlled with the same controller. The problem is that almost all tuning methods for controllers are adapted only for set point behavior, i.e., the disturbance behavior is usually controlled not optimal.

An example of Fig. 1.8 should illustrate the optimal CLC by set point behavior and the loss of control quality by disturbance behavior. The settling time T_s by the set point behavior after the set point step of $w=5$ by $t=0$ s is $T_s=20$ s and is optimal, while the settling time T_s by the disturbance behavior after the disturbance step of $z=5$ by $t=30$ s is $T_s=50$ s.

Let us note at this place that according to the Data Stream Management, introduced in this book, there are two controllers in the same CLC, one for the set point behavior and another for the disturbance behavior, both optimal adjusted.

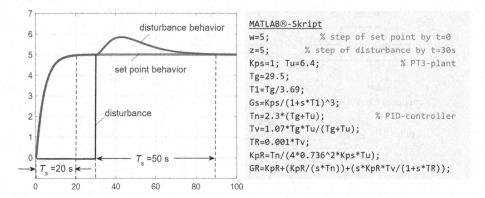

Fig. 1.8 Step responses by set point behavior and by disturbance behavior

Limitation of actuator
The capacity of real actuators is limited, i.e., only the maximum actuating value y could be reached by real control:

$$y_{min} < y < y_{max} \qquad (1.1)$$

The tuning of CLC upon classical control theory occurs usually without taking into account this condition. The consequences of neglecting the actuator limitation are shown in an example in Fig. 1.9. The settling time T_s of the real control under consideration of actuator limitation is $T_s = 100$ s, while the expected settling time, according to theoretical calculation, should be only $T_s = 20$ s.

Besides this, according to classical control theory, the expected step response without actuator limitation should run aperiodic, i.e., without overshooting over set point $w = 5$. In the reality, under consideration of actuator limitations $y_{max} = 10$ and $y_{min} = -10$ the percentage overshoot is 40%, which of course does not indicate good control quality

$$\text{percentage overshoot} = \frac{x_{max} - w}{w} 100\% = \frac{7 - 5}{5} 100\% = 40\% \qquad (1.2)$$

Looking ahead let us note at this place that with the Data Stream Management, introduced in this book, will be shown, how to tune controllers considering the actuator limitation.

1.3.2   Open Loop Control

Under certain conditions, a controlled variable can also be kept at a set point value, which can be constant or change over time, by controlling it without feedback. It is

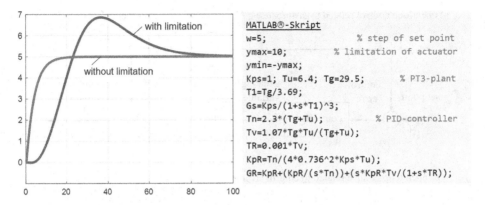

```
MATLAB®-Skript
w=5;                       % step of set point
ymax=10;                   % limitation of actuator
ymin=-ymax;
Kps=1; Tu=6.4; Tg=29.5;    % PT3-plant
T1=Tg/3.69;
Gs=Kps/(1+s*T1)^3;
Tn=2.3*(Tg+Tu);            % PID-controller
Tv=1.07*Tg*Tu/(Tg+Tu);
TR=0.001*Tv;
KpR=Tn/(4*0.736^2*Kps*Tu);
GR=KpR+(KpR/(s*Tn))+(s*KpR*Tv/(1+s*TR));
```

Fig. 1.9 Step responses by set point behavior with and without limitation of the actuator

called open loop control. In contrast to CLC there is no feedback by the open loop control.

There are known different kinds of the open loop control. One of them is shown in Fig. 1.10a. Not the controlled variable x is recorded continuously, as it is in the case by CLC, but the disturbance variable z. An actuating value y should be determined and saved in advance (either experimentally or theoretically) for each value of the disturbance variable z so that the disturbance will be fully compensated with the actuating value y. The disadvantage of the open loop control, however, is that only the measurable disturbance variables can be compensated.

The next option of an open loop control is shown in Fig. 1.10b. The controller gets the value of the control error and compares it with the desired, previously programmed, and saved set point value. Depending on the result the actuating value will be calculated and sent to the plant.

The advantages of this kind of open loop control include the possibility of programming any control algorithms, such as the Sequential Function Chart (SFC) i.e., a sequence of commands, with which various operations are carried out according to defined algorithms due to the technology. Also, the schedule control can be implemented by preprogramming the setpoint w so that the set point can be changed over time.

One more advantage of open loop control over CLC is that no instability can occur by controlling because of the missing feedback. In the ideal case, the set point can be exactly reached by open loop control, while in the case of CLC, when a disturbance occurs, there is at least a small temporary deviation of the controlled variable is needed to start the controller.

The disadvantage of the open loop control is the requirement that the behavior of the plant should be previously precisely known because there is no feedback, to recognize to correct possible errors.

a) open loop control related to disturbance z disturbance

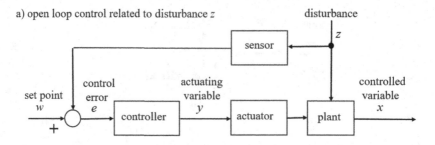

b) open loop control related to control error e

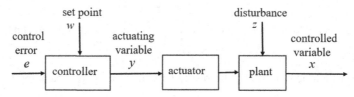

Fig. 1.10   Functional block diagrams of an open loop control

1.3.3   Examples of open and closed loop control

1.3.3.1 An antenna's angular positions control
Figure 1.11 is shown an example of the control of an antenna's angle:

a. Closed loop control
b. Open loop control

Closed loop control
The current angle α_x (controlled variable) is measured by a potentiometer (sensor) and converted into the voltage U_x, as shown in Fig. 1.11a. The voltage difference $U_e = U_w - U_x$ (control error) is formed by a comparison with the encoded set point value U_w.

If the set point $U_w = U_x$ then the error U_e is equal to zero, i.e., $U_e = 0$, the motor stops.

If the angle α_x increases, the voltage U_x also proportionally increases. Since the setpoint voltage U_w is constant, a negative voltage U_e is created. This voltage, amplified by two amplification stages (controller, power amplifier), results in the control of the motor U_A. The motor moves the antenna and the sliding contact of the potentiometer until it is $U_w = U_x$ or the angle α_x is equal to the setpoint angle α_w. Significant for the CLC is the feedback.

The functional block diagram of CLC in Fig. 1.11a is like the functional block diagram shown in Fig. 1.7b.

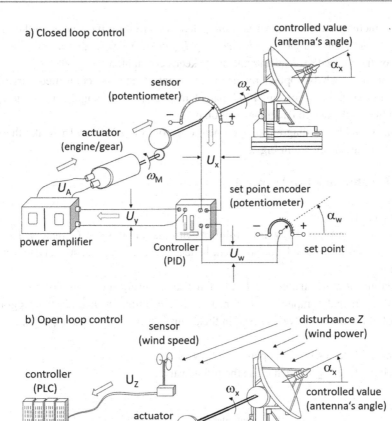

a) Closed loop control

controlled value
(antenna's angle)

α_x

sensor
(potentiometer)

ω_x

actuator
(engine/gear)

U_x

ω_M

U_A

set point encoder
(potentiometer)

α_w

U_y

power amplifier

Controller
(PID)

U_w

set point

b) Open loop control

sensor
(wind speed)

disturbance Z
(wind power)

controller
(PLC)

U_z

α_x

ω_x

controlled value
(antenna's angle)

actuator
(engine/gear)

U_A

ω_M

Fig. 1.11   Antenna's angular position control

Open loop control

The angular position α_x (controlled variable) of an antenna should be kept constant under the simplified assumption that there is only one relevant disturbance variable z, namely the fluctuation in wind power, which can easy be measured with a wind sensor, as shown in Fig. 1.11b.

The PLC-controller (programmable logical unit) is previously set in such a way that the antenna angle α_x (controlled variable) is equal to the setpoint value α_w and the control voltage $U_A = 0$ of the motor is equal to zero, i.e., $U_A = 0$. If the wind speed increases (disturbance variable z), the angular position α_x of the antenna would be changed.

But the increase in wind speed is immediately reported by the sensor to the control unit PLC, which controls the motor, changing U_A in such a way, that no change of the angular position α_x occurs. The antenna angle keeps constant, namely $\alpha_x = \alpha_w$.

Since no feedback is applied by open loop control and the controlled variable α_x is not measured, the control unit PLC cannot recognize if the angular position α_x has exactly achieved the desired angular position α_w.

The functional block diagram of the open loop control of Fig. 1.11b is like the functional block diagram shown in Fig. 1.10a.

1.3.3.2 A water tank level control

Open loop control

The tank, shown in Fig. 1.12, will be fulfilled with the inflow valve (actuator). The level sensor of the tank is installed, but not used. Instead of it, the flow sensor sends its signals to the controller.

A program, how to change the level of the tank during the technology process, is saved in the personal computer (set point w). Another program, namely the program of the control of the inflow valve, is saved in the controller. It is very simple program:

- If the actual level $x < w$, then open the inflow valve,
- If the actual level $x \geq w$, then close the inflow valve.

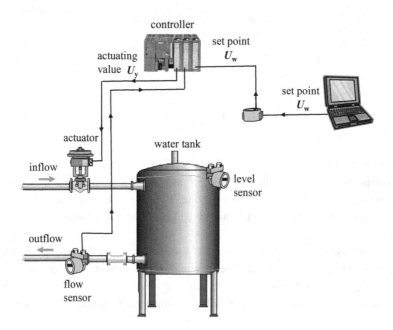

Fig. 1.12 Open loop control of tank level

To realize these conditions is needed only one logical block of the PLC. Generally, the PLC logical blocks have two inputs, In1 and In2, and one output OUT. Supposing In1 is connected to x and In2 to w, then the following logical block is enough for tank level control:

- LT (*less then*) for $x < w$ condition, i.e., if In1 < In2, then OUT is TRUE (inflow valve open). If the condition $x < w$ is not met, e.g., $x > w$ or $x = w$, than the inflow valve will be kept closed.

Such a program is an example of the open loop control according to the functional block diagram of Fig. 1.10a. It is clearly seen that the only advantage of the control option of Fig. 1.12 is its simplicity. The implementation of logical blocks LT (*less than*), LE (*less equal*), GT (*greater then*), etc. is possible with simple PLC (Programmable Logical Control), without relative expensive PID-controllers.

Closed loop control

Let us improve the control of the same tank, this time using the level sensor. As it is seen from Fig. 1.13 now it is a closed loop control with the functional block diagram given in Fig. 1.14.

As far as the plant has one controlled value and one actuating value it is often called SISO (Single Input Single Output).

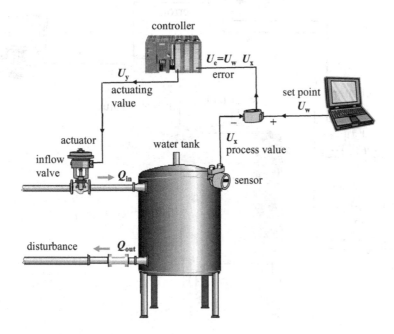

Fig. 1.13 Closed loop control of tank level

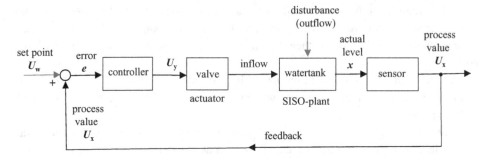

Fig. 1.14 Functional block diagram of tank level closed loop control

Closed loop control with two inputs

The best solution of control is shown in Fig. 1.15, with the corresponding functional block diagram of Fig. 1.16.

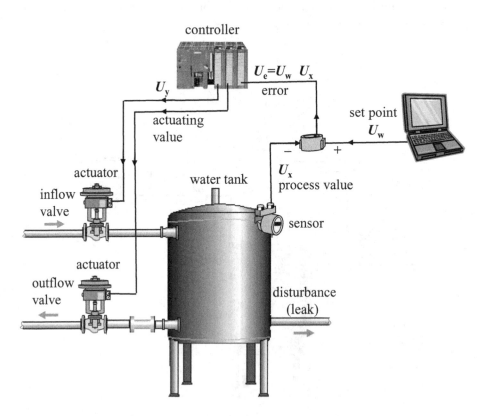

Fig. 1.15 Closed loop control of the tank level with two valves

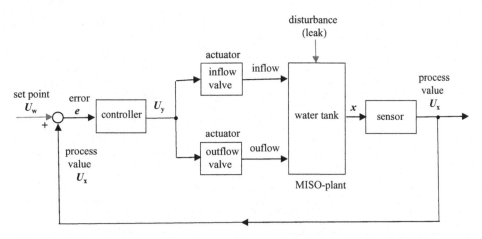

Fig. 1.16 Functional block diagram to the tank level closed loop control of Fig. 1.15 with two valves

It is again the closed loop control, but with two inputs of the plant, i.e., inflow and outflow, which are controlled with valves. As far as the plant has only one output (level *x*), it could be called MISO (Multi Input Single Output).

Looking ahead let us note that in the technology are known also SIMO (Single Input Multi Output) and MIMO (Multi Input Multi Output) plants which will be discussed in this book.

Finally, let us ask the question: Is it possible to combine the open and the closed loop controls, and if "yes" for what reason? An example of the answer to this question is given in the next section.

Combined open and closed loop control

Supposing the MISO tank shown above in Fig. 1.15 is equipped with two kinds of valves:

- an analog inflow valve, which lets the flow proportionally to each actuator's value e.g., between 0 and 100%.
- a binary outflow valve for outflow, which lets either 100% when its input is TRUE and is closed (0% flow) if its input is FALSE.

Of course, it is possible to use the closed loop control or open loop control equally for both valves. But considering the low costs and simplicity of the open loop control it is better to control only analog valve (inflow) with closed loop. In this case the binary valve (outflow) will be controlled with a logical block, i.e., with the open loop control.

The control system of a water tank level, consisting of the CLC algorithm (controller block PIDP1) and the open loop control (logical block GT), is shown in Fig. 1.17.

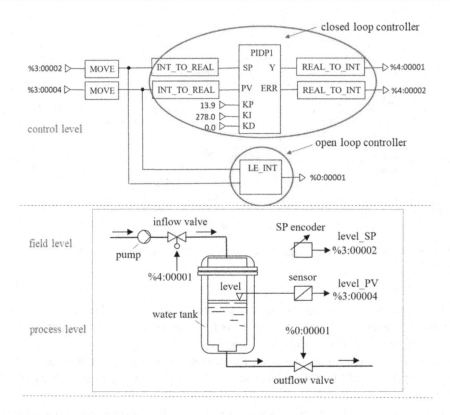

Fig. 1.17 Combined open and closed loop control of tank level

In the tank are implemented two actuators (valves):

- an analog valve for inflow feed with the PLC analog output address %4:00001
- a binary valve for outflow feed with the PLC binary output address %0:00001

There are also two controllers:

- closed loop controller PIDP1 function block with 4 input/output addresses and internal functional blocks:
  - %3:00002 analog input for set point w (desired level)
  - %3:00004 analog input for process variable x (actual level)
  - %4:00001 analog output for set point actuator value (inflow valve)
  - %4:00002 analog output for error $e = w - x$ calculated in PIDP1 function block
  - MOVE is used as an assign value instruction, which only copies the input data and sent its value unchanged to the output.

- INT_TO_REAL and REAL_TO_INT are data type convertors, which are needed because the PLC inputs are only of INT data type, while the PIDP1 control algorithm acts with REAL data type.
- KP, KI, KD are proportional-, integral and derivative coefficients of the PID-algorithm, which are considered in Chap. 2
- open loop controller, which is implemented as logical function block GT (*less then*) with following two inputs and one output:
  - In1 is connected to x,
  - In1 is connected to w,
  - OUT is connected to outflow binary valve.

Thus the algorithm implemented with LT comparison block is: if $x > w$, i.e., if the actual value x greater as the set point w, which is the same that the error $e = (w - x) < 0$, then open the outflow valve, i.e., the outflow valve = TRUE. The outflow valve will be opened until $x = w$.

The following comments are worth to make at this point, that the open loop control with logical blocks like LT, GT, etc.:

- can lead to oscillations of controlled variable x around set point w, which can be eliminate with so called hysteresis,
- is also known as two points closed loop controller,
- is an effective method to be used together with the closed loop control.

1.3.4 Bus

Without a bus is today no automation system imaginable. Since the functionality and structures of real buses are irrelevant to the goals of this book, they will further not be discussed. The mechanisms and algorithms of buses are quite well described in the literature.

Instead of real busses the virtual busses like bus-creator and bus-selector of MATLAB ® will be implemented in the closed loop control (CLC). The goal of such a bus implementation consists in a simple and, according to today's technical terms, comprehensible mapping of signal paths of a CLC as data streams of a bus system.

The controller and the plant will to be supposed as bus elements in segments of a virtual bus system. According to the rules of CLC the flow of information between bus segments will be considered only in one direction. The feedback of CLC will be integrated into the bus system. The *functional block diagrams* with busses will be called *signal flow diagrams*, they will replace the classical CLC block diagrams.

The CLC with virtual busses are detailed described in Chap. 4 and implemented in thereafter following chapters.

1.4   Management

The *Management*, as a subsystem of an industrial automation system, concerning planning and organizing the resources and activities of a business. should be very significant for the book conception, as it follows from the book's title.

But important for the new approach, called *Data Stream Management*, introduced in this and in the previous book [1] are not the definitions, functions, levels of management. Important for this book are the principals, and how the management itself is structured and organized.

In the following it will be shown without going into details what is management and how its principles will be applied to the goal of this book, i.e., to the closed loop control.

1.4.1   Management of Automation System

What is management of an industrial automation system? [23]

- Management oversees and supervises its target processes.
- Management works with human, financial, and physical resources.
- Management is a dynamic function and adapts to changes in its environment.
- effectiveness in management relates to the completion of tasks.
- Managers work closely with and provide guidance to the members of their team.
- To be effective, managers influence their team members to apply their unique strengths toward achieving the organization's goals.
- A manager considers a staff member both as an individual with diverse needs and as a component of the larger group.

There is no doubt, that the activities of a business of an industrial automation system is impossible to organize without management. In other words, the management is necessary condition for control function of an automation system.

Life cycle of products

"A product life cycle is the length of time from a product first being introduced to consumers until it is removed from the market. A product's life cycle is usually broken down into four stages; introduction, growth, maturity, and decline." (Quote: Source [24])

The life cycle of products of an automation system consists of following stages:

- Marketing
- Planning

- Development
- Test
- Commission
- Maintenance

1.4.2 Management of CLC

What is Management of CLC?
In this book is made a projection of management principles for an automation system on the whole life cycle of a CLC. The Management of CLC:

- considers not only one controller in the loop, but staff of control devices including open loop control, logical elements, and bus connection.
- oversees and supervises the processes of engineering and control in the CLC.
- threats mathematical and experimental resources of CLC.
- adapts itself to changes of plants parameters.

Life cycle of CLC with DSM
The closed loop control (CLC) is an academic subject of engineering and control theory. It is under no circumstances the academic subject of management.

Otherwise, the CLC has like all other products of an industrial company its own life cycle, which consists of the following stages (Fig. 1.18):

- Engineering
  - Planning
  - Development
    Plant identification
    Controller design
    Tuning of controller parameters
  - Supply of components
  - Installation
  - Test
- Commission
- Implementation
- Control of technological processes

Fig. 1.18 Life cycle of a closed loop controller

- Set point behavior
- Disturbance behavior
• Maintenance

1.4.3 Example of a CLC Life Cycle

To show the stages of CLC's life cycle and to explain how a CLC will be designed let us consider the simplest of all possible control loop namely one of the worldwide first feedback controllers [25], the water level controller of an oil lamp, shown in Fig. 1.1.

The stages of the life cycle of this controller are shown in Fig. 1.19.

Engineering
The first stage of engineering is the conception of control. As mentioned above we decide for CLC. The device consists of two connected to each other vessels filled with water. The controlled value is the water level of lower vessel (container). The container as a plant has two inputs:

• actuating value y, which is the water flow from the upper vessel (tank),
• disturbance z, which is leak through the hole in the container.

If the water level of the container is falling, it should be taken constantly by the amount of air entering the air pipe and led to the upper vessel (water tank).

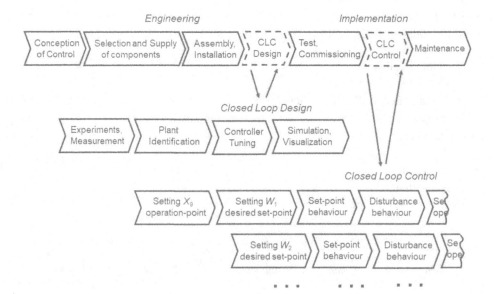

Fig. 1.19 Stages of CLC life cycle, treated in this book

The functional block diagram of the CLC according to this conception is shown in Fig. 1.20.

The thereafter following three engineering stages of the life cycle (Fig. 1.18) are given below without any comment:

- Selection and supply of components (Fig. 1.21),
- Assembly, installation (Fig. 1.22)
- Test, commission, implementation (Fig. 1.23)

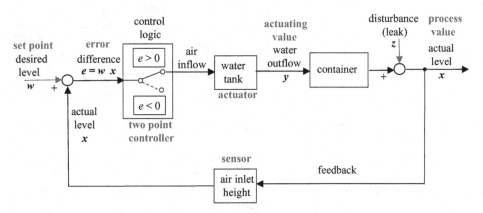

Fig. 1.20 Functional block diagram to the conception of the water level CLC

Fig. 1.21 Supply of components of the designed water level CLC

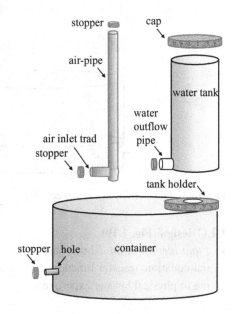

Fig. 1.22 Assembly of vessels
and pipes of water level CLC

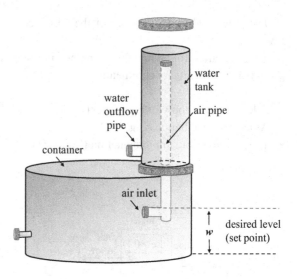

Fig. 1.23 Preparations to test and commission

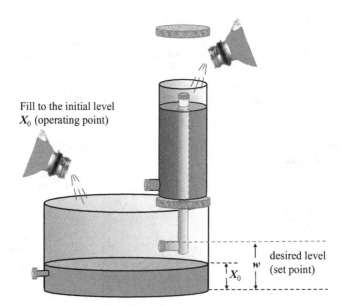

CLC design (Fig. 1.19)

- *Plant identification.* At this stage the mathematical description of the plant (differential equation, transfer function, or frequency response) will be defined either according to physical laws or experimentally upon measurements (see Chap. 2).

- *Controller tuning*. The controller parameter will be calculated for the identified mathematical model of the plant (see Chap. 6).
- *Simulation, Visualization*. These software tools are today well developed and widely applied to animate the behavior of dynamic systems inclusive CLC upon mathematical models, developed in previous stages of design. The simulation and visualization technique are realized in this book exclusively with MATLAB®-Scripts or MATLAB®/Simulink and implemented using several examples as well of SISO-loops (Single Input Single Output) as of loops with many inputs or outputs variables.

According to Fig. 1.19 the life cycle of a CLC has two kinds of behavior:

- set point behavior if no disturbance z acts. The controlled variable x is initially located by some operating point X_0 and should be changed to some given set point value w.
- disturbance behavior if the disturbance z acts but set point value w must not change.

Set-point behavior
(Siehe. Fig. 1.24)

a) Initial state: the controlled variable x is by operating point X_0, which is smaller as set point w. The leak hole is closed with the stopper so that no disturbance z acts. Because of the error $e = w-x$ the air inlet trad of the pipe (actuating value y) is open and the air pressures water from the water tank into the container.
b) Final state: the controlled variable x is equal to set point w, the error $e = w-x = 0$. The air inlet trad of the pipe is closed (actuating value $y=0$), and no water flows from the water tank into the container.

Disturbance behavior
(Siehe. Fig. 1.25)

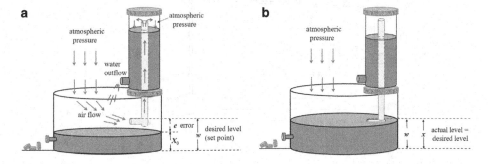

Fig. 1.24 Set point behavior of water level closed loop control: **a**) initial state x < w and e > 0; **b**) final state x = w and e = 0

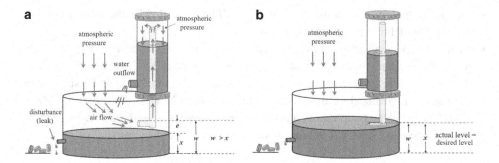

Fig. 1.25 Disturbance behavior of water level closed loop control

a) Initial state: the leak hole is opened (disturbance z acts) and the water level (controlled variable x) will be smaller as the set point w. And again, arises an error $e = w - x$ like above by set point behavior. The air inlet trad of the pipe (actuating value y) is open and the air pressures water from the water tank into the container.

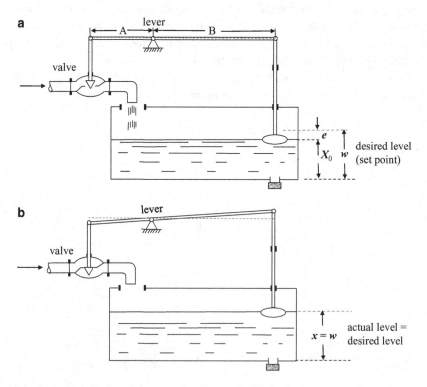

Fig. 1.26 Set point behavior of the water level control with the lever as controller: **a)** initial state, **b)** final state

b) Final state: the controlled variable $x = w$, the error $e = w - x = 0$. The air inlet trad of the pipe is closed (actuating value $y = 0$), and no water flows from the water tank into the container.

As far as the leak hole is opened the state (a) happened again after some time, then followed the state (b), etc. until the water tank will be empty.

Returning to Fig. 1.20, the question arises: where is the controller? It is no special device here, which can be called the controller. The whole construction including the water tank and air pipe serves as a controller.

Water level control with the lever as controller

Otherwise, it is in case of water level control with the float (bobber) as sensor and a valve as an actuator. The set point behavior and the disturbance behavior are shown in Fig. 1.26 and 1.27. The actons are the same as in the previous Fig. 1.24 and Fig. 1.25. The differences by construction are immediately seen:

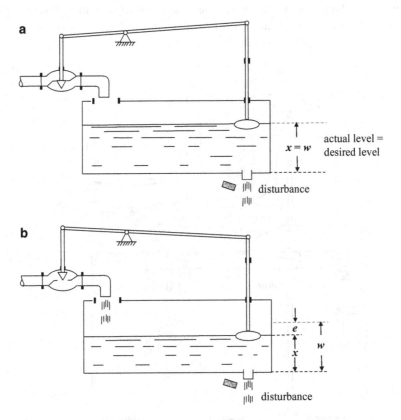

Fig. 1.27 Disturbance behavior of the water level control with the lever as controller: **a)** initial state, **b)** final state

- The valve is the actuator, through which the inflow water is pumped.
- The float is the sensor.
- The lever is served as a controller, which will be adjusted by a suitable ratio of lengths A and B.

Significant for this kind of CLC with the lever as controller is the stable connection between sensor (float) and actuator (valve), by which to every position of lever corresponds is only value of input flow. As result the lever cannot deliver the same set point by different disturbances, the control error e is not equal zero. Such a CLC is called proportional control.

Open loop water level control
Finally let us briefly discuss, how the device of Fig. 1.23, which originates from antique book [23], could be used for open loop control of another antique example, the automated opened door (Fig. 1.28).

The upper water tank is practically not used for control. As far as the hole in the container, which by the previous example was used as disturbance, will be opened, the process of filling the water bowl below starts and after some time the door opens.

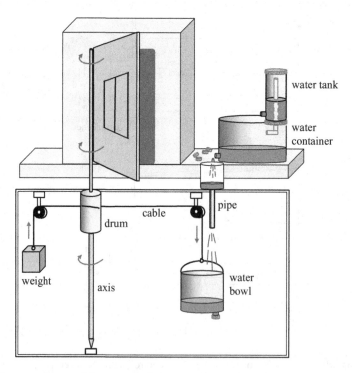

Fig. 1.28 Automated opened door as open loop control

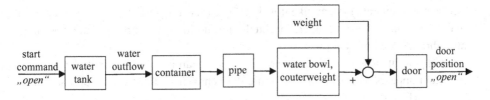

Fig. 1.29 Functional block diagram of the open loop control of automated door of Fig. 1.28

The functional block diagram of this open loop control is shown in Fig. 1.29.

At this place we finish the overview about life cycle of CLC. It only remains to list the conceptions, approaches, methods, used in this book, and to answer the most important question, what is a Data Stream Manager (DSM)?

1.5   Conceptions and Approaches, used in this book

1.5.1   Classical Control Theory

The basics of the classical theory of the CLC with linear time invariant (LTI) elements and standard PID-controllers, considering three domains (time-, Laplace-, and frequency domain) are briefly given in Chap. 2.

Significant for the classical control theory is that each controlled variable x is controlled with only one controller with its actuating variable y for corresponding set point w. The cascade control, the multivariable control, or the override control are no exceptions in this sense.

The next significant point of the classical control theory is that about 99% of known tuning rules are developed for set-point behavior, i.e., for steps of set point w. As a result, the disturbance behavior i.e., the reaction of a disturbance step z is stable, but not optimal.

Finally, it should be noted that the controller design is done by classical control theory without taking in account the restrictions y_{max} and y_{min} of the actuator or actuating variable y. Systems, which are optimally designed neglecting actuator restrictions, are often far from desired behavior by real applications with actuator's limitations.

All three omissions of the classical theory mentioned above are considered and partially improved in this book.

1.5.2   System-Approach

The classical systems approach, which was introduced in the fifties of the last century, is applicated in the book to the closed loop control (CLC). It leads to the consideration, that a CLC is not a separate system, like it is done by classical control theory, but is a part of an entire automation system, interconnected with the logical control and the open loop control through the bus system.

The important aspect of the system approach regarding CLC is that the automation system appears as *a macroworld* while the behavior of the closed loop itself is *its microworld* (Fig. 1.30).

Just as the management of macroworld (automation system) is distributed among many functions (MSE, ERP, CPM, etc.), the control functions of microworld (identification of the plant, tuning of controller, set-point behavior, disturbance behavior, etc.) are distributed among many Data Stream Managers (DSM).

1.5.3   Bus-Approach

The main idea of the bus-approach, introduced in [1–3] and [8–11] means that a CLC is considered not as a transfer of single variables in a classical functional block diagram (Fig. 1.31a), but rather as the data stream in a bus with the interconnected closed loop elements like the real bus in a real automation system. The application of bus-approach for a single control loop (SISO) with one controlled variable x and one actuating value

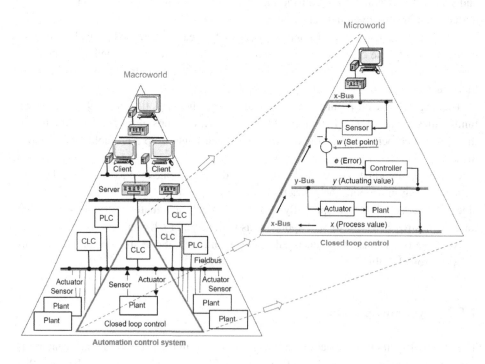

Fig. 1.30 CLC as part of an entire automation system

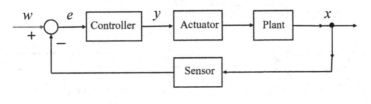

a) Classical functional block diagram of closed loop control

b) Bus-approach

Fig. 1.31 Bus-approach of a simple SISO closed loop control

y is shown in Fig. 1.31b. The example of multivariable control (MIMO) with two controlled variables x_1, x_2 and two actuating values y_1, y_2 is given in Fig. 1.32.

1.5.4 Symmetry and Antisymmetry

The branches of mathematics, such as topology and group theory, which deal with symmetry, have immense positive influenced for many science and engineering fields. A simple example of two triangles with the symmetry axis L1 shows in Fig. 1.33 without an explanation, what an antisymmetry means.

However, the system theory and control engineering are only slightly affected with symmetry and antisymmetry. One example of symmetric and antisymmetric CLCs with negative and positive poles and zeros is given in Fig. 1.34.

The application of the symmetry operations, introduced in [1–7], leads to new so called Antisystem-Approach (ASA) and resulted in various new control conceptions [12–19] like shadow-plant, ASA-controller, Turbo-controller, ASA-controller with bypass etc.

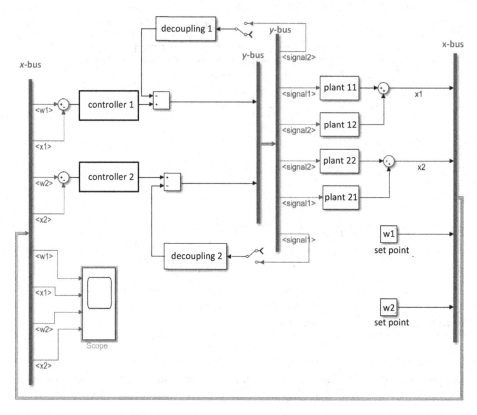

Fig. 1.32 Bus-approach of a MIMO closed loop control with two controlled variables

1.5.5 CLIMB & HOLD-modes

The idea of this conception is very simple: as far as the classical set point behavior occurs without disturbance and the disturbance behavior in its turn occurs without set point changes, two kinds of behavior are suggested in this book for the same CLC:

- CLIMB (setpoint behavior).
- HOLD (disturbance behavior).

The CLIMB-mode will be controlled by the main controller, also called CLIMB-controller, which is usually used in CLC. The task of this controller is to bring the controlled variable x possibly rush and without error to the setpoint value w.

As far as the set point is arrived, i.e., it will be $x=w$, the CLIMB-controller will be switched off. The other controller, called HOLD-controller will be switched on. Its task

Fig. 1.33 Simple example of symmetry and antisymmetry with L1 symmetry axis

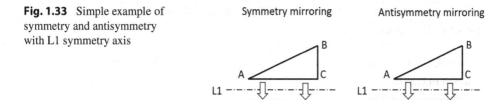

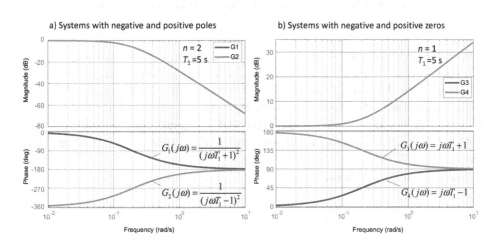

Fig. 1.34 Example of symmetry and antisymmetry in the CLC

is to keep constant not only the controlled variable but the actuating value y_w, which was needed for the CLIMB-controller, to arrive the set point w.

In other words, the HOLD-controller acts like open loop control while the CLIMB-controller is the classical CLC.

Such a two-controller approach has advantages against classical one-controller-approach especially by plants with dead time (inappropriate slow at the begin of control) and/or with derivatives (inappropriate fast at the begin of control).

1.6   Summary: Data Stream Management

One of the important tasks of an industrial automation systems is the management of the information flow between levels, shown in Fig. 1.4. As far as the CLC is considered as a part of the entire automated system, like shown in Fig. 1.30, the management of data stream between levels of CLC also becomes important.

The aim of this book is the design of Data Stream Managers (DSM) for following tasks of CLC-engineering:

- identification of plants,
- tuning of controllers.
- simulation of control.

These tasks are carried out in two domains:

- Real world, which is the world of physical devices and signals (Fig. 1.15).
- Virtual world, which is the world of mathematical variables and information (Fig. 1.16).

A Data Stream Manager (DSM) is a functional block, which consists of closed loop control, open loop control, and logical operators. It allows us to manage design and control as well for the real world of devices as for the virtual world of mathematics and simulations. The exchange of physical signals occurs in the real world with the real fieldbus, the information of virtual world is exchanged with the virtual bus, e.g., with simulated MATLAB®/Simulink bus-creator and bus-selector. The forerunners of DSM are software-agents ([26, 27]).

The DSM are divided in two groups:

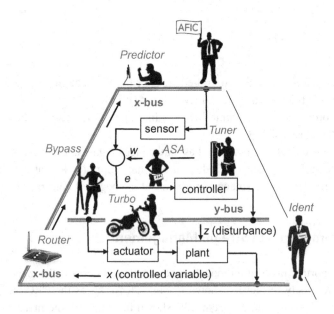

Fig. 1.35 Virtual world of CLC with Design Manager (offline DSM)

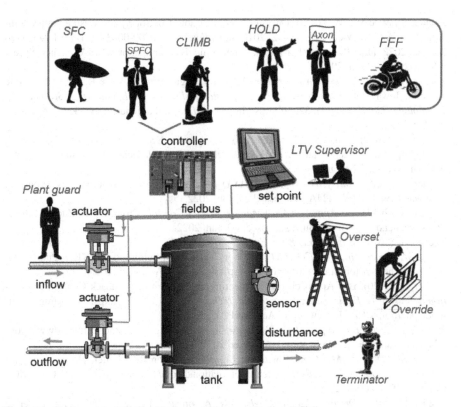

Fig. 1.36 Real world of CLC with Control Manager (online-DSM)

- *Design-DSM* (Fig. 1.35) for plant identification, controller tuning and structure of the CLC. They receive information from the real world or from operator and act in the virtual world. The Design-DSM are offline-managers.
- *Control-DSM* (Fig. 1.36) for management of control processes. They are placed in real controllers and act in the real world using models of virtual world, created by Design-DSM. The Control-DSM are online-managers.

References

1. Zacher, S. (2021). *Regelungstechnik mit Data Stream Management*. Verlag Springer Vieweg. https://link.springer.com/book/10.1007/978-3-658-30860-5. Accessed 20. Jan. 2022.
2. Zacher, S., & Reuter, M. (2022). *Regelungstechnik für Ingenieure*, (16th ed.). Verlag Springer Vieweg. https://link.springer.com/book/10.1007/978-3-658-36407-6. Accessed 20. Jan. 2022.
3. Zacher, S. (2020). *Drei Bode-Plots-Verfahren für Regelungstechnik*. Wiesbaden: Verlag Springer Vieweg. https://link.springer.com/book/10.1007/978-3-658-29220-1. Accessed 20. Jan. 2022.

4. Zacher, S., & Saeed, W. (2010). Design of multivariable Control Systems using Antisystem-Approach. AALE, Wien, Feb. 11–12, 2010, https://orcid.org/0000-0003-0021-7691 Accessed 20. Jan. 2022. https://www.zacher-international.com/wiss_Publikationen/ASA_Zacher_Saeed.pdf. Accessed 20 Jan. 2022.
5. Zacher, S. (2021) Antisystem-Approach (ASA) for Engineering of Wide Range of Dynamic Systems, Iowa State University, USA: *International Journal on Engineering, Science and Technology (IJonEST), 3*(1), 52–66 https://ijonest.net/index.php/ijonest/issue/view/4 Accessed 20. Jan. 2022.
6. Zacher, S. (2020) *Antisystem-Approach (ASA)*. Chicago, USA: IConEST, October 15–18, 2020 https://www.zacher-international.com/IJONEST_Journal/Zacher_Antisystem_Presentation.pdf. Accessed 20. Jan. 2022.
7. Zacher, S. (2020) *Antisystem-Approach (ASA) for Feedback Control*. https://www.youtube.com/watch?v=UhUrWrx24Ag. Accessed 20. Jan. 2022.
8. Zacher, S. (2014) *Bus-Approach for Feedback MIMO-Control*. Verlag Dr. S. Zacher https://www.szacher.de/my-Books/Bus/. Accessed: 20. Jan. 2022.
9. Zacher, S. (2019). *Bus-Approach for Engineering and Design of Feedback Control*. Denver, CO, USA: *Proceedings of ICONEST*, October 7–10, 2019, published by ISTES Publishing, pp. 26–27. https://ijonest.net/index.php/ijonest/issue/view/4. Accessed 20. Jan. 2022.
10. Zacher, S. (2020). Bus-Approach for Engineering and Design of Feedback Control. *International Journal of Engineering, Science and Technology, 2*(1), 6–24. https://www.ijonest.net/index.php/ijonest/article/view/9/pdf. Accessed 20. Jan. 2022.
11. Zacher, S. (2020). *Bus-Approach for Feedback Control*. Verlag Dr. S. Zacher https://www.youtube.com/watch?v=dXMXKQJtuIQ. Accessed 20. Jan. 2022.
12. Zacher, S. (2019). *MIMO-Contorl with Bus-Approach*. Automation-Letter, No 40, Stuttgart: Verlag Dr. S. Zacher https://www.zacher-international.com/Automation_Letters/40_MIMO_Control.pdf. Accessed 20. Jan. 2022.
13. Zacher, S. (2016). *Die unwirksame Regelstrecke*. Automation-Letter, No 06, Stuttgart: Verlag Dr. S. Zacher. https://www.zacher-international.com/Automation_Letters/06_Unwirksame_Strecke.pdf Accessed 20. Jan. 2022.
14. Zacher, S. (2016). *Schatten-Strecke*. Automation-Letter, No 07. Verlag Dr. S. Zacher. https://www.zacher-international.com/Automation_Letters/07_Schatten_Strecke.pdf. Accessed 20. Jan. 2022.
15. Zacher, S. (2016). *ASA Implementierung*. Automation-Letter, No 08. Verlag Dr. S. Zacher https://www.zacher-international.com/Automation_Letters/08_ASA-Implementierung.pdf. Accessed 20. Jan. 2022.
16. Zacher, S. (2016). *ASA-Regler für I -Strecke*. Automation-Letter, No 25. Verlag Dr. S. Zacher https://www.zacher-international.com/Automation_Letters/25_ASA_Regler_OSLO.pdf. accessed Jan 20, 2022
17. Zacher, S. (2016). *ASA-Regler mit Bypass*. Automation-Letter, No 29. Verlag Dr. S. Zacher https://www.zacher-international.com/Automation_Letters/29_ASA_Regler_mit_Bypass.pdf. Accessed 20. Jan. 2022
18. Zacher, S. (2017) *ASA-Regler: Test und Nachbesserung* Automation-Letter, No 32. Verlag Dr. S. Zacher https://zacher-international.com/Automation_Letters/32_ASA_Regler_Test.pdf. Accessed 20.Jan. 2022.
19. Zacher, S. (2017). *ASA-Bilanzregelung*. Automation-Letter, No 33. Verlag Dr. S. Zacher https://zacher-international.com/Automation_Letters/33_ASA_Bilanzregelung.pdf. Accessed 20. Jan. 2022.

20. Zacher, S. (2021). *Surf Feedback Control*. Automation-Letter, No 42. Verlag Dr. S. Zacher https://www.zacher-international.com/Automation_Letters/42_Surf_Control.pdf. Accessed 20. Jan. 2022.
21. Zacher, S. (2021). *Terminator im Regelkreis*. Automation-Letter, No 43. Verlag Dr. S. Zacher https://www.zacher-international.com/Automation_Letters/43_Terminator.pdf. Accessed 20. Jan. 2022.
22. Zacher, S. (2021) *Schubert-Terminator*. Automation-Letter, No 45. Verlag Dr. S. Zacher https://zacher-international.com/Automation_Letters/45_Schubert_Terminator.pdf. Accessed 20. Jan. 2022.
23. *What is Management? | Management Study HQ*. https://www.managementstudyhq.com/what-is-management.html. Accessed 20. Jan. 2022.
24. *What is the product life cycle? (Definition, Stages, Examples)* www.twi-global.com. Accessed 20. Jan. 2022.
25. *Hero of Alexandria*. https://en.wikipedia.org/wiki/Hero_of_Alexandria. Accessed 20. Jan. 2022.
26. Walzer, M. (2002) Softwareagenten in der Automatisierungstechnik. *atp- automatisierungstechnische Praxis, 44*(5), 26, 27.
27. Zacher, S. (2002) Softwareagenten für die Steuerung. Künstliche Intelligenz für Produktionssysteme. *Hessen TTN, Fachhochschule Wiesbaden*. Hannover Messe, Halle 18, Stand A16.

Control theory is the study of structure and behavior of control systems. A control system is interconnection of many functional subsystems, which are called "blocks" in this book. The aim of this book is the application of control theory to the management of the information and signal flow, which is called "Data Stream", between blocks of a control system.

Classes of control

There are known two classes of control:

- Open loop control, when the input of the system (actuating value y or control) is in no way influenced from its output (controlling variable x). The disadvantage of such control is that the output will not produce the desired value if uninspected disturbances act.
- Closed loop control, when the input of the system (actuating value y or control) is modified from in its output (controlling variable x), adjusting the input to the desired value even if uninspected disturbances act. The disadvantage of this class of control are the not desired oscillations of controlling variable x.

This chapter is assumed to encompass the questions to the classical mathematical description of dynamics of closed loop control systems (CLC). The following chapters built a bridge between the classical CLC and the management of data stream.

Kinds of closed loop control systems

The common classification of CLC is well known as:

© The Author(s), under exclusive license to Springer Nature Switzerland AG 2022   43
S. Zacher, *Closed Loop Control and Management*,
https://doi.org/10.1007/978-3-031-13483-8_2

- Continuous CLC, which consists of two classes of systems:
  - Linear systems
    with time-invariant blocks
    with time-varying blocks
  - Nonlinear systems
- Diskrete CLC

This chapter described linear systems with time-invariant blocks (LTI) and with one controlling variable (SISO-Single Input Single Output).

2.1   Three Levels of Mathematical Descriptions

2.1.1   Introduction

Generally, the feedback control systems analysis and design take several different forms:

- in the time-domain, when a control system is described with differential equations or with state space equations. The argument of differential equations in the time domain are t (time). For more precise identification, a block shows how the output variable reacts to input steps (sudden change in the input variable).
- in the s-domain, when the differential equations of the time domain are Laplace-transformed, and transfer functions of similar blocks are used. The argument of transfer functions in this domain are s (Laplace-operator). The advantage of this domain consists in the easy mathematical description by coupling of blocks into a system. It allows optimally adjust controllers changing the damping of the loop or shifting poles of TF to desired places.
- in the frequency domain, when frequency responses with arguments $s = j\omega$ are used for controller tuning (ω is angular frequency of the loop).

2.1.2   Time Domain

Steady state and dynamic behavior
By analysis of systems a distinction is made between the steady state (static behavior) and the dynamic (time) behavior. In the proportional systems if the input Y is constant, the output signal X is also constant. If the input variable Y changes by the time t_0 from Y_0 to Y_{End} as it is shown for example in Fig. 2.1 the output variable X also changes beginning from X_0 and will be usually set after a certain time to a constant value X_{End}.

It is also possible that a steady state X_{End} cannot be reached at all. The system is then unstable.

The relationships between the signals Y and X in the steady state are described with the help of static characteristics or functions $X = F(Y, Z)$, whereby Z is the disturbance.

Fig. 2.1 Step and step response. *Source* [16, 18, 19]

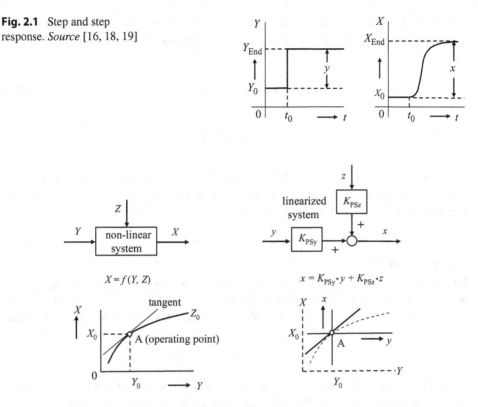

Fig. 2.2 Non-linear and linearized static characteristics. *Source* [16, p. 49]

It is useful to designate the small deviations ΔY, ΔZ, and ΔX simply by the small letters y, z, and x. The actual values are composed of the steady state operating point values and the time-dependent deviations:

$$Y_{End}(t) = Y_0 + y(t)$$
$$Z_{End}(t) = Z_0 + z(t) \tag{2.1}$$
$$X_{End}(t) = X_0 + x(t)$$

Linearization

Generally, the static characteristics are non-linear (Fig. 2.2) and should be discussed in the corresponding chapter "Non-linear control". In this book we will use the classical basics of the control theory and static characteristics of a controlled system will be linearized by the tangents at the operating point (X_0, Y_0, Z_0).

The function $X = F(Y, Z)$ becomes the following expression for differentials

$$dX = \left(\frac{\partial X}{\partial Y}\right)_0 \cdot dY + \left(\frac{\partial X}{\partial Z}\right)_0 \cdot dZ \tag{2.2}$$

The index 0 stands for the values X_0, Y_0, and Z_0 of the operation point A. The partial derivatives in Eq. 2.2 at the operating point A are denoted by the coefficients K_{PSy} and K_{PSz}

$$K_{PSy} = \left(\frac{\partial X}{\partial Y} \right)_0 \tag{2.3}$$

$$K_{PSz} = \left(\frac{\partial X}{\partial Z} \right)_0 \tag{2.4}$$

Denoting dX, dY and dZ in Eq. 2.2 and considering Eq. 2.1 due to small deviations x, y and z from the operating point A, the linearized description of the static behavior becomes the following expression:

$$x = K_{PSy} \cdot y + K_{PSz} \cdot z \tag{2.5}$$

Dynamic behavior

The dynamic behavior of output $x(t)$ and input $y(t)$ of linear or linearized systems is described in time domain with differential equations with constant coefficients. As an example, is below given the 1st order differential equation with coefficients T_1 (time delay) and K_p (gain):

$$T_1 \, \dot{x}(t) + x(t) = K_P y_0(t) \tag{2.6}$$

As an input y_0 is in this book usually a step signal used, like shown in Fig. 2.3, but the step y_0 will be applied by the time $t_0 = 0$, i.e., the differential equation will be solved by the zero initial conditions.

When setting up the differential equation of an unknown system, one must apply the physical laws e.g., the mechanical, hydraulic, pneumatic, electrical laws, etc. Otherwise,

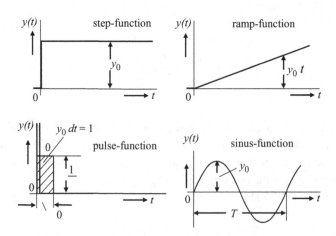

Fig. 2.3 Test input functions

one can define the coefficients T_1 and K_p of the differential equation experimentally, which is called *identification,* and which will be discussed later. Generally, the identification means, that the system differential equations will be determined as a system response to one of the possible test input functions according to Fig. 2.3.

Solution of differential equations
The solution of the differential equation Eq. 2.6 is shown graphically in Fig. 2.4 and corresponds to the exponential function

$$x(t) = K_P \cdot y_0 \cdot \left(1 - e^{-\frac{t}{T_1}}\right) \qquad (2.7)$$

The curve $x(t)$ has the greatest slope for $t=0$. If we put the tangent on the curve $x(t)$ at time $t=0$, then it intersects the steady-state value $x(\infty)$ for $t=T_1$.

Kinds of differential equations
There are known three kinds of differential equations and accordingly three kinds of dynamic elements:

• Proportional	or shortly P-elements, e.g., Eq. 2.6
• Integral	or I-elements, e.g., $x(t) = K_i \int y(t) dt$
• Derivativ	or D-elements, e.g., $x(t) = K_D \frac{dy(t)}{dt}$ or $x(t) = K_D \dot{y}(t)$

The same kind of differential equations, P. I or D, fit evenly to plants and to controllers. Only difference are inputs and outputs variables, which are noted in this book different, as it is shown below.

▶ **Definition**
The inputs of the plants are actuator values $y(t)$, while the outputs of the plant are controlled variables $x(t)$. The gains of the plants are indexed in book as K_{PS}. Accordingly, are considered also the gains and the variables of controllers. Inputs $e(t)$ of controller are the errors between set point $w(t)$ and controlled variable $x(t)$:

Fig. 2.4 Input step and plant's output as a solution of the differential equation of the 1st order. *Source* [17, p. 220]

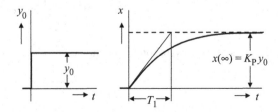

$$e(t) = w(t) - x(t) \tag{2.8}$$

The output of a controller is the actuating value $y(t)$. For example, so called P-plant and P-controller have the following differential equations:

$$x(t) = K_{PS} \cdot y_0 \quad \text{(plant)} \tag{2.9}$$

$$y(t) = K_{PR} \cdot e_0 \quad \text{(controller)} \tag{2.10}$$

Advantages and disadvantages of the time-domain
The main advantage of the time-domain is that the dynamic behavior of plants could be described with the physical laws e.g., the mechanical, electrical laws, etc. Or it could be defined experimentally from system responses to test input.

But there are following disadvantages:

- It is not easy to solve the differential equations with classic analytical methods,
- The handling with graphical solutions of differential equations is not clear and concise,
- The description of blocks connections, e.g., the series connection of a plant and a controller and further the building of a closed loop is possible, but very complicated.

These are reasons, why the analysis and design of closed loops is usually done in other domains, namely in the s-domain (also called Laplace-Domain) or in the frequency domain.

2.1.3 Laplace-Domain

In the case of linear systems, it is advantageous for the control theory not to solve differential equations in the time domain but using Laplace transformation.

Laplace-transform
The Laplace-transform of a continuous function $f(t)$ with the time t as argument and with the zero initial condition $f(t)=0$ by $t=0$ is the function $f(s)$ with the operator s as argument. The operator s, called Laplace-operator, is one of the exponents of the integral

$$f(s) = \int_{-\infty}^{\infty} f(t)e^{-st}dt$$

This operation is symbolically denoted as "L":

$$L[x(t)] = x(s)$$

According to the Laplace transform, shown above, the following transforms are obtained for derivatives and integrals, assuming that the initial condition is zero:

$$L[x(t)] = x(s)$$
$$L[\dot{x}(t)] = s \cdot x(s)$$
$$L[\ddot{x}(t)] = s^2 \cdot x(s)$$
$$\cdots \quad \cdots \quad \cdots$$
$$L\left[\int x(t)dt\right] = \frac{1}{s} \cdot x(s).$$

(2.11)

Solution of differential equations with Laplace-transformation
The differential equation of Eq. 2.6 will be transferred as follows:

$$T_1 \dot{x}(t) + x(t) = K_P y(t)$$
$$\Downarrow \qquad \Downarrow \qquad \Downarrow$$
$$T_1 \cdot s \cdot x(s) + x(s) = K\ y(s)$$

(2.12)

The Laplace-transform leads the differential equation to the algebraic equation without derivatives and integrals:

$$(1 + sT_1)\,x(s) = K_P y(s)$$

(2.13)

For the input step function $y_0(t)$ is the Laplace transform

$$L[y(t)] = y(s) = \frac{1}{s}y_0$$

(2.14)

Inserting Eq. 2.14 into Eq. 2.13, it follows the same solution of the differential equation, Eq. 2.12, but in Laplace domain:

$$x(s) = \frac{K_P}{1 + sT_1}\,y(s) = \frac{K_P}{1 + sT_1} \cdot \frac{1}{s}\,y_0$$

(2.15)

The back transformation of Eq. 2.15 into the time domain can be done by means of partial fraction decomposition, residue theorem, or correspondence table, which are not considered in this book. The solution Eq. 2.15 will be back transformed to the solution in the time domain Eq. 2.7.

Transfer function (TF)
The back Laplace transform from Laplace domain into time domain is as complicated as the solution of a differential equation in the time domain. That is why in the control theory it will be used another approach, namely it wil be defined so called transfer function (TF), which is nothing else as the relation between an input $y(s)$ and output $x(s)$ of Eq. 2.15:

$$G(s) = \frac{x(s)}{y(s)} = \frac{K_P}{1 + sT_1} \tag{2.16}$$

From the last expression follows the solution of the differential equation in Laplace domain:

$$x(s) = G(s)y(s) \tag{2.17}$$

The TF $G(s)$ is the significant achievement of Laplace transform, which brings many advantages to the analysis and design of dynamic systems in the s-domain:

a) It is very easy to describe the connections of blocks,
b) It is possible to handle dynamic systems upon zeros and poles of transfer functions in the s-domain without back transform into the time domain.

The detailed description of (a) and (b) follows.

Interconnection of blocks
Generally, there are known three kinds of connections, which are shown in Fig. 2.5 with corresponded transfer functions.

Transfer function of CLC
The typical closed control loop with the set point $w(s)$ as input is generally built of the following blocks:

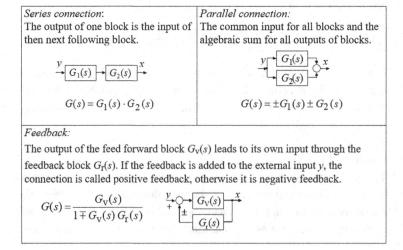

Fig. 2.5 Three kinds of blocks connection and the corresponding transfer functions. *Source* [17, p. 219]

- Plant with transfer function $G_S(s)$ including actuator
- Controller with transfer function $G_S(s)$
- Sensor $G_M(s)$

The closed loop is often influenced of a disturbance, which is unexpected external input $z(s)$. In is supposed that the set point and a disturbance do not appear at the same time. The dynamic of the closed loop, when only set point $w(s)$ appeared, is called set point behavior and is mathematically described with the TF

$$G_w(s) = \frac{x(s)}{w(s)} \tag{2.18}$$

Accordingly, the influence of the disturbance $z(s)$ is denoted as TF of disturbance behavior:

$$G_z(s) = \frac{x(s)}{z(s)} \tag{2.19}$$

To express the transfer functions of Eqs. 2.18 and 2.19 through blocks of the loop is considered so called TF of the open loop:

$$G_0(s) = \frac{x_r(s)}{e(s)} \tag{2.20}$$

The input of the TF of Eq. 2.20 is the error $e(s)$, the output is the feedback variable $x_r(s)$, i.e., the open loop consists of series of block in the loop.

The TF of the open loop shown in Fig. 2.6 is the series connection:

$$G_0(s) = G_R(s)G_P(s)G_M(s) \tag{2.21}$$

The other TF, which is needed to define the TF of the loop, is the feed forward transfer function $G_v(s)$. By the set point behavior, it is

$$G_{vw}(s) = \frac{x(s)}{w(s)} = G_R(s)G_P(s) \tag{2.22}$$

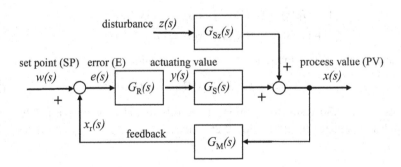

Fig. 2.6 Typical control closed loop. *Source* [16, p. 45]

By the disturbance behavior the feed forward transfer function is

$$G_{vz}(s) = \frac{x(s)}{z(s)} = G_{Sz}(s) \tag{2.23}$$

Finally, the TF of the closed loop will be defined as follows:

- for set point behavior:

$$G_w(s) = \frac{x(s)}{w(s)} = \frac{G_{vw}(s)}{1 + G_0(s)} \tag{2.24}$$

- for disturbance behavior:

$$G_z(s) = \frac{x(s)}{z(s)} = \frac{G_{vz}(s)}{1 + G_0(s)} \tag{2.25}$$

Devoted to Fig. 2.6 the TF of the closed loop are:

- for set point behavior, considering Eqs. 2.21, 2.22 and 2.24:

$$G_w(s) = \frac{x(s)}{w(s)} = \frac{G_R(s)G_P(s)}{1 + G_R(s)G_P(s)G_M(s)} \tag{2.26}$$

- for disturbance behavior, considering Eqs. 2.21, 2.23 and 2.25:

$$G_z(s) = \frac{x(s)}{z(s)} = \frac{G_{Sz}(s)}{1 + G_R(s)G_P(s)G_M(s)} \tag{2.27}$$

Zeros and poles of transfer functions

Significant is that the transfer functions of the loop of Eqs. 2.26 and 2.27 have the same denominator, which is called characteristic polynomial $P(s)$:

$$P(s) = 1 + G_R(s)G_P(s)G_M(s) \tag{2.28}$$

This designation is given to the polynomial $P(s)$, because it corresponds with the solution of the differential equation Eqs. 2.26 and 2.27 in Laplace-transform:

$$\begin{aligned} [1 + G_R(s)G_P(s)G_M(s)] \cdot x(s) = G_R(s)G_P(s) \cdot w(s) \\ [1 + G_R(s)G_P(s)G_M(s)] \cdot x(s) = G_{Sz}(s) \cdot z(s) \end{aligned} \tag{2.29}$$

Both equations above have the same characteristic equation:

$$1 + G_R(s)G_P(s)G_M(s) = 0 \tag{2.30}$$

The roots of Eq. 2.30 result to solution of the differential equations Eq. 2.29 and are called poles of transfer functions Eqs. 2.26 and 2.27. In other words, the poles characteristic equation Eq. 2.30 define the behavior of the loop. This behavior is the same for input $w(s)$ or $z(s)$.

The roots of the differential equations, built from numerators of Eqs. 2.26 and 2.27

$$G_R(s)G_P(s) = 0$$
$$G_{Sz}(s) = 0 \tag{2.31}$$

are called zeros of the corresponding TF Eqs. 2.26 or 2.27.

Example: Zeros and poles of a transfer function

Let us define the zeros and poles of the given TF of a closed loop:

$$G_w(s) = \frac{2(1 + 5s)}{(1 + 4s)(1 + 10s)}$$

The zero is: $s_{N1} = -0,2$ and the poles are $s_1 = -0,5$ and $s_2 = -0,1$. ◄

As far as the characteristic equation in Laplace-domain leads to the solution of the differential equation of a block in the time domain, the poles of the TF represent the behavior of the block. Generally, the roots of Eq. 2.30 are complex i.e., they consist of real part and imaginary parts. If e.g., the Eq. 2.30 is of the 2nd order, then poles $s_{1,2}$ have real parts α and imaginary parts ω:

$$s_{1,2} = \alpha \pm j\omega \tag{2.32}$$

Example: Characteristic equation of the 2nd order

Supposing the characteristic equation is given as below:

$$P(s) = 40s^2 + 3s + 1 = 0$$

Then the poles according to the solution with MATLAB® are complex.
p=[40 3 1];
roots (p)
ans =

$$-0.0375 + 0.1536i$$
$$-0.0375 - 0.1536i ◄$$

The relation between pole places in the Gaussian plane with axes (*Re*, *Im*), step responses $x(t)$ and damping ϑ are shown in Fig. 2.7.

Conclusion:

- If the real parts of poles are placed in the negative s-halfplane, the step response achieved some stable state. Such systems are called "stable".
- If at least one real part of poles is placed in the positive s-halfplane, the step response achieved no stable state and block is unstable.

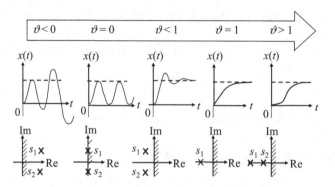

Fig. 2.7 Step responses $x(t)$ and damping ϑ dependent of pole places s_1, s_2. *Source* [19, p. 2]

- If at least one pole is placed directly on the imaginary axe, the step response is critical and has permanently oscillations.
- If all poles are real, i.e., the imaginary parts are equal to zero and poles are placed only on the real axe, the step response has no oscillations. One says the oscillations are damped with the damping factor of

$$\vartheta \geq 1$$

- If some of the poles are complex conjugate numbers with negative real parts, then the system is not enough damped and has oscillations:

$$0 < \vartheta < 1$$

- If at least one pair of poles are conjugate numbers with the positive real parts, then the system is not at all damped, has oscillations with growing magnitude and is unstable:

$$\vartheta < 0$$

2.1.4 Frequency Domain

The frequency domain results from the Fourier-transform of some function $f(t)$ with the real argument t into the function $f(j\omega)$ with the imaginary argument $j\omega$:

$$f(j\omega) = \int\limits_{-\infty}^{\infty} f(t)e^{-j\omega t}\,dt$$

It could be achieved simply changing the Laplace operator s of the Laplace transform (see Abschn. 2.1.3) through imaginary frequency ω:

$$s = j\omega$$

In this case for analysis and design of a CLC, mentioned in Abschn. 2.1.3, instead of TF $G(s)$ it will be used the function $G(j\omega)$, which is called frequency response.

Solution of differential equations with sinus input functions

The differential equation Eq. 2.6 was solved in time domain, when the input function is a step function, shown in Fig. 2.3. In the same figure are shown also ramp-, pulse- and sinus-functions, each of them could be used for the solution of differential equations. Especially great importance in the control theory won the sinus input function:

$$y(t) = y_0 \sin(\omega\, t) \tag{2.33}$$

with the magnitude y_0 and angular frequency ω.

If we put the exponential form of sinus function

$$\sin(\omega\, t) = \frac{e^{j(\omega\, t)} - e^{-j(\omega\, t)}}{2j} \tag{2.34}$$

into Eq. 2.18 and solve the differential equation Eq. 2.6 the solution will be seen as follows complicated expression

$$x(t) = \frac{K_P y_0}{A} \left[\sin(\omega\, t + \varphi) - \sin(\varphi) \cdot e^{-\frac{t}{T_1}} \right], \tag{2.35}$$

whereby there are:

$$A = \sqrt{1 + (\omega\, T_1)^2} \tag{2.36}$$

$$\tan \varphi = -\omega\, T_1 \text{ or } \varphi = -\arctan(\omega\, T_1) \tag{2.37}$$

This example of the differential equation of the 1st order has shown that after a sinusoidal input of Eq. 2.34 with frequency ω at the output occurs also the sinusoidal oscillation of the same frequency ω but with a phase shift φ. In general, in every linear system with the differential equation of any order, a harmonic oscillation Eq. 2.34 at the input also produces a harmonic oscillation like Eq. 2.36 at the output.

Frequency response

The application of sinusoidal input signals in so-called frequency domain has advantages over the step function in the time domain, especially for fast systems and is commonly used to examine closed control loops.

Similar to the step response in the time domain after application of a step input the output of a system after sinusoidal input is called frequency response. Let us show now, what relations are valid between frequency domain and Laplace-Domain. The input of Eq. 2.34 and the output of Eq. 2.36 could be written as

$$y(t) = |y| \cdot (\sin \omega t) = |y| \cdot e^{j\omega t} \tag{2.38}$$

$$x(t) = |x(\omega)| \cdot (\sin \omega t + \varphi) = |x(\omega)| \cdot e^{j(\omega t + \varphi)} \tag{2.39}$$

The ratio $G(t)$

$$\frac{x(t)}{y(t)}$$

of the output $x(t)$ to the input $y(t)$ variable, which is called the frequency response, is no longer a function of time, but of the $j\omega$:

$$G(j\omega) = \frac{|x(\omega)| \cdot e^{j(\omega t + \varphi)}}{|y| \cdot e^{j\omega t}} = \frac{|x(\omega)|}{|y|} \cdot e^{j\varphi} \tag{2.40}$$

The function of the magnitude

$$|G(\omega)| = \frac{|x(\omega)|}{|y|} \tag{2.41}$$

is called the amplitude response and the function $\varphi(\omega)$ is called phase response.

Pass from Laplace domain to frequency domain
Let us again consider the differential equation Eq. 2.6 of the 1st order

$$T_1 \dot{x}(t) + x(t) = K_P y_0(t),$$

which corresponds in the Laplace-domain with the transfer function of Eq. 2.16

$$G(s) = \frac{x(s)}{y(s)} = \frac{K_P}{1 + sT_1}$$

To pass from the Laplace domain into the frequency domain it is enough, as it is derived in the control theory and is given below without proof, to replace the Laplace operator (the complex variable s) with imaginary frequency $j\omega$:

$$s = j\omega \tag{2.42}$$

After such replacing it follows:

$$G(j\omega) = \frac{x(j\omega)}{y(j\omega)} = \frac{K_P}{1 + j\omega T_1} \tag{2.43}$$

The frequency response $G(j\omega)$ is the same complex function as Eq. 2.25.

Nyquist plot
It is useful, to represent the frequency response $G(j\omega)$ graphically in the Gaussian plane as real and imaginary parts. To separate Eq. 2.44 into real $Re(G)$ and imaginary $Im(G)$ parts, $G(j\omega)$ should be expanded with the complex conjugate expression of the denominator:

$$G(j\omega) = \frac{K_P}{1 + j\omega T} \cdot \frac{1 - j\omega T}{1 - j\omega T} = \frac{K_P(1 - j\omega T)}{1 + (\omega T)^2} = Re\,(G) + j \cdot Im\,(G) \tag{2.44}$$

Then follows:

$$\text{Re}\,(G) = \frac{K_P}{1+(\omega\,T)^2} \tag{2.45}$$

$$\text{Im}\,(G) = \frac{-K_P \omega\,T}{1+(\omega\,T)^2} \tag{2.46}$$

It useful to represent the frequency response $G(j\omega)$ as function of magnitude $|G(\omega)|$ and phase $\varphi(\omega)$:

$$G(j\omega) = |G(\omega)|e^{\varphi(\omega)} \tag{2.47}$$

$$[G(\omega)] = \sqrt{\text{Re}^2(\omega) + \text{Im}^2(\omega)} \tag{2.48}$$

$$\varphi(\omega) = \arctan\frac{\text{Im}(\omega)}{\text{Re}(\omega)} \tag{2.49}$$

An example of Nyquist plot for transform function of Eq. 2.43 is shown in Fig. 2.8.

Bode plot
Bode plot is the graphical representation of the frequency response $G(j\omega)$ of Eq. 2.47 in two separate charts:

- Magnitude $|G(\omega)|$ according to Eq. 2.48 but in logarithmic scale in decibels (dB):

$$|G\,(\omega)|_{dB} = 20 \cdot \lg |G\,(\omega)|$$

- Phase $\varphi(\omega)$ according to Eq. 2.49

Fig. 2.8 Nyquist plot

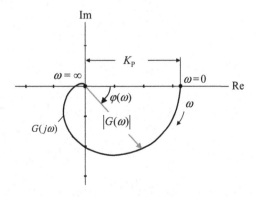

The angular frequency ω in both diagrams has logarithmic scale (dec). An example of bode plot for transform function of Eq. 2.43 is given in Fig. 2.9.

2.1.5 Summary

Overview of domains
Analysis and design of the closed loop control systems can take place in the one of the domains, described above and shown in Fig. 2.10. To each domain are in Fig. 2.10 shown also the in the praxis mostly used characteristics. The passes between domains are possible.

Limit conditions and retained error
The box shape in the middle of the Fig. 2.10 illustrates the limit values of arguments which lead to the initial and final static behavior. In between these conditions the system has dynamic behavior. These conditions followed from the Laplace-transform. Related to the CLC with the transfer function Eq. 2.24 the limit condition for the controlling variable looks like follows:

$$x(\infty) = \lim_{t \to \infty} x(t) = \lim_{s \to 0} G_{\mathrm{w}}(s) \cdot \hat{w} \tag{2.50}$$

Fig. 2.9 Bode plot

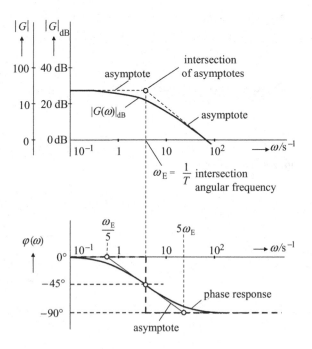

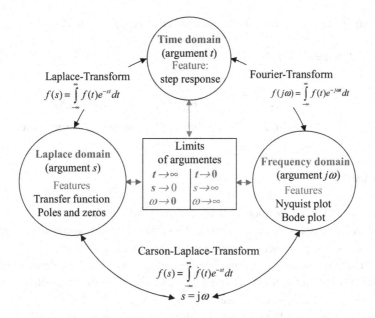

Fig. 2.10 Domains of analysis and design of CLC. *Source* [17, p. 20]

In this case the retained error $e(t)$ by $t = \infty$ in the final static state is equal to

$$e(\infty) = \hat{w} - x(\infty) \tag{2.51}$$

From Eqs. 2.50 and 2.51 follows that the CLC will have no retained error, i.e.,

$$e(\infty) = 0$$

if meets the condition below:

$$\lim_{s \to 0} G_w(s) = 1 \tag{2.52}$$

Example: Limits conditions and static behavior

Supposing the CLC is given by Eq. 2.26

$$G_w(s) = \frac{x(s)}{w(s)} = \frac{G_R(s)G_P(s)}{1 + G_R(s)G_P(s)G_M(s)}$$

with the transfer functions:

$$G_R(s) = K_{PR}$$

$$G_P(s) = \frac{K_{PS}}{1 + sT_1}$$

$$G_M(s) = 1$$

According to Eqs. 2.50 and 2.51 the retained error $e(\infty)$ is defined below:

$$x(\infty) = \lim_{s \to 0} G_w(s) \cdot \hat{w} = \frac{K_{PR}K_{PS}}{1 + K_{PR}K_{PS}} \cdot \hat{w}$$

$$e(\infty) = \hat{w} - x(\infty) = \hat{w} - \frac{K_{PR}K_{PS}}{1 + K_{PR}K_{PS}} \cdot \hat{w} = \frac{1}{1 + K_{PR}K_{PS}} \cdot \hat{w}$$

As far as condition Eq. 2.52 is not fulfilled the retained error $e(\infty)$ is not zero. ◀

2.2   Plants

Plants are parts of a closed loop control system in which the controlled variable x is influenced by the actuator value y. Simplified an actuator could be considered within the plant. In this case the plant gets its input directly from controller.

Plants treated below are linear time-invariant blocks (LTI) with one controlling variable (see SISO-Single Input Single Output). It is supposed that the parameters of the plant don't change by control. control device.

The classification of different kinds of plants occurs corresponding their dynamical behavior. Accordingly, the plants are represented in this book with:

- differential equations in time domain,
- transfer functions in Laplace domain,
- bode plots in frequency domain.

There are known three types of dynamical behavior:

- proportional or P-behavior,
- integral or I-behavior,
- derivative or D-behavior.

Relating to the time delay of step responses after the input steps there are following kinds of behavior:

- without time delay, i.e., P, I or D.
- with the time delay of the 1st order, i.e., P-T1, I-T1 or D-T1.
- with the time delay of the 2nd order, i.e., P-T2, I-T2 or D-T2 and so on.

2.2.1   Proportional P-plants

Significant for a P-plant is that its step response $x(t)$ achieves steady-state value $x(\infty) = $ const after some time, which is mathematically defined as $t = \infty$ and which depends on the time delay constant T of the plant. The overview of P-plants is given by Fig. 2.11.

P-T1 plant

A P-T1 plant has single pole and accordingly the pole is real, i.e., the step response proceeds aperiodic, without oscillations. If pole is negative, the damping of a P-T1 plant according to Fig. 2.7 is $\vartheta = 1$. If real pole is positive, the P-T1 plant is unstable, the step response achieved no steady state, and the damping is $\vartheta < 0$.

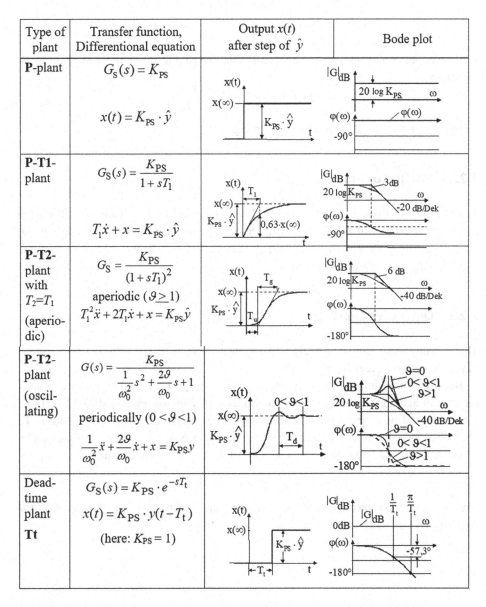

Type of plant	Transfer function, Differential equation	Output $x(t)$ after step of $\hat{y}$	Bode plot
P-plant	$G_S(s) = K_{PS}$ $x(t) = K_{PS} \cdot \hat{y}$		
P-T1-plant	$G_S(s) = \dfrac{K_{PS}}{1 + sT_1}$ $T_1\dot{x} + x = K_{PS} \cdot \hat{y}$		
P-T2-plant with $T_2 = T_1$ **(aperiodic)**	$G_S = \dfrac{K_{PS}}{(1 + sT_1)^2}$ aperiodic $(\vartheta \geq 1)$ $T_1^2\ddot{x} + 2T_1\dot{x} + x = K_{PS}\hat{y}$		
P-T2-plant (oscillating)	$G(s) = \dfrac{K_{PS}}{\frac{1}{\omega_0^2}s^2 + \frac{2\vartheta}{\omega_0}s + 1}$ periodically $(0 < \vartheta < 1)$ $\frac{1}{\omega_0^2}\ddot{x} + \frac{2\vartheta}{\omega_0}\dot{x} + x = K_{PS}y$		
Dead-time plant **Tt**	$G_S(s) = K_{PS} \cdot e^{-sT_t}$ $x(t) = K_{PS} \cdot y(t - T_t)$ (here: $K_{PS} = 1$)		

Fig. 2.11 Proportional behavior of plants. *Source* [19, p. 7]

P-T2 plant

The characteristic equation of a P-T2 plant is an algebraic quadratic equation and has two poles.

- If both poles are negative real, the step response is according to Fig. 2.7 aperiodic, without oscillations. The damping is $\vartheta > 1$.

$$G_S(s) = \frac{K_{PS}}{(1 + sT_1)(1 + sT_2)} \tag{2.53}$$

- If poles are conjugate with negative real parts, the step response proceeds with oscillations:

$$G_S(s) = \frac{K_{PS}}{a_2 s^2 + a_1 s + 1} \tag{2.54}$$

The parameters of oscillations are shown in Fig. 2.12.

2.2.2 Integral or I-plants

The overview of I-plants is shown in Fig. 2.13. Characteristic for an I-plant is that even a very small input will be accumulated, so its step response runs like a ramp without steady state, i.e., by $t = \infty$ there is $x(\infty) = \infty$. A steady state can be achieved only if the input y of the plant is equal to zero.

2.2.3 Derivative D-plants

The D-plants occur relatively rarely in the industrial practice. The ideal D-plants, shown in Fig. 2.14, are not realizable.

Because of the jumping output after a step by $t = 0$ the D-plants are relatively difficult for control and are not desired. The real plants with D-terms are called D-T1 or D-T2 plants, dependent of the number of delay time constants. Like P-T2-plants the D-T2 plant also can either have an aperiodic step response or produce oscillation like given by MATLAB® simulation in Fig. 2.15. The roots of the characteristic equation

$$a_2 s^2 + a_1 s + a_0 = 0$$

are conjugate with negative real parts, so the step response has oscillations with the damping $0 < \vartheta < 1$ according to Fig. 2.7.

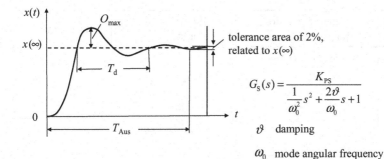

$$G_S(s) = \cfrac{K_{PS}}{\cfrac{1}{\omega_0^2}s^2 + \cfrac{2\vartheta}{\omega_0}s + 1}$$

ϑ damping

ω_0 mode angular frequency

$T_{Aus} = \dfrac{\ln 25}{\vartheta\omega_0} = \dfrac{3{,}22}{\vartheta\omega_0}$	settling time	$O_{max}\% = e^{-\vartheta\cdot\omega_0\frac{T_d}{2}}$	max. overshoot
$N = \sqrt{\dfrac{1}{\vartheta^2} - 1} \approx \dfrac{1}{\vartheta}$	number of halfways	$\omega_d = \omega_0\sqrt{1-\vartheta^2}$	oscillations angular frequency
$T_d = \dfrac{2\pi}{\omega_d}$	oscillation period	$\omega_D \approx \omega_d$	crossover angular frequency

Fig. 2.12 Oscillating P-T2 behavior. *Source* [20, p. 177]

Type of plant	Transfer function, Differentional equation	Output $x(t)$ after step of $\hat{y}$	Bode plot
I-plant	$G_S(s) = \dfrac{K_{IS}}{s}$ $x = K_{IS}\int y(t)dt$		
I-T1-plant	$G_S(s) = \dfrac{K_{IS}}{s(1+sT_1)}$ $T_1\dot{x} + x = K_{IS}\int y(t)dt$		

Fig. 2.13 Integral or I-plants. *Source* [19, p. 9]

2.3 Controllers

In this section are considered blocks, which are parts of the closed control loops, and which compare the controlled variable $x(t)$ with a specified, constant setpoint value w. These blocks, known as standard controllers, influence the plants via an actuator value

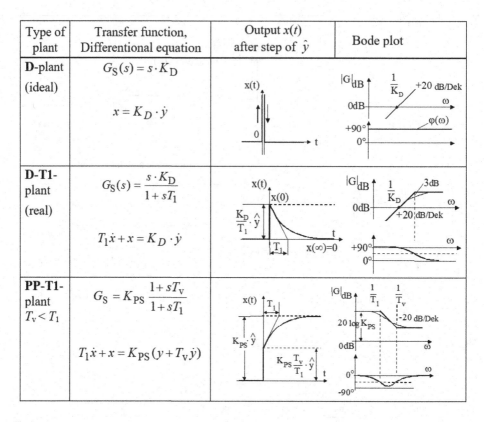

Type of plant	Transfer function, Differential equation	Output $x(t)$ after step of $\hat{y}$	Bode plot
D-plant (ideal)	$G_S(s) = s \cdot K_D$ $x = K_D \cdot \dot{y}$		
D-T1-plant (real)	$G_S(s) = \dfrac{s \cdot K_D}{1 + sT_1}$ $T_1\dot{x} + x = K_D \cdot \dot{y}$		
PP-T1-plant $T_v < T_1$	$G_S = K_{PS}\dfrac{1 + sT_v}{1 + sT_1}$ $T_1\dot{x} + x = K_{PS}(y + T_v\dot{y})$		

Fig. 2.14 Derivative D-plants with aperiodic behavior. *Source* [19, p. 8]

$y(t)$ in such a way that the control difference $e(t)$ is zero or as small as possible. A controller contains the following devices:

- to record the controlled variable $x(t)$,
- to compare the controlled variable $x(t)$ with the setpoint or the reference variable w,
- to form the controller output variable $y(t)$

The different kinds of controller according to the controlling algorithm are shown in Fig. 2.16:

- proportional or P-controllers are simple and act fast. But are not able to correct closed loop control and leave the retained error $e(\infty)$ by steady state.
- integral or I-controllers are very slow. But advantage of I-controllers is that they leave no retained error $e(\infty)$ when they are used in a closed loop control because they have the same features as I-plants described in Abschn. 2.2.2.

```
1 -    s=tf('s');                                    % Laplace operator
2 -    KdS=2;Td=5;                                   % plant parameters
3 -    a2=5;a1=1.7;a0=1;
4 -    Gs=KdS*(1+s*Td)/(a2*s^2+a1*s+a0);  % plant TF
5 -    y=4;                                          % input step
6 -    step(y*Gs,30); grid                           % step response
```

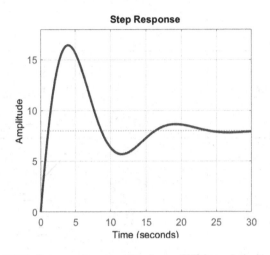

Fig. 2.15 MATLAB®-script of the oscillating D-T2 plant and step response x(t) after input step of y = 4

- proportional-integral or PI-controllers that are built as a sum of the P- and I-terms and consecutively have features both of P- and I-controllers, namely they are quite fast and at the same time they control correct, without retained error, i.e., $e(\infty) = 0$.
- proportional-derivative or PD-controllers that are built as a sum of the P- and D-terms. Because of D-terms the PD-controllers are very fast. But because of P-terms they are not able to correct control leaving the retained error $e(\infty)$. Like D-plants of Abschn. 2.2.3 there are ideal PD-controllers, which are practically not realizable, and PD-T1 or real PD-controllers. The necessary condition for fast control is that $T_v > T_R$. Otherwise it is no more PD-controller but PP-T1 plant, given in Fig. 2.14.

The universal PID-controller, which was patented 1922 by *Nicolas Minorsky*, contents all three terms and has their features, it is and leaves no retained error. Like PD-controllers there are ideal and real PID-controller (see Fig. 2.17).

Significant for PID-controllers are two forms of transfer functions:

- multiplicative form, which is used by calculations.
- additive form, which is used by real or simulated standard PID-controllers.

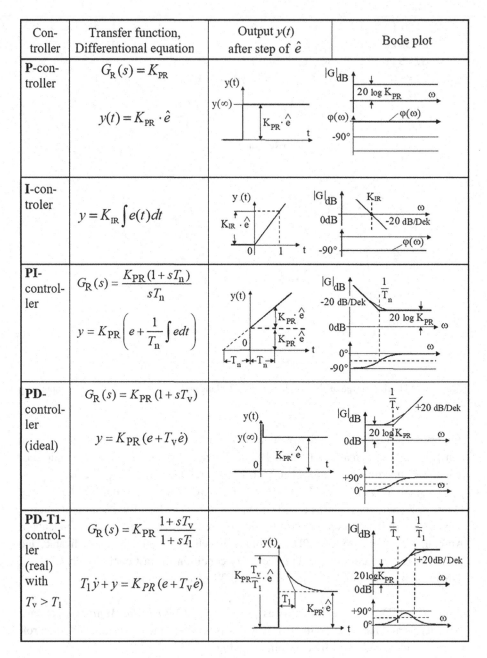

Con-troller	Transfer function, Differential equation	Output $y(t)$ after step of $\hat{e}$	Bode plot
P-con-troller	$G_R(s) = K_{PR}$ $y(t) = K_{PR} \cdot \hat{e}$		
I-con-troler	$y = K_{IR}\int e(t)dt$		
PI-control-ler	$G_R(s) = \dfrac{K_{PR}(1+sT_n)}{sT_n}$ $y = K_{PR}\left(e + \dfrac{1}{T_n}\int edt\right)$		
PD-control-ler (ideal)	$G_R(s) = K_{PR}(1+sT_v)$ $y = K_{PR}(e+T_v\dot{e})$		
PD-T1-control-ler (real) with $T_v > T_1$	$G_R(s) = K_{PR}\dfrac{1+sT_v}{1+sT_1}$ $T_1\dot{y} + y = K_{PR}(e+T_v\dot{e})$		

Fig. 2.16 Proportional, integral and derivative controllers

In Fig. 2.17 are given expressions to convert the calculated multiplicative form into the additive form. Without conversion a discrepancy between expected calculated and realized or simulated step responses takes place, as it is illustrated in the example of Fig. 2.18. It is clearly seen that step responses of the real PID-T1 controller in multiplicative form (blue) and in the additive form (red) with the same parameters differ one from another. After conversion of the parameters into additive form (green points) no difference is seen.

Controller	Transfer function, Differential equation	Output $y(t)$ after step of $\hat{e}$	Bode plot
PID-controller, ideal	multiplicative Form $$G_R = \frac{K_{PR}(1+sT_n)(1+sT_v)}{sT_n}$$ additive form $$G_R = K_{PR}^*\left(1+\frac{1}{sT_n^*}+sT_v^*\right)$$	multiplicative Form 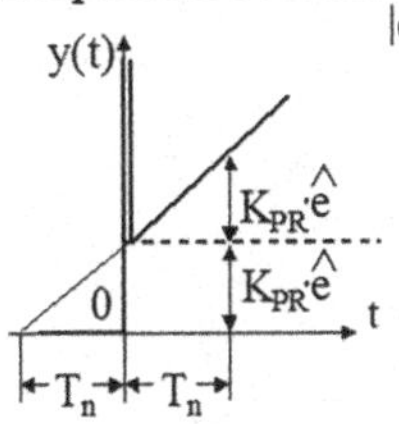	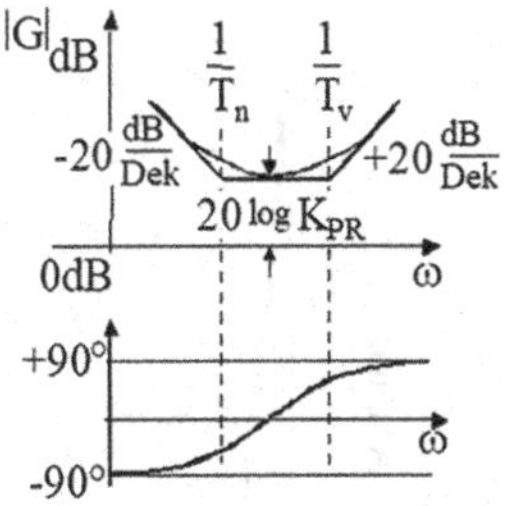
PID-controller, ideal	conversion multiplicative form into additive form $$K_{PR}^* = K_{PR}\left(1+\frac{T_v}{T_n}\right)$$ $$T_n^* = T_n + T_v$$ $$T_v^* = \frac{T_n T_v}{T_n + T_v}$$	multiplicative form $$G_R = \frac{K_{PR}(1+sT_n)(1+sT_v)}{sT_n}$$ into additive form $$G_R = K_{PR}^*\left(1+\frac{1}{sT_n^*}+sT_v^*\right)$$	
PID-T1-controller (real)	multiplicative form $$G_R = \frac{K_{PR}(1+sT_n)(1+sT_v)}{sT_n(1+sT_R)}$$ additive form $$K_{PR}^*\left(1+\frac{1}{sT_n^*}+\frac{sT_v^*}{1+sT_R}\right)$$	multiplicated form 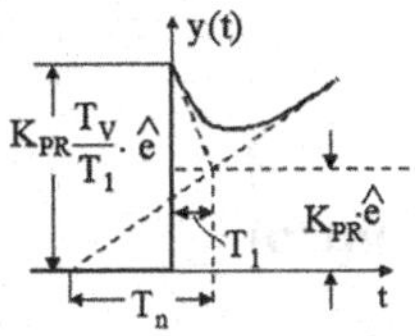	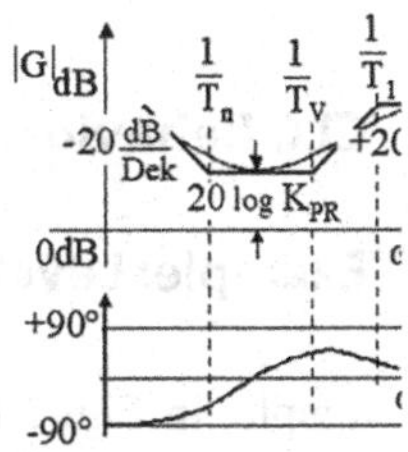

Fig. 2.17 PID (proportional-integral-derivative) controllers. *Source* [19, p. 9]

```
 1 -    s=tf('s');                    % Laplace operator
 2 -    KpR=0.1;Tn=30;Tv=5;TR=0.01*Tv; % parameters of PID-controller
 3 -    Rm=KpR*(1+s*Tn)*(1+s*Tv)/(s*Tn*(1+s*TR)); % multiplicative form
 4 -    Ra=KpR*(1+(1/s*Tn)+(s*Tv/(1+s*TR)));      % additive form
 5 -    KpR_c=KpR*(1+(Tv/Tn));         % conversion KpR into KpR_c
 6 -    Tn_c=Tn+Tv;                    % conversion Tn into Tn_c
 7 -    Tv_c=Tn*Tv/Tn_c;               % conversion Tv into Tv_c
 8 -    P=KpR_c;                       % P-Term
 9 -    I=KpR_c/(s*Tn_c);              % I-Term
10 -    D=s*KpR_c*Tv_c/(1+s*TR);       % D-Term
11 -    Rc=P+I+D;                      % multipl.form converted in additive form
12 -    step(Rm,Ra,Rc,1);             % step responses
13 -    grid; legend; title('PID: Rm-multiplicative, Ra-additive, Rc-converted')
```

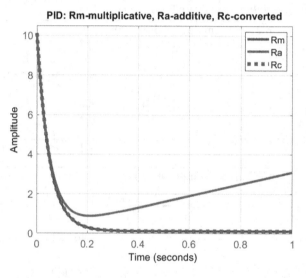

Fig. 2.18 Step responses of the real PID-T1 controller in multiplicative form (blue), in the additive form (red) with the same parameters and in the additive form (green points) with converted parameters

2.4   CLC Behavior

2.4.1   Example: Level Control

As an example, in Fig. 2.19 are shown two options of the level control. The tank as a plant has one input (valve V1) for the flow Q_{in}. The valve V2 for the flow Q_{out} is not measurable and not controllable disturbance. The tank on the left in Fig. 2.19 is controlled with the lever, that is used as actuator and as P-controller. The same tank with the same sensor and actuator (float) is shown in Fig. 2.19 on the right, but instead of lever here are an amplifier as controller and the motor M as actuator implemented. The functional block diagram of both options of level control is given below in Fig. 2.19.

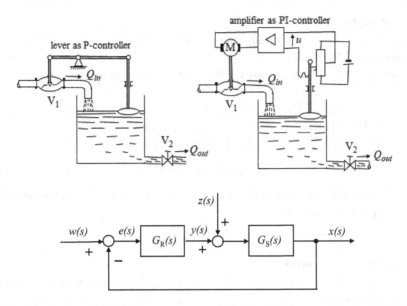

Fig. 2.19 Level control with the P-controller (on the left), with PI-controller (on the right) and functional block diagram of level control. *Source* [16, p. 99]

Supposing that the TF of the plant including sensor and actuator is known

$$G_S(s) = \frac{K_{PS}}{s(1 + sT_1)}, \tag{2.55}$$

let as discuss the reference behavior and the disturbance behavior of both control options.

2.4.2 Stabilty

The necessary and sufficient stability condition
The necessary and sufficient condition for stability of a CLC is that all poles of its TF, i.e., the roots of its characteristic equation have negative real parts (see Fig. 2.7). If the characteristic equation has higher order as 2, the stability can be proved upon so called *stability criterions* without defining poles. There are known many *stability criterions* in all three domains: in the time domain, Laplace domain and frequency domain. The stability criteria of the frequency domain are discusses in Chap. 6. The simplest and often used in the time domain and Laplace domain algebraic stability criterion of Hurwitz is presented below.

Hurwitz stability criterion

Without going into the exact formulation of the Hurwitz stability criterion let us define it for systems of 1st, 2nd, and 3rd order.

The systems of the 1st and 2nd order with the characteristic equations

$$a_1s + a_0 = 0 \qquad a_2s^2 + a_1s + a_0 = 0 \tag{2.56}$$

are stable, if all coefficients are positive, i.e., $a_2 > 0$, $a_1 > 0$ and $a_0 > 0$.

The system of the 3rd order with the characteristic equation

$$a_3s^3 + a_2s^2 + a_1s + a_0 = 0 \tag{2.57}$$

is stable, if all coefficients are positive, i.e., $a_3 > 0$, $a_2 > 0$, $a_1 > 0$, $a_0 > 0$ and besides of this the following condition is met:

$$a_2a_1 > a_3a_0 \tag{2.58}$$

Level control with P-controller

The TF $G_0(s)$ of the open loop and the TF $G_w(s)$ of the closed loop with P-controller by reference behavior are:

$$G_0(s) = G_R(s)G_S(s) = \frac{K_{PR}K_{PS}}{s(1 + sT_1)}$$
$$G_w(s) = \frac{G_0(s)}{1 + G_0(s)} = \frac{K_{PR}K_{PS}}{s(1 + sT_1) + K_{PR}K_{PS}} \tag{2.59}$$

The poles of TF $G_w(s)$ of the closed loop result from the characteristic equation

$$s(1 + sT_1) + K_{PR}K_{PS} = 0$$
$$s^2T_1 + s + K_{PR}K_{PS} = 0 \tag{2.60}$$

The CLC is stable, if the real part of roots $s_{1,2}$ of the last algebraic quadratic equation are negative:

$$s_{1,2} = \frac{-1 \pm \sqrt{1 - 4K_{PR}K_{PS}T_1}}{2T_1}$$

Example

Supposing there are given $K_{PR} = 2$; $K_{PS} = 0,8$ and $T_1 = 5$, then the CLC with poles $s_1 = -0,1 + j\,0,5568$ and $s_2 = -0,1 - j\,0,5568$ is stable. ◄

Level control with PI-controller

Repeating Eqs. 2.59 and 2.60 for the PI-controller it is easy to prove by what values of K_{PR} the CLC is stable:

$$G_0(s) = G_R Z G_S(s) = \frac{K_{PR}(1 + sT_n)}{sT_n} \cdot \frac{K_{PS}}{s(1 + sT_1)}$$

$$G_w(s) = \frac{G_0(s)}{1 + G_0(s)} = \frac{K_{PR}K_{PS}(1 + sT_n)}{s^2 T_n(1 + sT_1) + K_{PR}K_{PS}(1 + sT_n)}$$

$$s^3 T_n T_1 + s^2 T_n + s K_{PR}K_{PS}T_n + K_{PR}K_{PS} = 0$$

Considering that all coefficient of the characteristic equation by $K_{PR} > 0$ are positive and according to Hurwitz stability criterion, the CLC is stable if the condition Eq. 2.58 is fulfilled:

$$T_n K_{PR} K_{PS} T_n > T_n T_1 K_{PR} K_{PS} \quad \rightarrow \quad T_n > T_1$$

An example is given by Fig. 2.20. The critical case with the damping $\vartheta = 0$ of the stability boarder (oscillations with constant magnitude) is by $T_n = T_1$. By $T_n > T_1$ the CLC is stable, by $T_n < T_1$ is the CLC unstable.

2.4.3 Retained Error

Limit conditions

In this section we will define the retained error $e(\infty)$ corresponding to limit conditions given in Abschn. 2.1.5 by reference and disturbance behavior:

```
1 -    s=tf('s');                        % Laplace operator
2 -    Kps=0.8;T1=5;                     % parameters of plant
3 -    Gs=Kps/(s*(1+s*T1));             % plant
4 -    KpR=2;                            % KpR>0
5 -    Tn=T1;                            % critical case of Hurwitz stability condition
6 -    GR=KpR*(1+s*Tn)/(s*Tn);          % PI-controller
7 -    G0=GR*Gs;                         % open loop
8 -    Gw=G0/(1+G0);                     % closed loop
9 -    step(Gw,50);                      % step responses by input step w=1
10 -   grid;
11 -   title('Crirtical case Tn=T1')
```

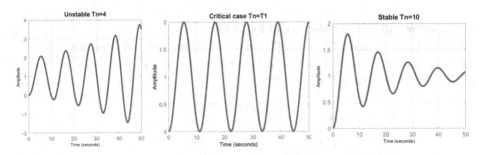

Fig. 2.20 Example of the use of the Hurwitz stability condition for the level control with PI-controller

- Reference behavior: reference value w is a step, and disturbance z is zero ($z = 0$).

$$x(\infty) = \lim_{s \to 0} G_w(s) \cdot \hat{w}$$
$$e(\infty) = \hat{w} - x(\infty) \tag{2.61}$$

- Disturbance behavior: reference value w is zero ($w = 0$), and disturbance z is a step.

$$x(\infty) = \lim_{s \to 0} G_z(s) \cdot \hat{z}$$
$$e(\infty) = \hat{w} - x(\infty) = 0 - x(\infty) = -x(\infty) \tag{2.62}$$

Level control with P-controller (left Fig. 2.19)

The TF $G_0(s)$ of the open loop and the TF $G_z(s)$ of the closed loop with P-controller by disturbance behavior are:

$$G_0(s) = G_R(s)G_S(s) = \frac{K_{PR}K_{PS}}{s(1 + sT_1)}$$
$$G_z(s) = \frac{G_S(s)}{1 + G_0(s)} = \frac{K_{PS}}{s(1 + sT_1) + K_{PR}K_{PS}} \tag{2.63}$$

From the limits condition considering Eq. 2.63 and taking in account that by disturbance behavior $w = 0$ follows:

$$x(\infty) = \lim_{s \to 0} G_z(s) \cdot \hat{z} = \frac{1}{K_{PR}} \cdot \hat{z} \tag{2.64}$$

$$e(\infty) = \hat{w} - x(\infty) = 0 - \frac{1}{K_{PR}}\hat{z} = -\frac{1}{K_{PR}}\hat{z}$$

From the last expression results that the retained error by disturbance behavior is $e(\infty) \neq 0$, i.e., the control with the P-controller leaves the retained error. To reduce the retained error, we should choose the maximal possible K_{PR}. However according to Eq. 2.64 the retained error could not be fully eliminated with a P-controller.

Level control with PI-controller (right Fig. 2.19)

Repeating Eqs. 2.63 and 2.64 for the PI-controller and supposing that the closed loop is stable it is easy to prove that the retained error $e(\infty)$ by steady state in case of PI-controller is fully eliminated i.e., $e(\infty) = 0$:

$$G_0(s) = G_R(s)G_S(s) = \frac{K_{PR}(1 + sT_n)}{sT_n} \cdot \frac{K_{PS}}{s(1 + sT_1)}$$
$$G_z(s) = \frac{G_S(s)}{1 + G_0(s)} = \frac{sK_{PS}T_n}{s^2T_n(1 + sT_1) + K_{PR}K_{PS}(1 + sT_n)}$$
$$x(\infty) = \lim_{s \to 0} G_z(s) \cdot \hat{z} = 0$$
$$e(\infty) = \hat{w} - x(\infty) = 0 - 0 = 0$$

2.5 Simulation of CLC with MATLAB®

2.5.1 Scripts

The transfer function of the CLC to be simulated will be entered as shown below in one of the three possible forms:

- polynomial form:
  - with Laplace-operator s,
  - with numerator/dominator form

$$G(s) = \frac{s^m + b_{m-1}s^{m-1} + \ldots + b_2 s^2 + b_1 s + b_0}{s^n + a_{n-1}s^{n-1} + \ldots + a_2 s^2 + a_1 s + a_0}$$

- pole/zero form

$$G(s) = K_0 \frac{(s - s_{N1})(s - s_{N2}) \ldots (s - s_{Nm})}{(s - s_{P1})(s - s_{P2}) \ldots (s - s_{Pn})}$$

- linear factor form

$$G(s) = K \frac{(1 + sT_{N1})(1 + sT_{N2}) \ldots (1 + sT_{Nm})}{(1 + sT_{P1})(1 + sT_{P2}) \ldots (1 + sT_{Pn})}$$

In the following examples is explained how to enter the transfer functions to each form.

Polynomial form with Laplace-operator s
The transfer function $G_0(s)$ of the open loop consisting of P-Tt plant (proportional plant with the time delay T_1 of the 1st order und the dead time T_t) controlling with the PID-T1 controller is given:

$$G_0(s) = G_R(s)G_S(s) = \frac{K_{PR}(1 + sT_n)(1 + sT_v)}{sT_n(1 + sT_R)} \cdot \frac{K_{PS}}{1 + sT_1} \cdot e^{-sT_t}$$

The MATLAB® script and the step response of the closed loop with transfer function $G_w(s)$ are shown in Fig. 2.21.

Polynomial form entering with numerator/denominator coefficients
The transfer function $G_0(s)$ of the open loop consisting of the PI-controller and P-T2 plant with parameters $Kps = 0{,}8$, $K_{PR} = 1{,}6$ and $T_n = 5$ is given below:

$$G_0(s) = G_R(s)G_S(s) = \frac{K_{PR}(1 + sT_n)}{sT_n} \cdot \frac{K_{PS}}{10s^2 + 7s + 1}$$

$$G_w(s) = \frac{G_0(s)}{1 + G_0(s)} = \frac{K_{PR}(1 + sT_n)K_{PS}}{sT_n(10s^2 + 7s + 1) + K_{PR}(1 + sT_n)K_{PS}}$$

$$G_w(s) = \frac{sK_{PR}K_{PS}T_n + K_{PR}K_{PS}}{10T_n s^3 + 7T_n s^2 + sT_n(1 + K_{PR}K_{PS}) + K_{PR}K_{PS}}$$

The script and the step response of the closed loop with transfer function $G_w(s)$ are shown in Fig. 2.22.

Pole/zero form

The transfer function $G_w(s)$ of the closed loop

$$G_w(s) = \frac{K_{PR}K_{PS}(s - s_{N1})}{(s - s_{p1})(s - s_{p2})(s - s_{p3})}$$

is given with its poles and zeros:

$$s_{N1} = -0{,}2$$

$$s_{p1} = -0{,}25 \quad s_{p2} = -0{,}25 \quad s_{p3} = -0{,}2$$

The gains of the plant and controller are also given as $Kps=0{,}8$, $K_{PR}=1{,}6$. The script and the step response of the closed loop with transfer function $G_w(s)$ are shown in Fig. 2.23.

```
1 -   Kps=2/33.5;T1=2.222;Tt=1.55;        % plant parameters
2 -   KpR=7.258; Tn=T1;Tv=0.2*Tt;         % controller parameters
3 -   s=tf('s');                          % Laplce operator
4 -   GR=KpR*(1+s*Tn)*(1+s*Tv)/(s*Tn);    % PID-T1 controller
5 -   Gs=Kps/(1+s*T1);                    % P-T1 part of the plant
6 -   Gt=exp(-s*Tt);                      % Tt (dead time) part of the plant
7 -   G0=GR*Gs*Gt;                        % TF of the open loop
8 -   Gw=G0/(1+G0);                       % TF of the clsoed loop
9 -   w=exp(-0.1*s);                      % input step by t=0.1 sec
10 -  step(Gw,w)                          % step response and step w=1
11 -  grid
```

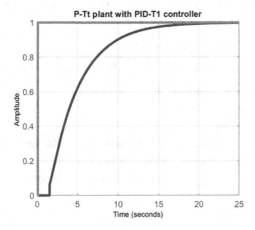

Fig. 2.21 Script of the CLC in polynomial form, entering with Laplace operator s

```
 1 –    Kps=0.8; KpR=1.6;Tn=5;                         % parameter
 2 –    num = [ 0        0   KpR*Kps*Tn    KpR*Kps]; % numerator of G0
 3 –    den = [ 10*Tn  7*Tn    1*Tn         0     ]; % denominator of G0
 4 –    G0=tf(num,den);                               % transfer function G0
 5 –    numGw=num;                                    % numerator of Gw
 6 –    denGw=num+den;                                % denominator of Gw
 7 –    Gw=tf(numGw,denGw)                            % transfer function Gw
 8 –    w=2;                                          % input step
 9 –    step (w*numGw, denGw)                         % step response
10 –    hold on
11 –    num_w=[w];                                    % numerator of input step
12 –    den_w=[1];                                    % denomitator of input step
13 –    step(num_w,den_w)                             % input step
14 –    grid
15 –    title('Polynomial form entering with (num, den)')
```

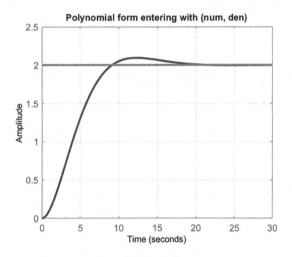

Fig. 2.22 Script of CLC with polynomial form, entering with (num, den)

Linear factor form

Instead of poles and zeros of previous example the CLC consisting of P-T2 plant and PI-controller is given in form of linear factors with time delays $T_1 = 1$; $T_2 = 5$; and $T_3 = 10$. The gain Kps of the plant is $Kps = 0{,}8$. The PI-controller is adjusted with reset time $T_n = T_3$ and gain $K_{PR} = 1{,}25$. The transfer function $G_w(s)$ of the closed loop is given below:

$$G_w(s) = \frac{K_{PR}K_{PS}(1 + sT_n)}{(1 + sT_1)(1 + sT_2)(1 + sT_3)}$$

The implementation of the corresponding script is given in Fig. 2.24.

```
 1 -    clearvars                           % cleare workspace
 2 -    z1=-0.2;                            % given zeros
 3 -    p1=-0.25; p2=-0.25;p3=-0.2;         % given poles
 4 -    z=[z1];                             % vector zeros
 5 -    p=[p1  p2  p3];                     % vector poles
 6 -    TN1=-1/z1;                          % time delay of zeros
 7 -    Tp1=-1/p1; Tp2=-1/p2; Tp3=-1/p3;    % time delay of poles
 8 -    k=TN1/(Tp1*Tp2*Tp3);                % gain
 9 -    [numGw, denGw] =zp2tf(z,p,k);       % conversion
10 -    Gw=tf(numGw,denGw)                  % transfer function Gw
11 -    w=3;                                % input step w=3
12 -    step (w*numGw, denGw)               % step response
13 -    hold on
14 -    num_w=[w];                          % numerator of input step
15 -    den_w=[1];                          % denomitator of input step
16 -    step(num_w,den_w)                   % input step
17 -    grid
```

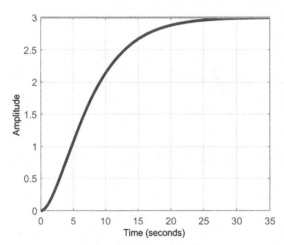

Fig. 2.23 Script and step response on a CLC by entering given poles and zeros

2.5.2 Simulink-Model

The level control of a tank is simulated and visualized in Figs. 2.25 and 2.26 similar to Fig. 2.19.

The plant has I-behavior with the integrator constant $Ki = 0,05$. The tank is fulfilled, if its input is positive, or it is emptied by the negative input. The actuator for the inflow is the pump, that is controlled with the two-point controller or relay. The actuator for output is the valve, that is controlled with the real PID-controller. The time delay T_R of the controller is $T_R = 0,4444$. To achieve this value the filter coefficient N in the configuration window of the PID controller, shown in Fig. 2.27, is adjusted by $N = 2,25$. It

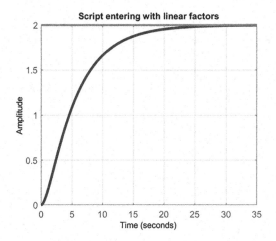

```
1 -    clearvars                          % cleare workspace
2 -    Kps=0.8; T1=1;T2=5;T3=10;          % plant parameters
3 -    KpR=1.25; Tn=T3;                   % controller parameters
4 -    z=[-1/Tn];                         % vector zeros
5 -    p=[-1/T1  -1/T2  -1/T3];           % vector poles
6 -    k=KpR*Kps*Tn/(T1*T2*T3);           % gain
7 -    [numGw, denGw] =zp2tf(z,p,k);      % conversion
8 -    Gw=tf(numGw,denGw)                 % transfer function Gw
9 -    w=2;                               % input step w=2
10 -   step (w*numGw, denGw)              % step response
11 -   hold on
12 -   num_w=[w];                         % numerator of input step
13 -   den_w=[1];                         % denomitator of input step
14 -   step(num_w,den_w)                  % input step
15 -   grid
16 -   title('Script entering with linear factors')
```

Fig. 2.24 Script with transfer function that was entered with linear factors

follows from the transfer function of PID controller in the parallel form, given in configuration window:

$$G_R(s) = P + I \cdot \frac{1}{s} + D \cdot \frac{N}{1 + N \cdot \frac{1}{s}}$$

This transfer function corresponds with the transfer function of the real PID controller in additive form, given in Abschn. 2.3:

$$G_R(s) = K_{PR} + \frac{K_{PR}}{T_n} \cdot \frac{1}{s} + K_{PR} \cdot \frac{sT_v}{1 + sT_R}$$

Compering both expressions it is to define the correlation between N and T_R, namely:

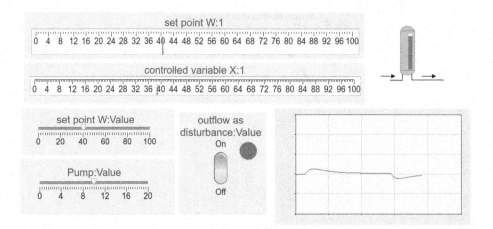

Fig. 2.25 Visualization of level control

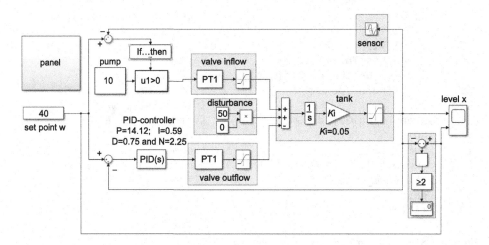

Fig. 2.26 Simulation of the level control

$$N = \frac{1}{T_R}$$

Significant is that the inflow and outflow valves have the same P-T1 dynamic behavior, but the outflow valve is configured with the negative gain, while the PID controller is configured with positive gain. In the reality a valve has only positive gain, and all three parameters P, I and D of the PID controller should be inversed. The actuating values of all valves and of the tank are limited with saturation blocks. The disturbance is simulated by the outflow, it could be manually switched on or off through the visualized panel.

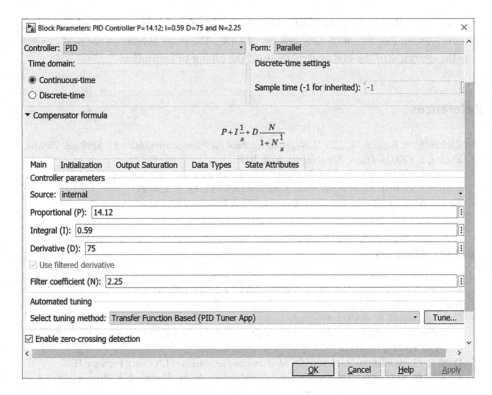

Fig. 2.27 Configuration of the PID-T1 controller

Summary

Finally let as summarize the subjects of this chapter.

- It is shown that the dynamic behavior of the closed loop control (CLC) is mathematically described in one of the domains: time domain, Laplace-domain and frequency domain.
- According to these domains the plants and controllers are classified by their P-, I- or D- behavior with differential equations, transfer functions and bode plots.
- The rules to define the transfer functions of series, parallel and feedback connections are given as far as the rules to define the transfer functions of CLC.
- The two kinds of behavior of CLC are discussed: the reference behavior, when no disturbance acts and only the set point is changed, and the disturbance behavior, when in opposite, the set point is constant. As performance criterions of the CLC behavior are considered the stability and the retained error.

This and the next chapter are a kind of introduction to main topics of this book, namely, to management of date streams of CLC. The main subject of the next chapter is the overview of the classical methods of the tuning of controllers.

References

16. Zacher, S., & Reuter, M. (2022). *Regelungstechnik für Ingenieure* (16th ed.). Springer Vieweg.
17. Zacher, S. (2003). *Duale Regelungstechnik*. VDE.
18. Zacher, S. (2017). *Übungsbuch Reglungstechnik* (6th ed.). Springer Vieweg.
19. Zacher, S. (2016). *Regelungstechnik Aufgaben* (4th ed.). Dr. S. Zacher.
20. Zacher, S. (Ed.). (2000). *Automatisierungstechnik kompakt*. Vieweg.

Further Reading

1. Brogan, W. L. (1991). *Modern control theory* (6th ed.). Prentice Hall.
2. Burl, J. B. (2000). *Linear optimal control*. Addison-Wesley.
3. Coughanowr, D. R. (1991). *Process systems analysis and control* (2nd ed.). McGrow-Hill.
4. DiStefano, J. J., Stubberud, A., & Williams, I. (1995). *Feedback and control systems* (2nd ed.). McCgaw Hill Professional.
5. Dorf, R. C., & Bishop, R. H. (2010). *Modern control systems* (12th ed.). Prentice Hall.
6. Francis, B. A. (1987). *A course in H∞ control theory*. In M. Thoma & A. Wyner (Eds.), *Lecture notes in control and information sciences*. Springer.
7. Franklin, G. F., Powell, J. D., & Emami-Naeini, A. (2010). *Feedback control of dynamic systems* (6th ed.). Pearson.
8. Gopal, M. (2002). *Control systems: Principles and design*. McGraw Hill Education.
9. Ogata, K. (2010). *Modern control engineering*. Prentice Hill.
10. Levine, W. S. (1999). *Control systems fundamentals*. CRC Press.
11. Lewis, P. H., & Yang, C. (1997). *Basic control engineering*. Prentice Hill.
12. Nise, N. S. (2008). *Control systems engineering* (5th ed.). Wiley.
13. Phillips, C .I., & Harbor, R. D. (2000). *Feedback control systems* (4th ed.). Prentice Hall.
14. Smith, C., & Corripio, A. (1985). *Principles and practice of automatic process control*. Wiley.
15. Westphal, L. C. (1995). *Sourcebook of control systems engineering*. Chapman & Hall. Inc.

Controller Tuning

3

3.1 Controller Tuning in Time Domain

3.1.1 Performance Criteria of Control

The step response and quality features of a closed loop control (CLC) are given in Fig. 3.1. There are:

- Retained error $e(\infty)$: it is the difference between the desired reference value w and the controlling variable $x(\infty)$ by the steady state $t \to \infty$ to the end of control.
- Damping ϑ is the feature of the CLC, which evaluates the oscillation of the controlling variable $x(t)$. Approximately the damping is reciprocal to the number N of hallways of the $x(t)$, the exact correlation between N and ϑ is given by Fig. 3.1. The following values of damping are recommended by the control:
  - $\vartheta = 1$ for control without oscillations,
  - $\vartheta = 0{,}707$ for control with one half wave and with overshooting of 4,3%, which is called "optimum magnitude",
  - $\vartheta = 0{,}5$ for control with two halfwaves.
- Maximum overshoot O_{max} is the greatest deviation of the controlled variable $x(t)$ from the setpoint w. The overshoot is expressed in percentage related to the steady state value $x(\infty)$.
- Settling time T_{aus} is the duration of the controlling process between initial state $x(0)$ to a final state $x(\infty)$. The controlling process is considered as finished when the

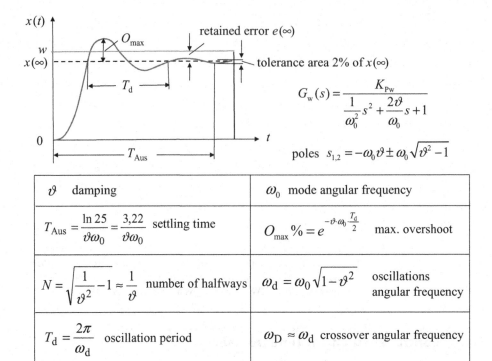

Equations shown with the figure:

$$G_w(s) = \cfrac{K_{Pw}}{\cfrac{1}{\omega_0^2}s^2 + \cfrac{2\vartheta}{\omega_0}s + 1}$$

poles $s_{1,2} = -\omega_0\vartheta \pm \omega_0\sqrt{\vartheta^2 - 1}$

ϑ damping	ω_0 mode angular frequency
$T_{Aus} = \dfrac{\ln 25}{\vartheta\omega_0} = \dfrac{3{,}22}{\vartheta\omega_0}$ settling time	$O_{max}\% = e^{-\vartheta\cdot\omega_0\cdot\frac{T_d}{2}}$ max. overshoot
$N = \sqrt{\dfrac{1}{\vartheta^2} - 1} \approx \dfrac{1}{\vartheta}$ number of halfways	$\omega_d = \omega_0\sqrt{1-\vartheta^2}$ oscillations angular frequency
$T_d = \dfrac{2\pi}{\omega_d}$ oscillation period	$\omega_D \approx \omega_d$ crossover angular frequency

Damping, step response, poles: real part $\alpha = -\omega_0\vartheta$; imaginary part $\beta = \omega_0\sqrt{\vartheta^2 - 1}$

$\alpha < 0$	$\alpha < 0$	$\alpha < 0$	$\alpha = 0$	$\alpha > 0$
$s_{1,2} = \alpha$	$s_{1,2} = \alpha$	$s_{1,2} = \alpha \pm j\beta$	$s_{1,2} = \pm j\beta$	$s_{1,2} = \alpha \pm j\beta$
$\vartheta > 1$	$\vartheta = 1$	$0 < \vartheta < 1$	$\vartheta = 0$	$\vartheta < 0$

Fig. 3.1 Step response and the quality features of a CLC

controlled variable enters the area of tolerance range $\pm 2\%$ of the final state $x(\infty)$ and remains in this area permanently.

An optimally tuned CLC should realize control with the smallest possible error $e(t)$ on the one hand and the greatest possible damping ϑ on the other. These demands contradict each other.

The optimal controller setting is achieved through a compromise solution, which, in turn, depends on the properties of the controlled systems. Thus, the success of the controller design essentially depends on the knowledge of the plant. The experimental determination of the mathematical description of the plant is referred to as identification. The more accurately the plant model is identified, the more accurate is the setting of controller parameters. Usually, the identification is done upon test signals (step, impulse, ramp, sinus) that are applied to the input of the plant. More about identification in Chap. 6.

3.1.2   Settability of PTn Plants

In cases where the mathematical description of the plant is not known or only approximately known, the empirical setting rules have proven to be successful. The advantage of empirical methods is that no mathematical efforts for plant identification are necessary. Nevertheless, the plant parameters like gain and delay time constants, which are dealt with in this section, are needed for empirical methods.

PTn plant as series connection of one PT1 plant with the dead time
Many industrial plants can be approximately represented as PT1, PT2 or commonly as PTn blocks.

> "PTn are in series connected PT1 (1st order) elements. Such plants are very common and can be found particularly in process technology, but also generally in mechanical engineering and electrical engineering. The series connection of such elements leads to step responses which are delayed. In particular, the dead times feared in control engineering can be approximated with linear models in this way" (Citation, source [6, p. 50])

From a step response, as shown in Fig. 3.2, the delay time T_u and recovery time T_g as well as proportionality coefficient K_{PS} of a plant can be determined by a rough approximation using the tangent drawn through the turning point t_w.

Fig. 3.2 Step response $x(t)$ of the P-Tn-plant after the input step y_0

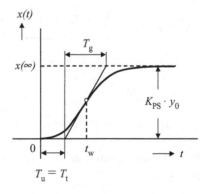

These three parameters are enough to approximate the plant as P-T1 block with dead time

$$G_S(s) = \frac{K_{PS}}{1 + sT_g} \cdot e^{-sT_u} \tag{3.1}$$

and to adjust the controller parameters upon empirical methods described in the following in this section.

The ratio T_g/T_u is called settability of the plant. The greater the settability, the greater the gain K_{PR} of the controller can be selected, as it follows from the *Ziegler-Nichols* condition:

$$K_{PRkrit} \approx \frac{1}{K_{PS}} \left(\frac{\pi}{2} \cdot \frac{T_g}{T_u} + 1 \right)$$

and from the condition of *Samal*

$$K_{PR} \approx \frac{1}{2K_{PS}} \cdot \left(\frac{\pi}{2} \cdot \frac{T_g}{T_u} \right)$$

The empirical values of settability and the corresponding step responses are summarized in Fig. 3.3.

"If the delay time T_u of the system is very short, the controller immediately recognizes a jump in the disturbance variable z and accordingly quickly eliminates the disturbance. It means good controllability. And vice versa, the greater the delay time T_u, the longer it takes for the disturbance to be transmitted to the controller input. In this case, the controller will react with the greater delay and will have to reduce a much larger control difference $e(t)$, which speaks for poor controllability." (Citation, Translation from source [16, p. 227])

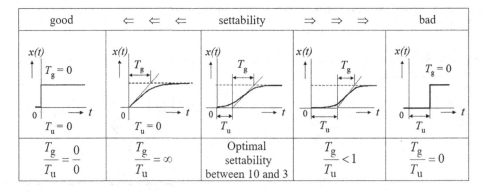

Fig. 3.3 Step responses $x(t)$ and settability T_g/T_u of P-plants with and without time delay. *Source* [2, p. 222]

PTn plant as series connection of the n number of PT1 plants

A PTn plant with step response shown in Fig. 3.2 can be identified as the number n of PT1 terms connected in series with only one for all terms identical time constants T_1.

$$G_S(s) = \frac{K_{PS}}{(1 + sT_1)^n} \tag{3.2}$$

One such methods was proposed by *Schwarze* (see [1, pp. 229, 230]) and is called the method of the time-percentage values. There are many other methods known to identify the PTn plant by means of settability times T_u and T_g. An example of MATLAB® script is given below. One can compare these methods by running the script and comparing the resulting step responses.

```
1.  clearvars; clc
2.  s=tf('s');                                    % Laplace operator
3.  Kps=1;                                         % gain of plant
4.  Tu=6.4; Tg=29.5;                               % settability times
5.  T1=Tg/3.69;                                    % upon [6] for Tg/Tu=4.61; n=3; T1=Tu/0.80;
6.  Gs=Kps/(1+s*T1)^3;                             % upon Eq. 3.2
7.  Gs1=Kps*exp(-s*Tu)/(1+s*Tg);                   % upon Eq. 3.1
8.  Gs2=Kps/((1+s*Tu)*(1+s*Tg));                   % Taylor row 1st order
9.  Gs3=Kps/((1+s*(Tu/3)^3)*(1+s*Tg));             % Taylor row 3rd order
10. Gs4=Kps*(1-s*0.5*Tu)/((1+s*0.5*Tu)*(1+s*Tg));  % Padé approximation 2nd order
11. step(Gs1,Gs,Gs2,Gs3,Gs4);                      % step responses of plant
12. title('Settablity'); grid
```

3.1.3 Empirical Tuning Methods

In the following is shown how controllers can be tunned directly in the time domain without defining the transfer function of the plant, based only upon parameters defined directly from step responses:

- the delay time T_u,
- recovery time T_g,
- gain K_{PS}.

There are empirical methods introduced by the following authors:

- Ziegler/Nichols (1942)
- Chien, Hrones, and Reswick (1952)
- Kuhn (1995)
- Strejc (1959)
- Latzel (1962)

Ziegler/Nichols method of controller tuning for damping $\vartheta = 0,2$ till $\vartheta = 0,3$ (Table 3.1)

Table 3.1 Ziegler/Nichols method

Parameter	P controller	PI controller	PID controller
$\frac{K_{PR}K_{PS}T_u}{T_g}$	1	0,9	1,2
$\frac{T_n}{T_u}$	–	3,3	2,0
$\frac{T_v}{T_u}$	–	–	0,5

Rules of Chien, Hrones, and Reswick for the controller in additive form (Table 3.2)
See Table 3.2.

Method of *Kuhn* (Table 3.3)
The time T_Σ (T-sum) is determined from the step response as a turning point that forms two equal areas F_1 and F_2, as shown in Fig. 3.4. Depending on the value of T_Σ, the controller parameter is to be calculated by Table 3.3.

Rule of *Strejc* for PI controller
The turning point t_w and the tangent drawn through the turning point are also the subject of the method of *Strejc* [1, p. 229]. From Fig. 3.5 we can define the delay time T_u, the recovery time T_g and the proportional coefficient K_{PS} of a plant. It is supposed that $T_2 = T_u$ and that $T_1 = T_2$.

The gain of PI controller K_{PR} and the reset time T_u are calculated as functions of the gain of plant K_{PS}, of the time T_u and of the factor k

$$K_{PR} = \frac{1}{K_{PS}} \cdot \frac{k^2 + 1}{2k}$$

$$T_n = \frac{(k^2 + 1)(k + 1)}{k^2 + k + 1} \cdot T_u$$

Table 3.2 Method of *Chien, Hrones, and Reswick*

Controller	Parameter	Without overshoot		With 20% overshoot	
		Reference behavior	Disturbance behavior	Reference behavior	Disturbance behavior
P	$K_{PR}K_{PS}\frac{T_u}{T_g}$	0,3	0,3	0,7	0,7
PI	$K_{PR}K_{PS}\frac{T_u}{T_g}$	0,35	0,6	0,6	0,7
	T_n	$1,2 \cdot T_g$	$4 \cdot T_u$	$1,0 \cdot T_g$	$2,3 \cdot T_u$
PID	$K_{PR}K_{PS}\frac{T_u}{T_g}$	0,6	0,95	0,95	1,2
	T_n	$1,0 \cdot T_g$	$2,4 \cdot T_u$	$1,35 \cdot T_g$	$2,0 \cdot T_u$
	$\frac{T_v}{T_u}$	0,5	0,42	0,47	0,42

Table 3.3 Tuning rules of *Kuhn* (T-sum method)

Parameter	P controller	PD controller	PI controller	PID controller
$K_{PR}\,K_{PS}$	1	1	0,5	2
T_n	–	–	$0,5\,T_\Sigma$	$0,8\,T_\Sigma$
T_v	–	$0,33\,T_\Sigma$	–	$0,194\,T_\Sigma$

Fig. 3.4 Step response with
the two equal areas F_1 and F_2

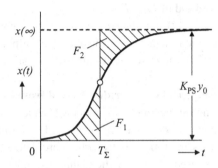

Fig. 3.5 Approximation of the
step response with the turning
point tangent supposing
T1 > T2

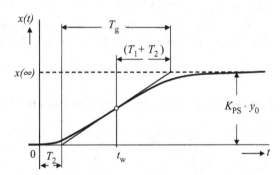

The parameter k was introduced by *Strejc* (see e.g., [1, p. 229]). It is the ratio of time
constants

$$k = \frac{T_1}{T_u}$$

while T_1 is the solution of the nonlinear equation

$$T_1 = T_g - \frac{T_1 T_u}{T_1 - T_u} \ln \frac{T_1}{T_u}$$

If we approximate T_1 as

$$T_1 = 2T_u$$

then it leads (as it is shown in [16, p. 229]) to

$$T_1 = T_g - 1{,}386 T_u$$

and following to

$$k = \frac{T_g}{T_u} - 1{,}386.$$

Method of *Latzel*

Latzel [7] introduced the tuning rules which are given by Table 3.4 based on the identification method of *Schwarze* for PTn plants as the number n of PT1 terms connected in series with only one for all terms identical time constants T_1 like given by Eq. 3.2.

Empirical tuning rules derived from exactly identified transfer function of the plant
The tuning rules based upon a wide range of exactly known transfer functions of the plant are described in [1, 4, 5]. These rules are applied in Table 3.5 for PTn plants with the same time delay T, which are defined according to Eq. 3.2 of *Schwarze*. In this case the need to identify the plant is eliminated, and the tuning will be done only upon settability times defined by Sect. 3.1.2.

The reset time T_n and the rate time T_v of PI and PID controller are supposed to be chosen according to compensating rules, i.e., $T_n = T_1$ and $T_v = T_1$. The desired settling

Table 3.4 Tuning rules of *Latzel* for control with 10% overshoot for different order n of PT-plants

Parameter	$n=3$		$n=5$		$n=10$	
	PI controller	PID	PI controller	PID	PI controller	PID
$K_{PR} K_{PS}$	0,877	2,543	0,543	1,109	0,328	0,559
$\frac{T_n}{T_1}$	1,96	2,47	2,59	3,31	3,73	4,80
$\frac{T_v}{T_1}$	–	0,66	–	0,99	–	1,57

Table 3.5 Tuning rules of *Zacher* [5] for control without overshooting or with $\vartheta = 0{,}5$ till $\vartheta = 0{,}7$

Controller type	P	PI	PID
$n=1$			
K_{PR}	$\frac{1}{K_{PS}}\left(\frac{5T_1}{T_{aus}} - 1\right)$ with $\vartheta = 1$	$\frac{5T_1}{K_{PS}T_{aus}}$ with $\vartheta = 1$	*Not recommended*
$n=2$			
K_{PR}	$\frac{1}{K_{PS}}$ with $\vartheta = 0{,}7$	$\frac{1}{4K_{PS}}$ with $\vartheta = 0{,}7$	$\frac{5T_1}{K_{PS}T_{aus}}$ with $\vartheta = 1$
$n=3$			
K_{PR}	$\frac{1}{K_{PS}}$ with $\vartheta = 0{,}5$	$\frac{1}{4K_{PS}T_1}$ with $\vartheta = 0{,}7$	$\frac{1}{4K_{PS}}$ with $\vartheta = 0{,}7$

time of the control is referred to T_{aus}. It supposed the condition $T_{aus} < T_1$. The use of P controller is not recommended due to big retained error $e(\infty)$. To reduce the retained error given in Table 3.5 the K_{PR}-values which lead to overshoot or low damping, namely the control occur by $n = 2$ with one halfwave ($\vartheta = 0{,}7$) and by $n = 3$ with two halfwaves ($\vartheta = 0{,}5$). The use of the PID controller for plants with $n = 1$ causes D-terms by control, so the controller should be carefully adjusted.

3.1.4   Tuning of Controller Minimizing Integral Criterion

Instead of many quality features of control, mentioned in Sect. 3.1.1, the control theory proposed only one criterion. This criterion is called the "integral criterion" because it estimates the control based on the integral of the error $e(t)$

$$Q = \int_0^\infty e(t)dt \qquad (3.3)$$

The optimal control is supposed to be achieved when the integral Eq. 3.3 achieved its minimum, i.e., if the area delimited by the step response, as shown in Fig. 3.6, is minimal.

Since the resulting area of Fig. 3.6 would have an infinitely large value for CLC with the retained error $e(\infty)$, the difference $[e(t) - e(\infty)]$ should be introduced instead of $e(t)$. The resulting quality criterion is referred to as the linear integral criterion Q_{lin}

$$Q_{lin}(t) = \int_0^\infty [e(t) - e(\infty)]dt \qquad (3.4)$$

Linear integral criterion
The classic method to minimize the linear integral criterion Q_{lin} is shown below as it is described in [3, pp. 234 and 235]. From the Laplace transform of the integral of Eq. 3.4

$$Q_{lin}(s) = L[Q_{lin}(t)] = \int_0^\infty [e(t) - e(\infty)]\,e^{-st}\,dt = \int_0^\infty e(t)e^{-st}\,dt - \int_0^\infty e(\infty)e^{-st}\,dt$$

Fig. 3.6 Integral criterion as the area delimited by the step response. *Source* [1, p. 223]

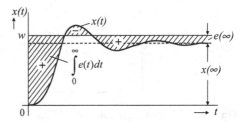

and the limits condition for steady state

$$e(\infty) = \lim_{t \to 0} e(t) = \lim_{s \to 0} s \cdot e(s)$$

follows:

$$Q_{\text{lin}}(s) = e(s) - \lim_{s \to 0} s \cdot e(s) \tag{3.5}$$

Example: Tuning of P-controller minimizing linear integral criterion Q_{lin}

The PT2 plant with the transfer function $G_S(s)$ with given parameters $K_{PS} = 0{,}8$ and $T_1 \neq T_2$ is to be controlled without overshoot, i.e., with the damping $\vartheta \geq 1$, with a PI controller $G_R(s)$, as shown in Fig. 3.7

$$G_S(s) = \frac{K_{PS}}{(1 + sT_1)(1 + sT_2)} \qquad G_R(s) = \frac{K_{PR}(1 + sT_n)}{sT_n}$$

The gain K_{PR} of the PI controller is defined upon minimization of the linear integral criterion Q_{lin} of Eq. 3.5.

To define the integral criterion, first of all, we define the transfer function $G_w(s)$ of the closed loop by disturbance behavior

$$G_z(s) = \frac{x(s)}{z(s)} = \frac{G_S(s)}{1 + G_R(s)G_S(s)} = \frac{K_{PS}}{s^2 T_1 T_2 + s(T_1 + T_2) + K_{PR}K_{PS} + 1} \tag{3.6}$$

After that, we define the error as the difference between setpoint, which is zero by disturbance behavior, and controlling variable $x(s)$

$$e(s) = \hat{w} - x(s) = 0 - G_z(s)\hat{z} = -\frac{K_{PS}}{s^2 T_1 T_2 + s(T_1 + T_2) + K_{PR}K_{PS} + 1} \hat{z}$$

Applying the limits condition for $t \to \infty$ or $s \to 0$, we define the returned error $e(\infty)$

$$e(\infty) = \hat{w} - \lim_{s \to 0} s \cdot e(s) = 0 - \lim_{s \to 0} G_z(s)\hat{z} = 0 - \frac{K_{PS}}{K_{PR}K_{PS} + 1} \hat{z}$$

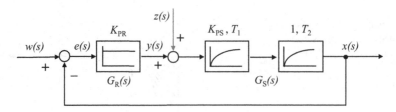

Fig. 3.7 CLC with P controller and PT2 plant

Finally, using all previous expressions and considering Eq. 3.5, we determine the linear integral criterion Q_{lin}

$$Q_{\text{lin}} = -\frac{K_{\text{PS}}\hat{z}}{s^2 T_1 T_2 + s(T_1 + T_2) + K_{\text{PR}} K_{\text{PS}} + 1} + \frac{K_{\text{PS}}\hat{z}}{K_{\text{PR}} K_{\text{PS}} + 1} \tag{3.7}$$

The minimum of Eq. 3.7 is achieved when the derivative of Q_{lin} is equal to zero

$$\frac{\partial Q_{\text{lin}}}{\partial K_{\text{PR}}} = -2K_{\text{PS}}\frac{K_{\text{PS}}(T_1 + T_2)}{(1 + K_{\text{PR}} K_{\text{PS}})^3}z_0 = 0 \tag{3.8}$$

The solution of Eq. 3.7, i.e., the condition of minimizing the linear integral criterion Q_{lin} is very simple, namely, the gain of P controller should be chosen as the possible big value

$$K_{\text{PR}} \rightarrow \infty$$

This recommendation is not new and is well known in the control theory also without linear integral criterion. It only confirms that the controllers without I-terms are not able to bring the error $e(t)$ to zero and left the retained error $e(\infty)$, which can be reduced by increasing the gain of P or PD controller.

The tuning of P-controller for the given PT2 plant should be done referring to the desired damping of the closed loop. Comparing the characteristic polynomial of the given closed loop of Eq. 3.6 and with the common PT2 behavior shown in Fig. 3.1,

$$s^2 T_1 T_2 + s(T_1 + T_2) + K_{\text{PR}} K_{\text{PS}} + 1 = \frac{1}{\omega_0^2}s^2 + \frac{2\vartheta}{\omega_0}s + 1$$

we get the ration between damping ϑ and controller gain K_{PR}

$$\vartheta = \frac{T_1 + T_2}{2\sqrt{T_1 T_2(1 + K_{\text{PR}} K_{\text{PS}})}} \tag{3.9}$$

Supposing the desired damping of $\vartheta = 1$, we define the gain K_{PR} of the controller

$$K_{\text{PR}} = \frac{1}{K_{\text{PS}}} \cdot \frac{(T_1 - T_2)^2}{4T_1 T_2} \tag{3.10}$$

In [3] was calculated the value of the linear integral criterion, setting Eqs. 3.10 in 3.7, as below

$$Q_{\text{lin}} = K_{\text{PS}}\frac{(4T_1 T_2)^2}{(T_1 + T_2)^2}\hat{z}$$

In the step responses, the controller gains K_{PR} and the values of the linear integral criterion Q_{lin} calculated with MATLAB® according to Eq. 3.9 for different values of damping of ϑ are given by Fig. 3.8. It confirms again that if K_{PR} is larger, then the Q_{lin} will be smaller. It is also seen, that minimizing Q_{lin} leads to an increase in oscillations. ◄

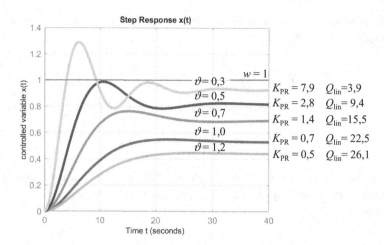

Fig. 3.8 Step responses and linear integral criterion to the example of PT2 plant controlled with the P controller with gain according to Eq. 3.10 for different dampings

3.1.5   ITAE Criterion

The disadvantage of the linear integral criterion Q_{lin} is that the areas occur with different signs, so the positive and the negative areas compensate each other and falsify the real value of the integral. To improve it, the quadratic integral criterion Q_{lsqr} or quadratic integral criterion Q_{labs} are proposed (Fig. 3.9).

These two criteria Q_{lsqr} and Q_{labs} are also not free from disadvantages. The amplitudes by the initial part of the step response hardly affect the integral value and decrease over time. The whole criterion is dependent only on the initial integral area.

By multiplying the integral criterion Q_{labs} with the time variable t, the initial area will be flatted out and small areas will be increased. Such an integral quality criterion is given in Fig. 3.9 and is known as the ITAE criterion (integral of time multiplied by absolute value of error).

ITAE controller test bench
The calculation of the ITAE value is usually carried out with MATLAB® toolboxes or Simulink models. One of such models, developed by the author of this book, is called "ITAE controller test bench" and is given by Fig. 3.10. By the calculation of ITAE, it is supposed that the control achieved without retained error is like the case by PI or PID controller.

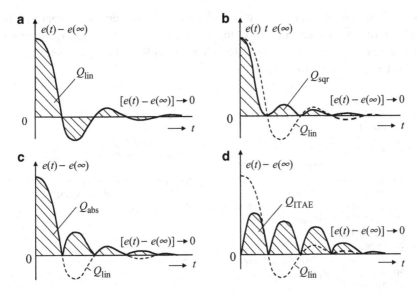

Fig. 3.9 Integral criteria: **a** linear Q_{lin}; **b** quadratic Q_{sqr}; **c** absolute Q_{abs}; **d** integral of time multiplied absolute value of error Q_{ITAE}

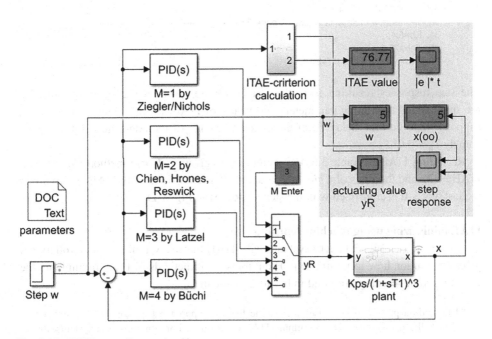

Fig. 3.10 ITAE controller test bench

The test bench consists of 4 closed control loops with the same PT3 plant and different PID controllers. A user can choose the desired controller by entering the index M. But first, the user should initialize the DOC block to implement the parameters. The content of the DOC block is given below.

```
13. clearvars; clc
14. s=tf('s');                              % Laplace operator
15. Kps=1;                                  % given gain of plant
16. Tu=6.4; Tg=29.5;                        % given settability times
17. T1=Tg/3.69;                             % calculated according to [6], page 50
18. Gs=Kps/(1+s*T1)^3;                      % PT3 plant with time delay T1
19. w=5;                                    % w: step of set point, here w=5
20. W=w*s/s;                                % set point as transfer function
21. %-----M=1: Ziegler-Nichols method (⊛ Table 3.1) -----------------------------
22. Kz=1.2*Tg/(Kps*Tu);
23. Tnz=3*Tu;
24. Tvz=0.5*Tu;TRz=0.01*Tvz;
25. % -----M=2: Chien,Hrones, Reswick method (⊛ Table 3.2) --------------------
26. Kch=0.6*Tg/(Kps*Tu);
27. Tnch=2*Tg;
28. Tvch=0.5*Tu;TRch=0.01*Tvch;
29. % -----M=3: Latzel rules (⊛ Table 3.4) ------------------------------------
30. Kla=2.543/Kps;
31. Tnla=2.47*T1;
32. Tvla=0.66*T1; TRla=0.01*Tvla;
33. % -----M=4: method of Büchi [6], page 52 ------------------------------------
34. Kb=5.4;
35. Tnb=75.2;
36. Tvb=5.6; TRb=0.01*Tvb;
```

The step responses resulting from the test bench are given in Fig. 3.11. In Table 3.6 are summarized the tuning parameters and values of ITAE. By all tuning methods, it is supposed that the time delay T_R of the controller is 1% of the controller's rate time T_v, namely $T_R = 0,01T_v$.

Regarding ITAE of Table 3.6, the preference is given to *Latzel*-method (M = 3 with ITAE = 77), and the rules of *Chien, Hrones, and Reswick* (M = 2 with ITAE = 926) are in the last place. It is confirmed with the step responses of Fig. 3.11.

ITAE minimizing using machine learning

The test bench of Fig. 3.10 lets evaluate with ITAE the ready tuning of controllers but gives no solution how to minimize it. From the known methods of ITAE minimizing the use of machine learning proposed in [6] attracts attention.

"The method presented here uses a machine learning approach to automatically find the optimal PID parameters of the minimum ITAE criterion ... For general stable systems, the parameters could even be found directly on the system. However, many systems can be described directly with PTn elements by measuring step responses. For these, the paper provides calculated table values of the minimized ITAE criterion with different control signal

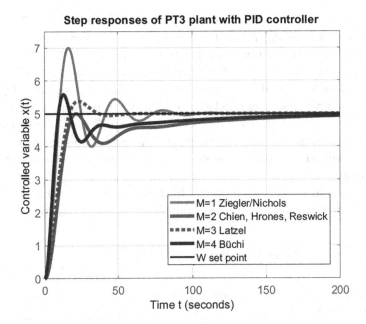

Fig. 3.11 Step responses and ITAE criteria of CLC with PID controllers tuned by Ziegler-Nichols (M = 1), by Chien, Hrones, and Reswick (M = 2), by Latzel (M = 3) and by Büchi (M = 4)

Table 3.6 ITAE values and PID controller tuning of the CLC with the PT3 plant as series connection of the $n = 3$ number PT1 plants with $K_{ps} = 1$ and $T_1 = 8$ s, approximated with $T_u = 6,4$ s and $T_g = 29,5$ s

M	Tuning method	ITAE value	K_{PR}	T_n	T_v
1	Ziegler-Nichols	260	5,53	19,2	3,2
2	Chien, Hrones, and Reswick	926	2,77	59	3,2
3	Latzel [7]	77	2,54	19,8	5,2
4	Büchi [6]	662	5,40	75,2	5,6

limitations. These are verified in practice using the example of a thermal system. The table values are already successfully in use in the control theory course for mechanical engineers at Zurich University of Applied Sciences, School of Engineering" (Citation [6, p. 50])

Significance is that the minimizing of ITAE is done in [6] considering the limitations of actuating value $y(t)$ which is limited between y_{max} and y_{min}. The ITAE as a function of the controller parameters is built and minimized under consideration of limitation $y_{min} < y < y_{min}$ through parameter search. Repeating the optimization of ITAE for different combinations of settability parameters T_u and T_g or converting these parameters to the time delay constant T_1 of Eq. 3.2 the new tuning recommendations are obtained and offered in [6] as a table.

As an example, an excerpt from these tuning rules is given in Table 3.7. This example was already used in discussing the ITAE test bench of Figs. 3.10 and 3.11.

Let us apply the ITAE test bench of Fig. 3.10 for tuning rule of Table 3.7 considering the limitation of the actuating value. For this purpose, the Simulink model of Fig. 3.10 is supplemented with the saturation block as limitation y_{max} and y_{min} as given in Fig. 3.12.

Step responses resulting after simulation of ITAE test bench considering the limitation of the actuating value, shown in Fig. 3.13, of course, differ from the step responses of Fig. 3.11. Significance is that the minimum ITAE achieves by tuning the rules of *Büchi* (M = 4 with ITAE = 122), while the method of *Latzel* (M = 3 with ITAE = 336) loses its priority.

The possibility to consider the actuator's limitation, as it is done in [6] opens a new chapter by controller tuning, which is important for practical applications.

Table 3.7 Tuning rules of *Büchi* for $n = 3$. *Source* [6, p. 52]

			Actuator limitation factor $\pm y_{max}/w$			
n: Number of series connected PT1 plants	Type of the plant	Parameters	± 2	± 3	± 5	± 10
3	PT3	$K_{PR}K_{PS}$	5,4	7	8,2	10
		T_n	$9,4\ T_1$	$10\ T_1$	$9,6\ T_1$	$9,7\ T_1$
		T_v	$0,7\ T_1$	$0,7\ T_1$	$0,7\ T_1$	$0,7\ T_1$

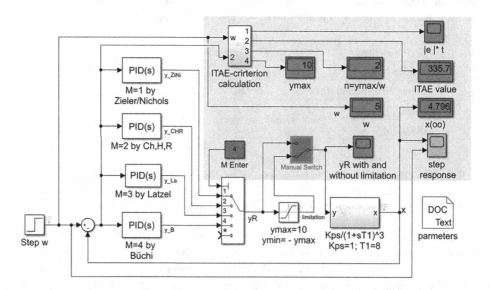

Fig. 3.12 ITAE controller test bench considering actuator's limitation

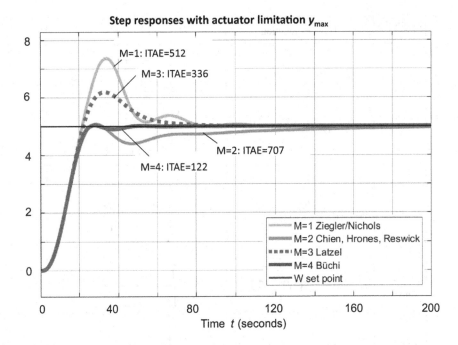

Fig. 3.13 Step responses and ITAE criteria considering actuator's limitation

3.2   Controller Tuning in Laplace Domain

3.2.1   Standard Controller

For controller design in Laplace domain is needed the transfer function $G_S(s)$ of the plant (see [1, 8–10]). The transfer function $G_R(s)$ of the controller will be chosen from one of the known standard controller types (Figs. 3.14 and 3.15). In Fig. 3.14 is seen the difference between ideal and real PD controllers. The last option is also called PD-T1 because it contains the time delay T_R of the controller. The ideal option ignored the time delay T_R of the controller and is practically not realizable while the order m of the numerator ($m = 1$) is bigger than the order n of the denominator ($n = 0$). Nevertheless, the ideal form is used by control theory to simplify the calculations.

The same difference is seen in Fig. 3.15 regarding ideal and real PID controllers. Besides this, there are two forms of transfer functions, the additive and the multiplicative.

The first worldwide PID controller, which was patented in 1922 by *Nicolas Minorsky*, was realized in multiplicative form as an analog amplifier with feedback. The parameters of multiplicative form are denoted in Fig. 3.15 as K^*_{PR}, T^*_n and T^*_v.

Controller type	Transfer function	Step response
P	$G_R(s) = K_{PR}$	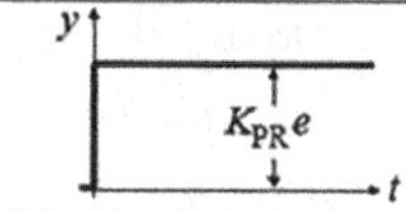
I	$G_R(s) = \dfrac{K_{IR}}{s}$	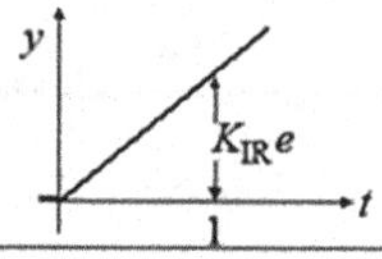
PI	$G_R(s) = \dfrac{K_{PR}(1 + sT_n)}{sT_n}$	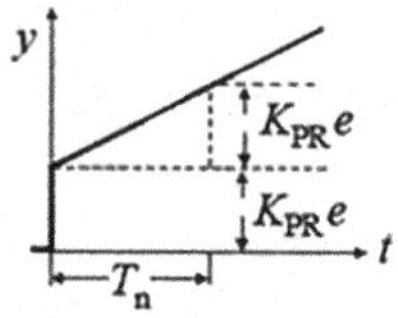
PD ideal	$G_R(s) = K_{PR}(1 + sT_v)$	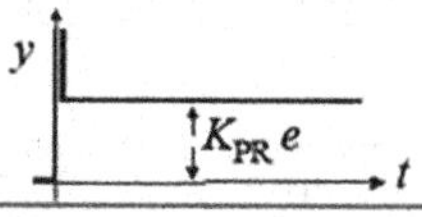
PD real PD-T$_1$	$G_R(s) = K_{PR}\dfrac{1 + sT_v}{1 + sT_1}$	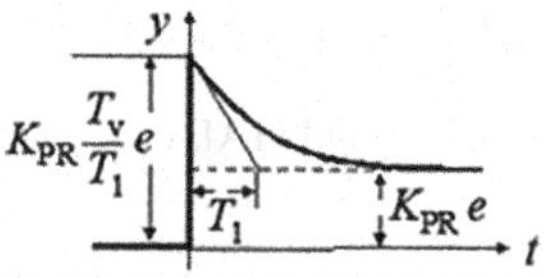

Fig. 3.14 Transfer function $G_R(s)$ and step responses of standard P, I, PI, and PD controllers

Today's PID controller are digital programmable devices that are implemented as the sum of three terms, P, I, D; and denoted in Fig. 3.15 with parameters K_{PR}, T_n, and T_v. The numerator of additive form

$$s^2 + s\frac{1}{T_v} + \frac{1}{T_n T_v}$$

differs from the numerator of multiplicative form and has the following zeros:

$$s_{1,2} = \frac{1}{2T_v} \pm \frac{1}{2T_v}\sqrt{1 - \frac{4T_v}{T_n}} \tag{3.11}$$

From Eq. 3.11 follows the limitation for time constants for the additive form of PID controller to avoid the output oscillations

$$T_n \geq 4T_v$$

Controller type	Transfer function	Step response
PID ideal	additive form $$G_R(s) = K_{PR}\left(1 + \frac{1}{sT_n} + sT_v\right)$$	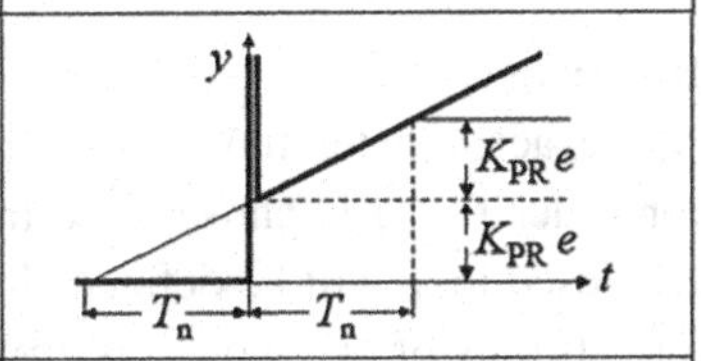
	multiplicative form $$G_R = \frac{K_{PR}^*(1+sT_n^*)(1+sT_v^*)}{sT_n^*}$$	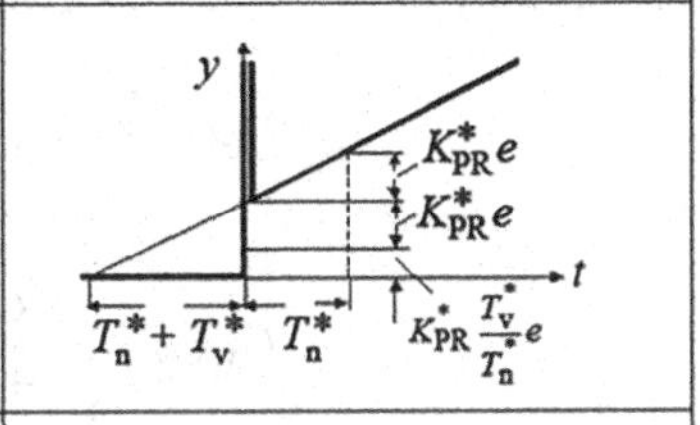
PID real PID-T_1	additive form $$G_R(s) = K_{PR}\frac{1 + \frac{1}{sT_n} + sT_v}{1 + sT_1}$$	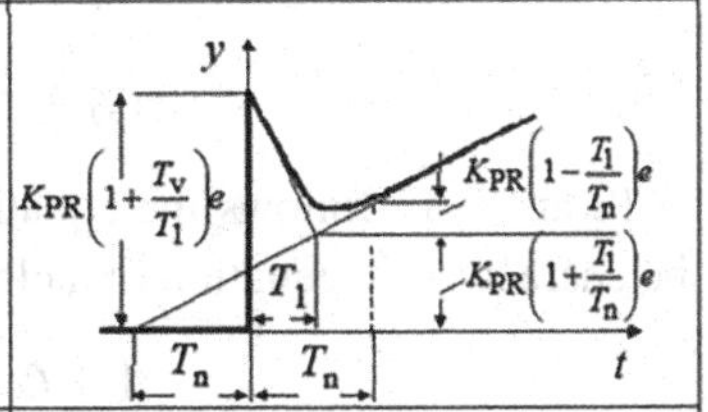
	multiplicative form $$G_R = \frac{K_{PR}^*(1+sT_n^*)(1+sT_v^*)}{sT_n^*(1+sT_1)}$$	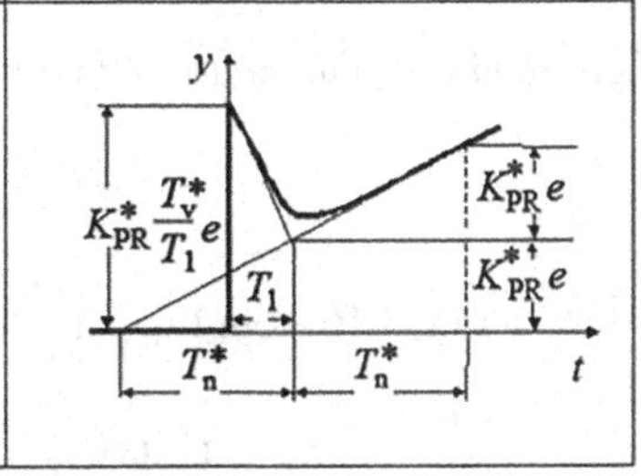

Fig. 3.15 Transfer function $G_R(s)$ and step responses of PID controller

The multiplicative form is suitable for calculations because it enables compensation of terms of numerator and denominator by the transfer functions $G_0(s)$ of the open loop. But as far as the realization of PID controller is done in the additive form, the conversion from the multiplicative form to the additive form is needed as shown below for ideal PID

$$T_n = T_n^* + T_v^*$$

$$T_v = \frac{T_n^* T_v^*}{T_n^* + T_v^*}$$

$$K_{PR} = K_{PR}^*\left(1 + \frac{T_v^*}{T_n^*}\right)$$

3.2.2 Open and Closed Loop

Transfer functions

As far as the exact transfer function of the plant $G_S(s)$ is known and the type of the standard controller $G_R(s)$ is chosen, the transfer functions $G_0(s)$ of the open loop and $G_w(s)$ of the closed loop will be defined. To simplify the calculations, it is supposed that the CLC consists only of two blocks as given in Fig. 3.16.

$$G_0(s) = G_R(s)G_S(s) = \frac{x(s)}{e(s)} \tag{3.12}$$

$$G_w(s) = \frac{G_0(s)}{1 + G_0(s)} = \frac{x(s)}{w(s)} \tag{3.13}$$

$$x(s) = G_w(s)w(s) \tag{3.14}$$

From Eqs. 3.12 and 3.13 follows the equation, that originated by $w(s)=0$ and which is known as the characteristic equation of a closed loop

$$1 + G_0(s) = 0 \tag{3.15}$$

If $G_0(s)$ is given with its numerator $N_0(s)$ and denominator $D_0(s)$

$$G_0(s) = \frac{N_0(s)}{D_0(s)} \tag{3.16}$$

then according to Fig. 3.16 from Eq. 3.12 directly follows

$$G_w(s) = \frac{N_0(s)}{N_0(s) + D_0(s)} \tag{3.17}$$

and from Eq. 3.14 follows the characteristic equation of a closed loop

$$N_0(s) + D_0(s) = 0 \tag{3.18}$$

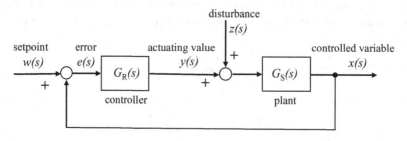

Fig. 3.16 Functional block diagram of a simplified CLC

Compensation rules

According to Eq. 3.12, the transfer function of open loop is defined as the multiplication of the transfer functions of the plant $G_S(s)$ and controller $G_R(s)$. From Figs. 3.14 and 3.15 is seen that all standard controllers in multiplicative form, except of P and I, contain at least one binomial $(1+sT_k)$ in the numerator, i.e., they contain at least one zero $s_k = -1/T_k$. On the contrary, most industrial plants contain binomials in denominators, i.e., they contain poles $s_p = -1/T_p$. It makes sense for the zeros of controllers to choose equal to the poles of the plant. In this case, the order of transfer function will be reduced.

Example: compensation plant time delay constants through controller reset time T_n and/or rate time T_v

The plant is given as

$$G_S(s) = \frac{K_{PS}}{(1+sT_1)(1+sT_2)(1+sT_3)} \tag{3.19}$$

The ideal PID controller in the multiplicative form

$$G_R(s) = \frac{K_{PR}(1+sT_n)(1+sT_v)}{sT_n} \tag{3.20}$$

is tunned with $T_n = T_1$ and $T_v = T_2$.

In this case, poles of the plant are compensated with zeros of PID controller

$$G_0(s) = \frac{K_{PR}(1+sT_n)(1+sT_v)}{sT_n} \cdot \frac{K_{PS}}{(1+sT_1)(1+sT_2)(1+sT_3)} = \frac{K_{PR}K_{PS}}{sT_n(1+sT_3)} \tag{3.21}$$

As a result, the time delay constants of the plant disappear from Eq. 3.21 and lead to faster control. ◄

Considering limitation $T_n > 4T_v$ for the additive form of PID controller, there are common compensation rules:

- T_n compensates the biggies time delay constant of plant,
- T_v compensates the next biggies time delay constant of plant.

As an example, let us suppose in Eq. 3.19 the values $T_1 = 7$ s, $T_2 = 10$ s, and $T_3 = 3$ s. Then the compensation $T_n = T_1$ and $T_v = T_2$ by Eq. 3.21 is false and should be corrected: $T_n = T_2$ and $T_v = T_1$ resulting instead in Eq. 3.21 to

$$G_0(s) = \frac{K_{PR}K_{PS}}{sT_n(1+sT_3)} \tag{3.22}$$

3.2.3   Model Based Compensating Controller

Another option of compensation known as *model-based compensating controller* is possible if the transfer function $G_R(s)$ of the controller is a reverse of the transfer function $G_S(s)$ of the plant

$$G_R(s) = \frac{1}{G_S(s)} \tag{3.23}$$

It follows to open loop

$$G_0(s) = G_R(s)G_S(s) = 1$$

and to closed loop

$$G_w(s) = \frac{x(s)}{\hat{w}} = \frac{G_0(s)}{1 + G_0(s)} = 0{,}5$$

Such compensation seems to be not applicable because of the 50% error

$$e(s) = \hat{w} - x(s) = \hat{w} - G_w(s)\hat{w} = 0{,}5\hat{w}$$

It could be improved by including the transfer function $G_M(s)$ of desired CLC behavior in the transfer function $G_R(s)$ of the controller:

$$G_R(s) = \frac{1}{G_S(s)} \cdot \frac{G_M(s)}{1 - G_M(s)} \tag{3.24}$$

Example: model based compensating controller

If the plant is given with Eq. 3.19 and the desired behavior is given as

$$G_M(s) = \frac{1}{1 + sT_M} \tag{3.25}$$

then the model-based compensating controller is programmed according to Eq. 3.24:

$$G_R(s) = \frac{(1 + sT_1)(1 + sT_2)(1 + sT_3)}{K_{PS}} \cdot \frac{1}{sT_M} \tag{3.26}$$

It is easy to prove that the controller of Eq. 3.26 with the plant of Eq. 3.19 leads according to Eq. 3.13 to the desired transfer function Eq. 3.25 of the closed loop. ◄

The disadvantage of the above-described tuning method is that the transfer function Eq. 3.26 is not realizable. To solve this problem one put small time constants in Eq. 3.26 as shown below,

$$G_R(s) = \frac{(1 + sT_1)(1 + sT_2)(1 + sT_3)}{K_{PS}(1 + 0{,}01s)^2} \cdot \frac{1}{sT_M} \tag{3.27}$$

The controller of Eq. 3.27 is realizable, but the desired behavior Eq. 3.25 cannot be exactly arrived. Nevertheless, the big D-parts with rate times T_1, T_2 and T_3 of Eq. 3.27 cause disturbances by control. Besides this, a small abbreviation of the real plant from its identified model Eq. 3.19 leads to unexpected changes by control.

3.2.4 Tuning of Typical Simple Loops

In this section are summarized some CLCs with standard controllers which often occur by practical applications. The quality features of a CLC are commonly known as:

- retained error $e(\infty)$: it is desired to minimize or fully eliminate $e(\infty)$,
- damping ϑ or D (see ratio between damping and oscillations of step response in Fig. 3.1),
- settling time T_{aus} (see Fig. 3.1).

The aid of this section is to describe how the quality features depend on controller parameters K_{PR}, T_n, and T_v.

Retained error $e(\infty)$
P and PD controller are not able to eliminate the retained error. Although one exception is possible, namely, if the plant contains an I-term. It is discussed later in this section.

In the case of P controller with ṖT1 plant (Fig. 3.17), the retained error is minimized by the maximum of controller gain K_{PR} without causing oscillations, i.e., the damping is $D = 1$. The same is if a PD controller is connected to the PT1 plant.

But in the case of the PD controller with PT2 plant (Fig. 3.18), the increasing of controller gain K_{PR} leads to a decrease in damping D and oscillations. The tuning is the compromise between minimizing retained error and setting acceptable damping.

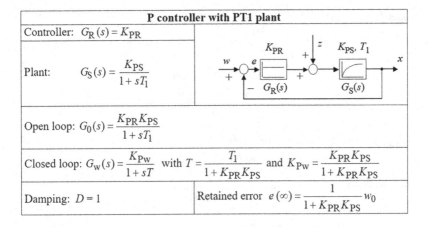

P controller with PT1 plant	
Controller: $G_R(s) = K_{PR}$	
Plant: $\quad G_S(s) = \dfrac{K_{PS}}{1 + sT_1}$	
Open loop: $G_0(s) = \dfrac{K_{PR}K_{PS}}{1 + sT_1}$	
Closed loop: $G_w(s) = \dfrac{K_{Pw}}{1 + sT}$ with $T = \dfrac{T_1}{1 + K_{PR}K_{PS}}$ and $K_{Pw} = \dfrac{K_{PR}K_{PS}}{1 + K_{PR}K_{PS}}$	
Damping: $D = 1$	Retained error $e(\infty) = \dfrac{1}{1 + K_{PR}K_{PS}} w_0$

Fig. 3.17 P controller with PT1 plant

Damping (referred to as ϑ or D)

The standard controllers with I-term such as I, PI, and PID fully eliminate the retained error. The disadvantage of the I controller is that the controlling is very slow. Oscillations with damping D are possible by control of a PT1 plant with the I controller (Fig. 3.19). Instead of it, the P controller causes no oscillation with the same plant (Fig. 3.17). Besides this, the I controller is not able to control the I plant because the loop is unstable by every value of controller gain K_{IR} (Fig. 3.19).

The best solution to improve the control of I plant is to use a PI controller (Fig. 3.20). The retained error disappeared ($e(\infty)=0$), and the desired damping D can be achieved by choosing the appropriate values of controller gain K_{PR} and reset time T_n.

Settling time T_{aus}

In Fig. 3.21 are shown two cases to adjust the reset time T_n if the PI controller is set to control the PT1 plant.

- without compensation (the damping depends on K_{PR} and T_n),
- with compensation (if $T_n = T_1$, then the damping is $D=1$ and is not dependent on K_{PR}).

In the last case, the loop has PT1 behavior with the time delay T_w.

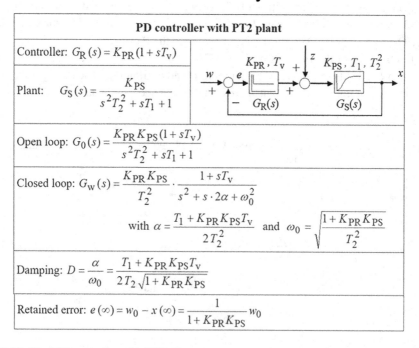

Fig. 3.18 Ideal PD controller with PT2 plant

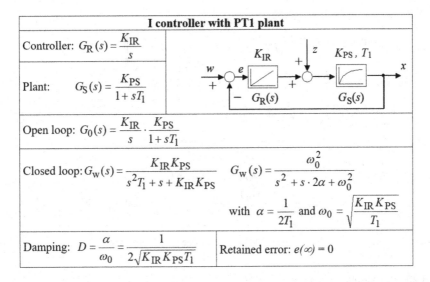

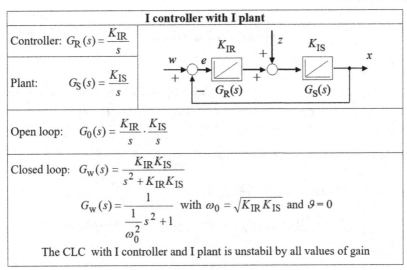

Fig. 3.19 I controller with PT1 plant and with I plant

By properly adjusting T_w, one can adjust the desired settling time T_{aus} of the step response of the closed loop without oscillations. It follows from the known ratio between T_{aus} and T_w by PT1 plants

$$T_{aus} = 3,9T_w$$

This ratio follows from the solution

$$x(t) = K_w\left(1 - e^{-\frac{t}{T_w}}\right)w_0$$

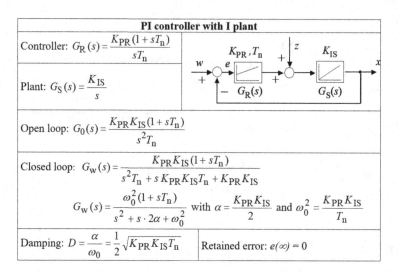

PI controller with I plant

Controller: $G_{\mathrm{R}}(s) = \dfrac{K_{\mathrm{PR}}(1+sT_{\mathrm{n}})}{sT_{\mathrm{n}}}$

Plant: $G_{\mathrm{S}}(s) = \dfrac{K_{\mathrm{IS}}}{s}$

Open loop: $G_0(s) = \dfrac{K_{\mathrm{PR}}K_{\mathrm{IS}}(1+sT_{\mathrm{n}})}{s^2 T_{\mathrm{n}}}$

Closed loop: $G_{\mathrm{w}}(s) = \dfrac{K_{\mathrm{PR}}K_{\mathrm{IS}}(1+sT_{\mathrm{n}})}{s^2 T_{\mathrm{n}} + s\,K_{\mathrm{PR}}K_{\mathrm{IS}}T_{\mathrm{n}} + K_{\mathrm{PR}}K_{\mathrm{IS}}}$

$$G_{\mathrm{w}}(s) = \frac{\omega_0^2(1+sT_{\mathrm{n}})}{s^2 + s\cdot 2\alpha + \omega_0^2} \quad \text{with } \alpha = \frac{K_{\mathrm{PR}}K_{\mathrm{IS}}}{2} \text{ and } \omega_0^2 = \frac{K_{\mathrm{PR}}K_{\mathrm{IS}}}{T_{\mathrm{n}}}$$

Damping: $D = \dfrac{\alpha}{\omega_0} = \dfrac{1}{2}\sqrt{K_{\mathrm{PR}}K_{\mathrm{IS}}T_{\mathrm{n}}}$ | Retained error: $e(\infty) = 0$

Fig. 3.20 PI controller with I plant

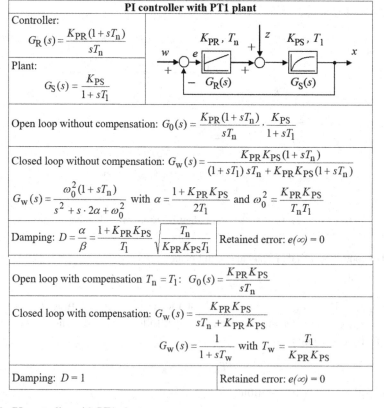

PI controller with PT1 plant

Controller:

$$G_{\mathrm{R}}(s) = \frac{K_{\mathrm{PR}}(1+sT_{\mathrm{n}})}{sT_{\mathrm{n}}}$$

Plant:

$$G_{\mathrm{S}}(s) = \frac{K_{\mathrm{PS}}}{1+sT_1}$$

Open loop without compensation: $G_0(s) = \dfrac{K_{\mathrm{PR}}(1+sT_{\mathrm{n}})}{sT_{\mathrm{n}}} \cdot \dfrac{K_{\mathrm{PS}}}{1+sT_1}$

Closed loop without compensation: $G_{\mathrm{w}}(s) = \dfrac{K_{\mathrm{PR}}K_{\mathrm{PS}}(1+sT_{\mathrm{n}})}{(1+sT_1)\,sT_{\mathrm{n}} + K_{\mathrm{PR}}K_{\mathrm{PS}}(1+sT_{\mathrm{n}})}$

$$G_{\mathrm{w}}(s) = \frac{\omega_0^2(1+sT_{\mathrm{n}})}{s^2 + s\cdot 2\alpha + \omega_0^2} \quad \text{with } \alpha = \frac{1+K_{\mathrm{PR}}K_{\mathrm{PS}}}{2T_1} \text{ and } \omega_0^2 = \frac{K_{\mathrm{PR}}K_{\mathrm{PS}}}{T_{\mathrm{n}}T_1}$$

Damping: $D = \dfrac{\alpha}{\beta} = \dfrac{1+K_{\mathrm{PR}}K_{\mathrm{PS}}}{T_1}\sqrt{\dfrac{T_{\mathrm{n}}}{K_{\mathrm{PR}}K_{\mathrm{PS}}T_1}}$ | Retained error: $e(\infty) = 0$

Open loop with compensation $T_{\mathrm{n}} = T_1$: $G_0(s) = \dfrac{K_{\mathrm{PR}}K_{\mathrm{PS}}}{sT_{\mathrm{n}}}$

Closed loop with compensation: $G_{\mathrm{w}}(s) = \dfrac{K_{\mathrm{PR}}K_{\mathrm{PS}}}{sT_{\mathrm{n}} + K_{\mathrm{PR}}K_{\mathrm{PS}}}$

$$G_{\mathrm{w}}(s) = \frac{1}{1+sT_{\mathrm{w}}} \quad \text{with } T_{\mathrm{w}} = \frac{T_1}{K_{\mathrm{PR}}K_{\mathrm{PS}}}$$

Damping: $D = 1$ | Retained error: $e(\infty) = 0$

Fig. 3.21 PI controller with PT1 plant

of the differential equation of PT1 behavior by the input step w_0

$$T_w \dot{x}(t) + x(t) = K_w w_0.$$

Namely, the 98% of the steady state value

$$x(\infty) = K_w w_0$$

will be arrived by the time $t = T_{aus}$ fulfilling the condition

$$x(T_{aus}) = 0{,}98 K_w w_0 = \left(1 - e^{-\frac{T_{aus}}{T_w}}\right).$$

In Fig. 3.22 is shown how the PT2 plant is controlled with a PID controller. It is seen that CLC with the fully compensated PID controller has the PT1 behavior which is shown in Fig. 3.21 with the compensated PI controller with PT1 plant. In this case, the tuning of a PID controller with the PT2 plant to achieve the desired settling time is the same as given above the tuning of a PI controller with the PT1 plant.

P or PD controller with plants containing I-term
Let us finish this section with one special case of the use of P and PD controllers. As it was written above the P and PD controllers are not able to eliminate the retained error. However, one exception is possible, namely, if the plant contains an I-term. By the setpoint behavior (no disturbance) that means not the controller, but the plant brings the I term in the loop and eliminates the retained error. By the disturbance

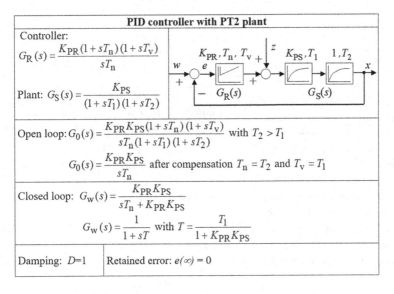

Fig. 3.22 Ideal PID controller with PT2 plant

behavior (no setpoint) are different cases possible depending upon, where is an external input applied (Fig. 3.23):

- before the I-term of the plant (disturbance z_1): the retained error is not eliminated.
- after the I-term of the plant (disturbance z_2), the retained error is fully eliminated.

3.2.5 Optimum Magnitude

In this section is discussed the tuning of standard controllers in the common case of the transfer function $G_0(s)$ of the open loop

- with I-term,
- without I-term.

Transfer function $G_0(s)$ with I-term
The PID controller with the PTn plant leads to the transfer function of the open loop with I-term, e.g.

$$G_0(s) = \frac{K_{PR} K_{PS} K_{IS} (1 + sT_n)(1 + sT_v)}{sT_n(1 + sT_1)(1 + sT_2)(1 + sT_3)(1 + sT_4)(1 + sT_5)} \qquad (3.28)$$

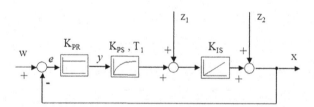

Behavior of the closed loop	Controlled variable by steady state	Retained error by steady state
Setpoint behavior (no disturbance) step $w \neq 0$; $z_1 = z_2 = 0$	$x(\infty) = \lim\limits_{s \to 0} G_w(s) \cdot \hat{w}$	$e(\infty) = 0$
Disturbance z_1 acts bevor I-term: step $w = 0$; $z_1 \neq 0$ and $z_2 = 0$	$x(\infty) = \lim\limits_{s \to 0} G_{z1}(s) \cdot \hat{z}_1$	$e(\infty) = -\dfrac{1}{K_{PS} \cdot K_{PR}}$
Disturbance z_2 acts after I-term: step $w = 0$; $z_1 = 0$ and $z_2 \neq 0$	$x(\infty) = \lim\limits_{s \to 0} G_{z2}(s) \cdot \hat{z}_2$	$e(\infty) = 0$

Fig. 3.23 P or PD controller with the plant containing I-term

Supposing $T_5 > T_4 > T_3 > T_2 > T_1$ the recommended steps of tuning are:

1. *Compensation*: $T_n = T_5$ and $T_v = T_4$
2. *Equivalent time delay constant T_E*. The transfer function Eq. 3.28 will be simplified through the building of equivalent time delay constant $T_E = T_1 + T_2 + T_3$. The simplification is allowed only if the following condition meets: one of the time delay constants is 5 times greater than the sum of the rest time delay constants, e.g., $T_1 > 5(T_2 + T_3)$. If this condition is not fulfilled, simplification is not allowed and some other tuning methods should be used. Supposing the condition meets, and the equivalent time delay constant is defined, then the transfer function of the open loop of Eq. 3.28 is simplified

$$G_0(s) = \frac{K_{PR} K_{PS} K_{IS}}{s T_n (1 + s T_E)} \tag{3.29}$$

As far as $T_n = T_5$ was set above by step 1 (compensation), only the gain K_{PR} is to be defined. In this section, the gain will be defined to achieve the desired damping.

3. *Transfer function $G_w(s)$*: From Eq. 3.29, considering Eqs. 3.18 and 3.19, the transfer function and the characteristic equation of the closed loop are defined

$$G_w(s) = \frac{K_{PR} K_{PS} K_{IS}}{s T_n (1 + s T_E) + K_{PR} K_{PS} K_{IS}} \tag{3.30}$$

$$s^2 T_n T_E + s T_E + K_{PR} K_{PS} K_{IS} = 0 \tag{3.31}$$

4. *Gain of controller*: The ratio between damping ϑ and gain K_{PR} of controller is defined by comparing Eq. 3.31 with the transfer function of PT2 behavior given in Fig. 3.1

$$K_{PR} = \frac{T_n}{4 \vartheta^2 K_{PS} K_{IS} T_E} \tag{3.32}$$

The gain K_{PR} of controller will be calculated setting in Eq. 3.32 the desired value of damping ϑ according to Fig. 3.1, the known gains of plant $K_{PS} K_{IR}$, the equivalent time delay T_E from step 2, and the reset time of controller from step 1.

The recommendation for the choice of the desired damping was mentioned in Sect. 3.1.1:

- $\vartheta = 1$ for control without oscillations,
- $\vartheta = 0,707$ for control with overshooting of 4,3%, which is called "optimum magnitude",
- $\vartheta = 0,5$ for control with two halfwaves,

Transfer function $G_0(s)$ without I-term

As an example, is given below the transfer function of the PTn plant with PD controller, with the same condition between time delay constants of the plant $T_3 > T_2 > T_1$

$$G_0(s) = \frac{K_{PR}K_{PS}(1 + sT_v)}{(1 + sT_1)(1 + sT_2)(1 + sT_3)} \tag{3.33}$$

Compensating $T_v = T_3$ and repeating steps 3 and 4 given above for Eq. 3.28, one gets the transfer function

$$G_w(s) = \frac{K_{PR}K_{PS}}{(1 + sT_1)(1 + sT_2) + K_{PR}K_{PS}} = \frac{K_{PR}K_{PS}}{s^2 T_1 T_2 + s(T_1 + T_2) + K_{PR}K_{PS}} \tag{3.34}$$

and the ratio between damping ϑ and gain K_{PR} of the controller

$$K_{PR} = \frac{(T_1 + T_2)^2}{4\vartheta^2 K_{PS} T_1 T_2} - \frac{1}{K_{PS}} \tag{3.35}$$

Both cases of Eqs. 3.28 and 3.33, referred to as Type A and Type B, are shown with quality features in Fig. 3.24.

3.2.6   Symmetrical Optimum

If the plant with I-term is controlled with the controller that also contend the I-term, like given by Fig. 3.25, then the transfer function of the open loop is represented as follows:

$$G_0(s) = \frac{K_{PR}K_{PS}K_{IS}}{s^2 T_n} \cdot \frac{(1 + sT_n)}{(1 + sT_1)} \tag{3.36}$$

The compensation of $T_n = T_1$ is not allowed because it leads to an unstable loop given in Fig. 3.19. Instead of compensation proposed *Kessler* (1958) to increase the reset time of controller:

$$T_n = kT_1 \tag{3.37}$$

The transfer function of the closed loop given by Eq. 3.36 by $k = 4$ is shown below

$$G_w(s) = \frac{1 + sT_n}{(1 + s \cdot 2T_1)(s^2 \cdot 2^2 T_1^2 + s \cdot 2T_1 + 1)} \tag{3.38}$$

The CLC has three poles:

$$s_1 = -\frac{1}{2T_1} \qquad s_{2,3} = \frac{-1 \pm j\sqrt{3}}{4T_1} \tag{3.39}$$

Type A (with I-term)	Type B (without I-term)
a) Transfer function of an open loop	
$G_0(s) = \dfrac{K_{PR}K_{PS}K_{IS}}{s\,(1+sT_1)}$	$G_0(s) = \dfrac{K_{PR}K_{PS}}{(1+sT_1)(1+sT_2)}$
b) Ratio between controller gain K_{PR} and damping ϑ	
$K_{PR}K_{PS}K_{IS} = \dfrac{1}{4\vartheta^2 T_1}$	$K_{PR}K_{PS} = \dfrac{(T_1+T_2)^2}{4\vartheta^2 T_1 T_2} - 1$
c) Tuning of controller with the desired damping $\vartheta = 0{,}707$	
$K_{PRopt} = \dfrac{1}{2K_{PS}K_{IS}T_1}$	$K_{PRopt} = \dfrac{(T_1+T_2)^2}{2K_{PS}T_1T_2} - \dfrac{1}{K_{PS}}$
d) Performance criteria and step response by reference behavior	

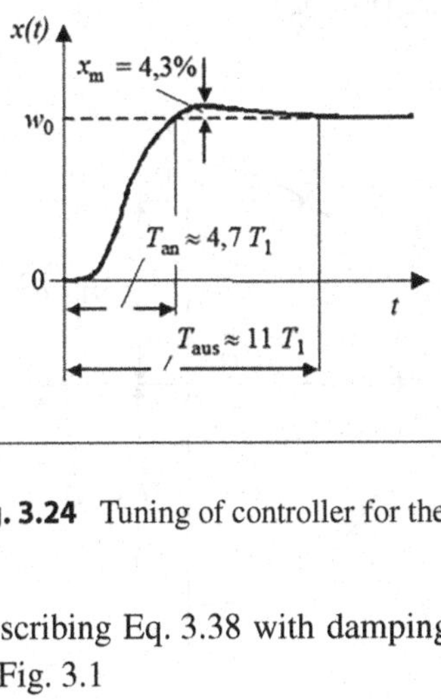

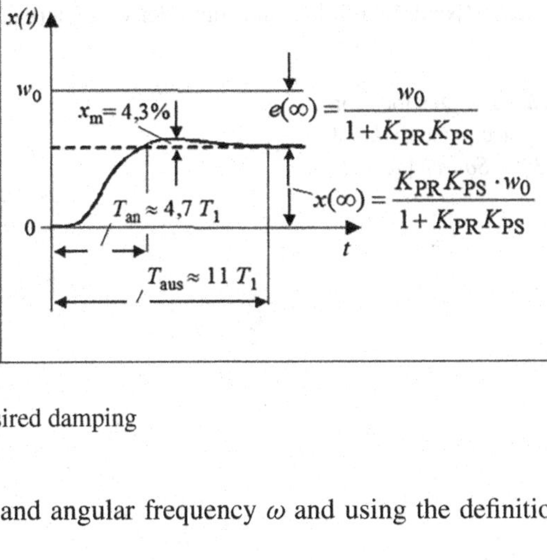

Type A plot: $x(t)$, $x_m = 4{,}3\%$, w_0, $T_{an} \approx 4{,}7\,T_1$, $T_{aus} \approx 11\,T_1$.

Type B plot: $x(t)$, w_0, $x_m = 4{,}3\%$, $T_{an} \approx 4{,}7\,T_1$, $T_{aus} \approx 11\,T_1$, $e(\infty) = \dfrac{w_0}{1+K_{PR}K_{PS}}$, $x(\infty) = \dfrac{K_{PR}K_{PS}\cdot w_0}{1+K_{PR}K_{PS}}$.

Fig. 3.24 Tuning of controller for the desired damping

Describing Eq. 3.38 with damping ϑ and angular frequency ω and using the definitions of Fig. 3.1

$$\alpha = \frac{1}{4T_1} \qquad \beta^2 = \frac{1}{2^2 T_1^2}$$

the poles of Eq. 3.39 are represented as

$$s_1 = -\omega_m \qquad s_{2,3} = -\beta(\vartheta \pm \sqrt{\vartheta^2 - 1})$$

The poles are placed on the circle of the radius ω_m (Fig. 3.26). The symmetrical location of poles in Fig. 3.26 is one of the reasons why the method of *Kessler* is called "symmetrical optimum".

	Transfer functions: plant $G_S(s)$, controller $G_R(s)$	condition	The resulting transfer function of an open loop
a)	$G_S(s) = \dfrac{K_{PS}K_{IS}}{s(1+sT_1)}$ $G_R(s) = \dfrac{K_{PR}(1+sT_n)}{sT_n}$	$\Rightarrow$	$G_0(s) = \dfrac{K_{PR}K_{PS}K_{IS}}{s^2 T_n}\dfrac{(1+sT_n)}{(1+sT_1)}$
b)	$G_S = \dfrac{K_{PS}K_{IS}}{s(1+sT_1)(1+sT_2)}$ $G_R(s) = \dfrac{K_{PR}(1+sT_n)}{sT_n}$	$T_1 \geq 5 \cdot T_2$ $\Rightarrow$ $T_E = T_1 + T_2$	$G_0(s) = \dfrac{K_{PR}K_{PS}K_{IS}}{s^2 T_n}\dfrac{(1+sT_n)}{(1+sT_E)}$
c)	$G_S = \dfrac{K_{PS}K_{IS}}{s(1+sT_1)(1+sT_2)}$ $G_R = \dfrac{K_{PR}(1+sT_n)(1+sT_v)}{sT_n}$	$T_1 > T_2 > T_3$ $\Rightarrow$ $T_v = T_2$	$G_0(s) = \dfrac{K_{PR}K_{PS}K_{IS}}{s^2 T_n}\dfrac{(1+sT_n)}{(1+sT_1)}$

Fig. 3.25 Examples of plant and controller with I-terms. *Source* [1, p. 242]

Fig. 3.26 Poles location in Laplace plane by $k=4$ in Eq. 3.37. *Source* [1, p. 246]

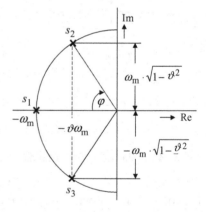

The step response of the CLC of Eq. 3.38 with $k=4$ is shown in Fig. 3.27. By $k=1$ is $\vartheta=0$ and the CLC is unstable. By $k=9$, all three poles are real $s_1=s_2=s_3$ and the damping is $\vartheta=1$, with no oscillations.

3.2.7   Pole Placing

The PT1 plant, given as

$$G_S(s) = \frac{K_{PS}}{1+sT_1},$$

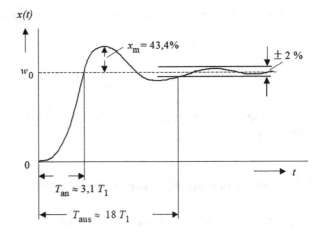

Fig. 3.27 Step response with damping $\vartheta = 0{,}5$ (two half waives) by $k = 4$

is to be controlled with the PI controller

$$G_R(s) = \frac{K_{PR}(1 + sT_n)}{sT_n}. \tag{3.40}$$

The desired poles of the CLC are given:

$$p_{1,2} = -\alpha \pm j\beta \tag{3.41}$$

According to the pole placing method the following steps are needed for tuning of the controller:

1. Transfer function of the closed loop

$$G_w(s) = \frac{G_R(s)G_S(s)}{1 + G_R(s)G_S(s)} = \frac{K_{PR}K_{PS}(1 + sT_n)}{s^2 T_n T_1 + sT_n(1 + K_{PR}K_{PS}) + K_{PR}K_{PS}}$$

2. Characteristic equation of the given closed loop

$$s^2 T_n T_1 + sT_n(1 + K_{PR}K_{PS}) + K_{PR}K_{PS} = 0 \tag{3.42}$$

3. Characteristic equation of the desired closed loop

$$T_n T_1(s - p_1)(s - p_2) = 0$$

$$s^2 + s(-p_1 - p_2) + p_1 p_2 = 0 \tag{3.43}$$

4. Comparison of coefficient of Eqs. 3.42 and 3.43

$$\begin{cases} T_n(1 + K_{PR}K_{PS}) = -p_1 - p_2 \\ K_{PR}K_{PS} = p_1 p_2 \end{cases} \tag{3.44}$$

5. Tunning parameters are the solution of the equations system Eq. 3.44

$$K_{PR} = \frac{p_1 p_2}{K_{PS}}$$

$$T_n = \frac{-p_1 - p_2}{p_1 p_2}$$

As an example of tuning with pole placing according to Eq. 3.44 is shown in the MATLAB® script in Fig. 3.28.

3.2.8 Algebraic Stability Criterion of Hurwitz

A system is stable if its step response monotony decreases with time until it achieves some steady state. Or if the magnitude of its oscillations goes to zero after some time as shown in Fig. 3.1 by the damping of $\vartheta > 0$. If the damping ϑ is null or negative or if the real parts α of the system poles $s_k = \alpha \pm j\beta$ have positive real parts, as shown in Fig. 3.1, then the system is unstable.

To check the stability, one can solve the characteristic equation and define the poles. Or one can use the so-called stability criteria. The most known and simple criterion is the algebraic criterion of *Hurwitz*. However, the practical use of this criterion is limited to the 3rd order of the characteristic equation. By the systems of higher order the conditions of Hurwitz criterion became to be so complicated that it is easier directly to solve the characteristic equation with numerical methods. In the following is given the formulation of the Hurwitz stability criterion without going into detail and only for systems of low order.

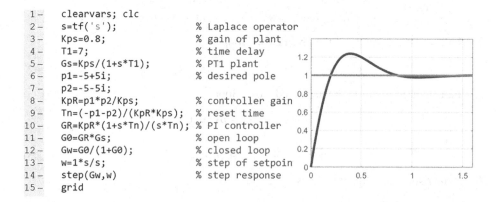

```
1 -    clearvars; clc
2 -    s=tf('s');              % Laplace operator
3 -    Kps=0.8;                % gain of plant
4 -    T1=7;                   % time delay
5 -    Gs=Kps/(1+s*T1);        % PT1 plant
6 -    p1=-5+5i;               % desired pole
7 -    p2=-5-5i;
8 -    KpR=p1*p2/Kps;          % controller gain
9 -    Tn=(-p1-p2)/(KpR*Kps);  % reset time
10 -   GR=KpR*(1+s*Tn)/(s*Tn); % PI controller
11 -   G0=GR*Gs;               % open loop
12 -   Gw=G0/(1+G0);           % closed loop
13 -   w=1*s/s;                % step of setpoin
14 -   step(Gw,w)              % step response
15 -   grid
```

Fig. 3.28 MATLAB® script of pole placing and step response to the example of the PT1 plant with the PI controller

The systems of the 1st and 2nd order with the characteristic equations

$$a_1s + a_0 = 0 \tag{3.45}$$

$$a_2s^2 + a_1s + a_0 = 0 \tag{3.46}$$

are stable, if all coefficients a_k are positive. If at least one coefficient is equal to zero or negative, the system is unstable.

Besides this, there is an additional condition by Hurwitz stability criterion for the system of the 3rd order, namely: the system with the characteristic polynomial

$$p(s) = a_3s^2 + a_2s^2 + a_1s + a_0 \tag{3.47}$$

Is stable, if all coefficients a_k are positive and the following condition meets:

$$a_2 \cdot a_1 > a_3 \cdot a_0 \tag{3.48}$$

Example: Controller tuning with Hurwitz criterion

Given is the CLC consisting of the PT2 plant with $T_4 > T_3 > T_2 > T_1$ and the PID controller

$$G_S(s) = \frac{K_{PS}}{(1 + sT_1)(1 + sT_2)(1 + sT_3)(1 + sT_4)} \qquad G_R(s) = \frac{K_{PR}(1 + sT_n)(1 + sT_v)}{sT_n}$$

The controller is to be tuned in such a way that the closed loop will be stable. It is done below with the following steps:

1. Open loop

$$G_0(s) = G_R(s)G_S(s) = \frac{K_{PR}(1 + sT_n)(1 + sT_v)}{sT_n} \cdot \frac{K_{PS}}{(1 + sT_1)(1 + sT_2)(1 + sT_3)(1 + sT_4)}$$

2. Compensation: $T_n = T_4$ and $T_v = T_3$.

$$G_0(s) = \frac{K_{PR}K_{PS}}{sT_n(1 + sT_1)(1 + sT_2)}$$

3. Characteristic polynomial of closed loop

$$p(s) = sT_n(1 + sT_1)(1 + sT_2) + K_{PR}K_{PS}$$
$$p(s) = s^3T_nT_1T_2 + s^2T_n(T_1 + T_2) + s + K_{PR}K_{PS}$$

4. Hurwitz stability criterion: as far as all values of time delays T_n, T_1, T_2, and gains K_{PR}, K_{PS} are supposed positive, the CLC is stable, if the condition Eq. 3.48 meets

$$T_n(T_1 + T_2) \cdot 1 > T_nT_1T_2 \cdot K_{PR}K_{PS} \tag{3.49}$$

5. The tuning rule for stable CLC results from Eq. 3.49

$$T_n(T_1 + T_2) \cdot 1 > T_n T_1 T_2 \cdot K_{PR} K_{PS}$$

The CLC is stable by controller gain

$$K_{PR} < \frac{T_1 + T_2}{T_1 T_2 K_{PS}}$$

If controller gain is

$$K_{PR} = \frac{T_1 + T_2}{T_1 T_2 K_{PS}},$$

the damping $\vartheta = 0$, the step response has oscillations with a constant magnitude as shown in Fig. 3.1. Such unstable systems are often called *critical*. ◀

3.3   Controller Tuning in Frequency Domain

The frequency response describes how the system transmits a sinusoidal input. The frequency domain has long established itself as an important tool for illustrating the dynamic behavior of the system and has the following advantages compared to time and Laplace domains:

- The graphical representation of the frequency response can be determined experimentally by measuring input and output oscillations for stable controlled systems.
- The check of stability and the controller tuning with the graphical representation are simpler as it is by methods of time and Laplace domains.

3.3.1   Nyquist Stability Criterion

The frequency response of an open loop $G_0(j\omega)$ follows from its transfer function $G_0(s)$ after substitution $s = j\omega$.

$$G_0(s) = \frac{x(s)}{e(s)} \quad \rightarrow \quad s = j\omega \quad \rightarrow \quad G_0(j\omega) = \frac{x(j\omega)}{e(j\omega)}$$

According to the Nyquist stability criterion the frequency response $G_0(j\omega)$ is represented graphically with real and imaginary parts $Re(\omega)$ and $Im(\omega)$ or with magnitude $|G_0(\omega)|$ and phase $\varphi(\omega)$

$$G_0(j\omega) = Re(\omega) + j Im(\omega) = |G_0(\omega)| e^{\varphi(\omega)}$$

This graphical representation of $G_0(j\omega)$ is called the *Nyquist plot*. Comparing $G_0(j\omega)$ of the open loop with the characteristic polynomial of closed loop $1 + G_0(j\omega)$, *Nyquist* proposed the condition for stability as it is shown in Fig. 3.29, namely: the point $(-1, j0)$ of

Fig. 3.29 Nyquist plots G0 for stable, unstable and critical closed loops Gw. *Source* [2, p. 254]

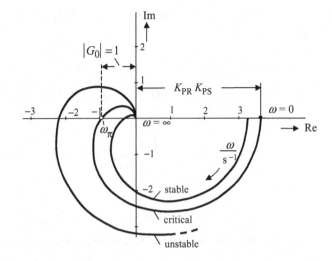

the intersection of the plot $G_0(j\omega)$ by the frequency ω_π with the real axis $Re(\omega)$ is decisive for the stability or instability of a CLC. This point is, therefore, referred to as the *critical point* in connection with the Nyquist criterion. In the following is described the simplified Nyquist criterion applied only to the stable open loops. The definition of the Nyquist criterion for unstable open lops is given in [1–4].

▶ **Definition of simplified Nyquist stability criterion**
A closed loop is stable if the Nyquist plot of the open loop $G_0(j\omega)$ intersecting the real axis by the frequency ω_π is placed on the right side of the critical point $(-1, j0)$, i.e.

$$|G_0(\omega_\pi)| < 1$$

This position of the point of intersection of the $G_0(j\omega)$ is determined graphically by drawing the plot (Fig. 3.29) or can be calculated from the following system of equations:

$$\begin{cases} |G_0| = 1 \\ \varphi = -180° \end{cases} \quad \text{or} \quad \begin{cases} \text{Im } G_0 = 0 \\ \text{Re } G_0 = -1 \end{cases}$$

Example: Controller tuning according to Nyquist stability criterion (Source [2, p. 255])

The PT1 plant with parameters K_{PS}; T_1 and T_t is controlled with a P controller

$$G_S(s) = \frac{K_{PS}}{1 + sT_1} e^{-sT_t} \qquad G_R(s) = K_{PR}$$

By what values of the controller gain K_P is the CLC stable?
   The frequency response of open loop

$$G_0(j\omega) = G_R(j\omega)\, G_S(j\omega) = K_{PR} \frac{K_{PS}}{1 + j\omega T_1} e^{-j\omega T_t}$$

consists of three series connected blocks, which are represented with their amplitude and phase responses:

- P controller

$$|G_1| = K_{PR} \quad \text{and} \quad \varphi_1 = 0$$

- PT1 plant

$$|G_2| = \frac{K_{PS}}{\sqrt{1 + (\omega T_1)^2}} \quad \text{and} \quad \varphi_2 = -\arctan(\omega T_1)$$

- Dead time

$$|G_3| = 1 \quad \text{and} \quad \varphi_3 = -\omega T_t$$

The amplitude of the whole system is the product, and the phase is the sum of each

$$|G_0| = |G_1| \cdot |G_2| \cdot |G_3| = \frac{K_{PR} K_{PS}}{\sqrt{1 + (\omega T_1)^2}} \tag{3.50}$$

$$\varphi_0 = \varphi_1 + \varphi_2 + \varphi_3 = -\arctan(\omega T_1) - \omega T_t$$

The frequency ω_π of the critical point $(-1, j0)$ is defined as the solution of the following equation:

$$\varphi(\omega_\pi) = -\arctan(\omega_\pi T_1) - \omega_\pi T_t = -\pi$$

$$\omega_\pi = 4 \text{ s}^{-1} \tag{3.51}$$

Setting Eqs. 3.51 in 3.51 results in the amplitude of $G_0(j\omega)$ at a critical point

$$|G_0(\omega_\pi)| = \frac{K_{PR} K_{PS}}{\sqrt{1 + (\omega_\pi T_1)^2}} \tag{3.52}$$

The closed loop is stable if meets the condition

$$|G_0| < 1. \tag{3.53}$$

From Eqs. 3.52 and 3.53 follows the tuning of the P controller

$$K_{PR} < \frac{\sqrt{1 + (\omega_\pi T_1)^2}}{K_{PS}} \quad \blacktriangleleft$$

3.3.2   Controller Tuning with Bode-Plot

Stability check and controller tuning with the Bode diagram have even more advantages than with the Nyquist plot, described in the previous section:

- The Bode plot can be determined experimentally by measuring input and output oscillations, but it can be sketched by hand with much less effort than Nyquist plot.
- The transfer function G(s) can easily be obtained from a Bode diagram or read out directly.
- The logarithmic scaling of the abscissa axis allows for easy handling of large frequency ranges, e.g., from $10^{-2}\,\mathrm{s}^{-2}$ to $10^5\,\mathrm{s}^{-2}$.
- Because of the logarithmic representation, one has only to add the amplitude and phase of frequency responses of series connected single elements to obtain the Bode diagram of the entire system.
- The amplitude and phase responses of the inverse system $1/G(j\omega)$ are obtained by reflecting the corresponding diagrams of the original system $G(j\omega)$ on the frequency axis. Other symmetry operations such as shifts up/down, or right/left are easy to use as it is described in [4].

The relation between Nyquist and Bode plots is shown in Fig. 3.30. The Fig. 3.31 illustrates the terms amplitude margin A_R and phase margin φ_R. According to this the Nyquist criterion is formulated not through the frequency ω_π of the intersection of phase response with the $(-180°)$-line, as in Fig. 3.29, but through the frequency ω_D of the intersection of amplitude response with the abscissa axis, i.e., with the 0 dB line.

According to Figs. 3.30 and 3.31, the Nyquist criterion can be formulated regarding to the frequency ω_D as far as regarding to the frequency ω_π:

▶ **Definition of simplified Nyquist stability criterion in Bode plot**
A closed loop is stable if the amplitude response of the open loop $G_0(j\omega)$ intersected the abscissa axis by the frequency ω_D and the phase $\varphi(\omega_D)$ is placed over $(-180°)$-line, i.e.

$$\varphi(\omega_D) > -180°$$

Example: Controller tuning with Bode plot according to Nyquist stability criterion

The PT3 plant with parameters $K_{PS}=3{,}5$ and $T_1=2$ s is controlled with a PI controller

$$G_S(s) = \frac{K_{PS}}{(1+sT_1)^3} \qquad G_R(s) = \frac{K_{PR}(1+sT_n)}{sT_n}$$

The PI controller is to be tuned to achieve the given phase margin $\varphi_R=60°$.
First, we define the frequency response of open loop

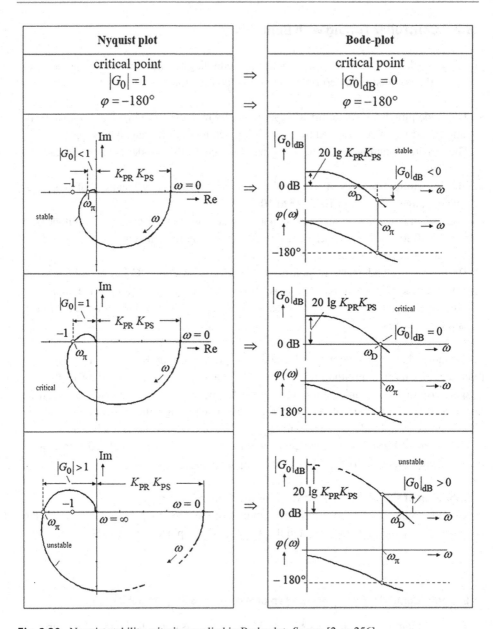

Fig. 3.30 Nyquist stability criterion applied in Bode plot. *Source* [2, p. 256]

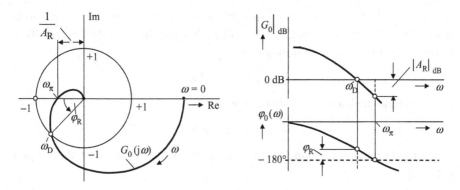

Fig. 3.31 Amplitude and phase margin by Nyquist plot and Bode plot. *Source* [2, p. 257]

$$G_0(s) = G_R(s)G_S(s) = \frac{K_{PR}(1 + sT_n)}{sT_n} \cdot \frac{K_{PS}}{(1 + sT_1)^3} \qquad (3.54)$$

Then after compensation $T_n = T_1 = 2$ s and setting the initial value of gain $K_{PR} = 1$, we define the Bode-plot using MATLAB® script shown in Fig. 3.32. The closed loop by the initial value of $K_{PR} = 1$ is unstable according to the Bode plot of Fig. 3.33 because by ω_D of the intersection of amplitude response with the 0-dB line the phase is $\varphi_{(\omega_D)} < -180°$, i.e., the Nyquist stability condition is not fulfilled. It is confirmed with the simulated step response of Fig. 3.32 for $K_{PR} = 1$.

In order to achieve the given phase reserve, according to the Nyquist stability criterion, the 0-dB line must intersect the amplitude response at the crossover frequency

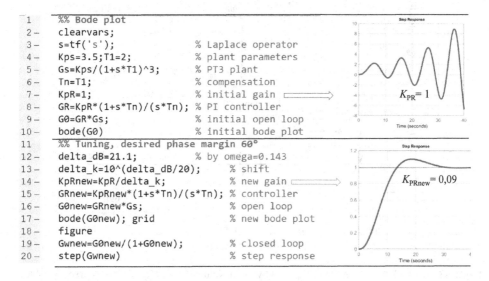

```
1    %% Bode plot
2 -  clearvars;
3 -  s=tf('s');              % Laplace operator
4 -  Kps=3.5;T1=2;           % plant parameters
5 -  Gs=Kps/(1+s*T1)^3;      % PT3 plant
6 -  Tn=T1;                  % compensation
7 -  KpR=1;                  % initial gain
8 -  GR=KpR*(1+s*Tn)/(s*Tn); % PI controller
9 -  G0=GR*Gs;               % initial open loop
10 - bode(G0)                % initial bode plot
11   %% Tuning, desired phase margin 60°
12 - delta_dB=21.1;          % by omega=0.143
13 - delta_k=10^(delta_dB/20);   % shift
14 - KpRnew=KpR/delta_k;     % new gain
15 - GRnew=KpRnew*(1+s*Tn)/(s*Tn); % controller
16 - G0new=GRnew*Gs;         % open loop
17 - bode(G0new); grid       % new bode plot
18 - figure
19 - Gwnew=G0new/(1+G0new);  % closed loop
20 - step(Gwnew)             % step response
```

Fig. 3.32 MATLAB® script to Eq. 3.54

$\omega_D = 0,143$ s^{-1}. To do this, either the amplitude response must be shifted up by $\Delta dB = 21,1$ dB or the 0-dB line must be shifted down by the same $\Delta dB = 21,1$ dB. To do this, we define the new value of gain K_{PRnew}

$$K_{PRnew} = K_{PR} \cdot \Delta K$$

going out from the known logarithmic expression

$$\Delta_{dB} = 20 \lg(\Delta K),$$

which results in the following calculation, given by rows 12–14 of the script in Fig. 3.32:

$$\Delta K = 10^{\frac{\Delta_{dB}}{20}} \tag{3.55}$$

$$K_{PRnew} = K_{PR} \cdot 10^{\frac{\Delta_{dB}}{20}} = 0,09 \tag{3.56}$$

The resulting Bode plot is shown in Fig. 3.33 as red curve. The phase reserve φ_R as the distance between the phase response at the crossover frequency ω_D and the $(-180°)$ line is $\varphi_R = 58°$ and thus corresponds to the desired value. It is confirmed with the step response shown in Fig. 3.32 for $K_{PRnew} = 0,09$.

Note: If the amplitude response is either shifted down or the 0-dB line is shifted up, the controller gain K_{PR} is reduced as follows:

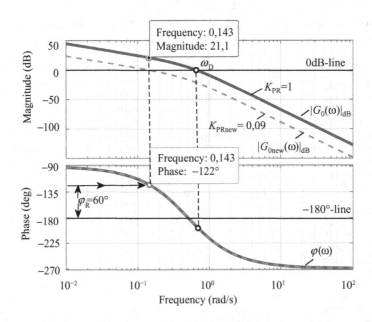

Fig. 3.33 Bode plots to example of Eq. 3.54 by initial value of gain $K_{PR} = 1$ (blue curve, unstable) and by $K_{PRnew} = 0,09$ (red curve, stable, with desired phase margin of 58°)

$$K_{\text{PRneu}} = K_{\text{PRalt}} \cdot \frac{1}{\Delta K} \qquad\qquad (3.57)$$

◀

References

1. Zacher, S., & Reuter, M. (2022). *Regelungstechnik für Ingenieure* (16th ed.). Springer Vieweg.
2. Zacher, S. (2003). *Duale Regelungstechnik*. VDE.
3. Zacher, S., & Reuter, M. (2021). *Regelungstechnik für Ingenieure* (15th ed.). Springer Vieweg.
4. Zacher, S. (2020). *Drei Bode-Plots-Verfahren für Regelungstechnik*. Springer Vieweg.
5. Zacher, S. (2021). *Regelungstechnik mit Data Stream Management*. Springer Vieweg.
6. Büchi, R. (2021) Optimal ITAE criterion PID parameters for PTn plants found with a machine learning approach: Luxembourg, 11th–14th November. In *Proceedings of the 9th IEEE international conference on control, mechatronics and automation (ICCMA2021)* (pp. 50–54).
7. Latzel, W. (1995). *Einführung in die digitalen Regelungen*. VDE.
8. Zacher, S. (2017). *Übungsbuch Reglungstechnik* (6th ed.). Springer Vieweg.
9. Zacher, S. (2016). *Regelungstechnik Aufgaben* (4th ed.). Dr. S. Zacher.
10. Zacher, S. (Ed.). (2000). *Automatisierungstechnik kompakt*. Vieweg.

Bus-Approach

4

4.1 Definition of Virtual Bus

4.1.1 Curse of Dimensionality

Simple closed loop control

The simple closed control loop (CLC) consists of plant, sensor, controller, actuator, as it is shown in Fig. 4.1. The engineering of such loops is well known, and it belongs to the basics of all technical university's courses. The main tool for analysis and design of CLC is the functional block diagram (Fig. 4.2), each element of this diagram is mathematically described as a transfer function $G(s)$:

- plant $G_S(s)$
- sensor $G_{sensor}(s)$
- actuator $G_{actuator}(s)$
- controller $G_R(s)$

The CLC, given in Fig. 4.1, is very simple. It has one controlled variable x and one actuating value y and could be called SISO (Single Input Single Output). Such loops are often simplified by neglecting sensors and actuators and leaving only plants and controllers (Fig. 4.3).

Single-loops MIMO control of separated MIMO plants

Generally, it is no problem to represent the MIMO control systems (Multi-Input Multi-Output), if the plants are separated, i.e., are not coupled together, like shown in Fig. 4.4 for $n = 4$ control variables x_1, x_2, x_3, and x_4. Such block diagrams are still understandable. But it's quite different to follow the signal ways when the loops are coupled.

© The Author(s), under exclusive license to Springer Nature Switzerland AG 2022   125
S. Zacher, *Closed Loop Control and Management*,
https://doi.org/10.1007/978-3-031-13483-8_4

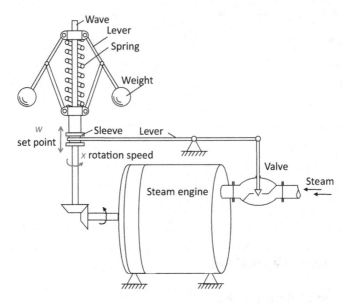

Fig. 4.1 Feedback control of steam engine, known since seventeenth century

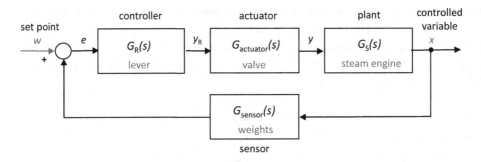

Fig. 4.2 Functional block diagram for CLC of steam engine, shown in Fig. 4.1

Fig. 4.3 Simplified functional
block diagram of SISO of the
Fig. 4.2

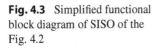
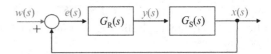

Separated MIMO control of coupled MIMO plants

The classical functional block diagram of coupled MIMO-loops is possible only for
$n = 2$ variables x_1 and x_2. An example of the MIMO of the 3rd order with control variables x_1, x_2, and x_3 is given in Fig. 4.5. The plants with controlled variables x_1, x_2, x_3 and
actuating values y_1, y_2, y_3 are designated with transfer functions $G_{11}(s)$, $G_{22}(s)$, $G_{33}(s)$

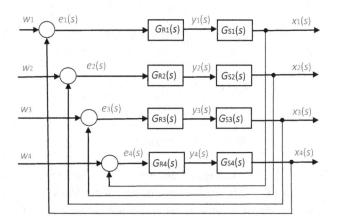

Fig. 4.4 Separated MIMO CLC

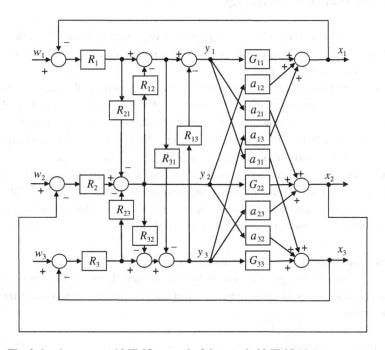

Fig. 4.5 The 3rd order separated MIMO control of the coupled MIMO plant

and called main plants. The plants with the controlled variable x_1, which are actuating by y_2 and y_3, are called coupling plants. Their transfer functions are designated as $a_{12}(s)$ and $a_{13}(s)$. The same for the coupling plants $a_{21}(s)$, $a_{23}(s)$ and $a_{31}(s)$, $a_{32}(s)$:

$$G_{11}(s) = \frac{x_1(s)}{y_1(s)} \quad G_{22}(s) = \frac{x_2(s)}{y_2(s)} \quad G_{33}(s) = \frac{x_3(s)}{y_3(s)} \tag{4.1}$$

$$a_{12}(s) = \tfrac{x_1(s)}{y_2(s)} \quad a_{13}(s) = \tfrac{x_1(s)}{y_3(s)} \tag{4.2}$$

$$a_{21}(s) = \tfrac{x_2(s)}{y_1(s)} \quad a_{23}(s) = \tfrac{x_2(s)}{y_3(s)} \tag{4.3}$$

$$a_{31}(s) = \tfrac{x_3(s)}{y_2(s)} \quad a_{32}(s) = \tfrac{x_3(s)}{y_2(s)} \tag{4.4}$$

Such function block diagrams like Fig. 4.5 are not clear and not understandable. As far as it is very difficult to follow the signal ways of functional block diagrams of MIMO CLC of the 3rd order and higher order, they are barely used in the control theory.

4.1.2   Course of Dimensionality

Richard Bellman called the problem caused by the exponential increase of calculations, which occurs with adding extra new variables to a matrix of a mathematical model, as "Curse of Dimensionality " ([4–6]).

The only known way to describe the MIMO coupled systems is the use of vectors and matrices. The single variables, e.g., x_1, x_2 and x_3 of Fig. 4.5, will be replaced through vectors **X**. In this case, the single transfer functions $G_{11}(s)$, $a_{12}(s)$, $a_{13}(s)$, $a_{21}(s)$, $G_{22}(s)$, … etc., of the plant will build the matrix $\mathbf{G_S}(s)$. The single transfer functions $R_1(s)$, $R_{12}(s)$, … etc., of the controller will build the matrix $\mathbf{R}(s)$ of the controller.

Although such replacement does not reduce the scope of calculations, it lets to simplify the representation of function block diagram of control of coupled MIMO plants (Fig. 4.6). The more variables are in the matrix the more calculations are needed and more computer time requires the engineering of MIMO.

Industrial solution for curse of dimensionality
In the information technology, the solution for the problem *Curse of Dimensionality* has long been found and is well known, namely, instead of *point-to-point* connection it will be used a fieldbus (Fig. 4.7).

Virtual bus for virtual world
The same solution is proposed in [1] for CLC, which is called virtual bus. An example of the $n = 4$ separated CLC is shown in Fig. 4.8 without explanation how a virtual bus is constructed. The detailed explanation follows in Sect. 4.2.1

Already in Fig. 4.8 are seen advantages of the virtual bus. However, its advantages are fully manifested for MIMO control, which was shown in Fig. 4.5. The virtual bus for the $n = 3$ coupled controlled variables is given in Fig. 4.9.

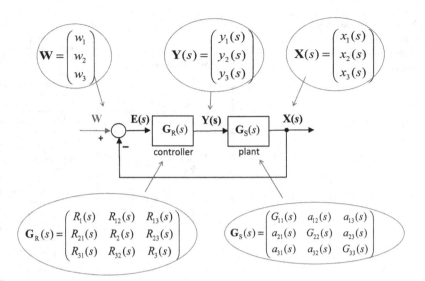

Fig. 4.6 MIMO of the Fig. 4.5 represented with matrices and vectors

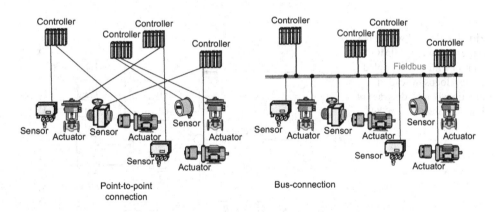

Fig. 4.7 Real world: fieldbus instead of point-to-point connection

„The signal ways are clearly seen on Fig. 4.9, and it is easy to track. The signal way 1-2 through block $a_{12}(s)$ will be compensated with the signal way 1-3-4 through block $R_{12}(s)$, then further through blocks $G_{11}(s)$ with the signal way 4-5, which results in first decoupling controller:

$$R_{12}(s)G_{11}(s) = a_{12}(s) \tag{4.5}$$

$$R_{12}(s) = \frac{a_{12}(s)}{G_{11}(s)} \tag{4.6}$$

Fig. 4.8 Virtual world: bus of separated MIMO with $n=4$ control loops instead of functional block diagram of Fig. 4.4

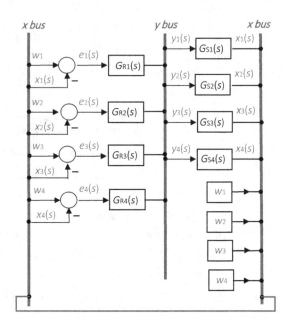

Fig. 4.9 Virtual world: bus representation of MIMO control with $n=3$ coupled control loops instead of functional block diagram of Fig. 4.5. *Source* [1, p. 84]

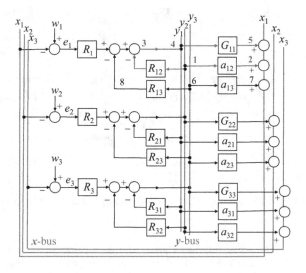

Similarly, the signal way 6-7 through block $a_{21}(s)$ will be compensated with the signal way 6-8 through block $R_{13}(s)$, then further through block $G_{22}(s)$ upon the signal way 8-3-4-5, which results in the transfer function of second decoupling controller:

$$R_{13}(s)G_{11}(s) = a_{13}(s) \qquad (4.7)$$

$$R_{13}(s) = \frac{a_{13}(s)}{G_{11}(s)}\text{''},\qquad(4.8)$$

(Quote: Source [1, p. 85])

4.2 Basics of Bus-Approach

4.2.1 Construction of Virtual Bus

The idea of bus-approach is that blocks of a closed control loop (CLC), which picture the transfer function $G_S(s)$ of a plant and the transfer function $G_R(s)$ of a controller, are connected through a bus. The bus is of course not a real fieldbus, but also a pictured connection, which links the controller to the plant like shown in Fig. 4.10.

Such a bus is called in the following *virtual bus*. It consists of three wires, namely,

- x-bus on the right, also called *bus-creator*, like it is done by MATLAB®,
- x-bus on the left, called again like in MATLAB® *bus-selector*,
- y-bus in the middle, which is assigned for actuating variables.

In Fig. 4.10 is shown the connection between two wires $x(s)$ on the left and on the right sides, which mentioned the feedback of the closed loop.

Transfer function of an open loop $G_0(s)$
The transfer function $G_0(s)$ of the open loop in the classical functional block diagram is the same for the virtual bus, namely:

$$G_0(s) = \frac{x(s)}{e(s)} = G_R(s)G_S(s)\qquad(4.9)$$

Fig. 4.10 Virtual bus-system of a simple CLC

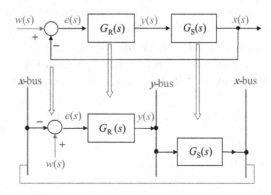

Fig. 4.11 Definition of
the transfer function of an
open loop as series of blocks
between to x-bus-wires

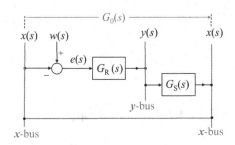

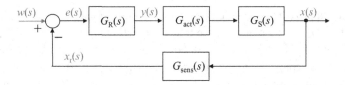

Fig. 4.12 Example of a CLC considering a sensor and an actuator

Transfer function of the closed loop $G_w(s)$ by reference behavior

As a result, the classical transfer function $G_w(s)$ of the closed loop by reference behavior
is the same for the virtual bus:

$$G_w(s) = \frac{x(s)}{w(s)} = \frac{G_0(s)}{1 + G_0(s)} \qquad (4.10)$$

It is very easy to define the transfer function $G_0(s)$ of the open loop directly from the virtual bus according to the following definition, explained in Fig. 4.11.

u **Transfer function of open loop $G_0(s)$** The series of function blocks $G(s)$, which are in
the way from the left bus wire $x(s)$ to the right bus wire $x(s)$, is the transfer function of
the open loop $G_0(s)$.

How should the transfer function of the open loop, given in Eq. 4.9, and the transfer
function of the closed loop by the reference behavior of the Eq. 4.10 be changed if the
CLC consists not only of a plant $G_S(s)$ and a controller $G_R(s)$ but also of a sensor $G_{sens}(s)$
and of an actuator $G_{act}(s)$ like shown in Fig. 4.12?

The answer to this question is given in Fig. 4.13, i.e., Eq. 4.9 remains the same, but
instead of Eq. 4.10 the following equation should be applicated:

$$G_w(s) = \frac{G_{vw}(s)}{1 + G_0(s)} \qquad (4.11)$$

Fig. 4.13 Virtual bus of CLC of Fig. 4.12

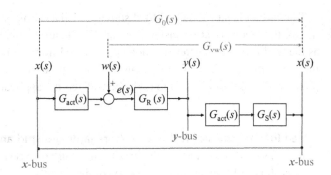

Fig. 4.14 Functional block diagram of a CLC by disturbance behavior

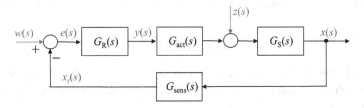

The transfer function $G_{vw}(s)$ is known in the classical control theory as the feed-forward transfer function from the set point $w(s)$ to the controlled variable $x(s)$.

⊔ **Feed-forward transfer $G_{vw}(s)$ regarding set point W(s)** The series of function blocks $G(s)$, which are in the way from set point $w(s)$ to the right bus wire $x(s)$, is the feed-forward transfer function $G_{vw}(s)$ of the set point $w(s)$.

According to this definition and considering Eq. 4.11, the reference behavior of the closed loop of Fig. 4.12 and 4.13 will be described as follows:

$$G_{vw}(s) = G_R(s)G_{act}(s)G_S(s) \tag{4.12}$$

$$G_0(s) = G_{sens}(s)G_R(s)G_{act}(s)G_S(s) \tag{4.13}$$

$$G_w(s) = \frac{G_R(s)G_{act}(s)G_S(s)}{1 + G_{sens}(s)G_R(s)G_{act}(s)G_S(s)} \tag{4.14}$$

Transfer function of the closed loop $G_z(s)$ by disturbance behavior
An example of the functional block diagram of a CLC by disturbance behavior is given in Fig. 4.14. It is assumed that only the disturbance $z(s)$ acts, the set point $w(s)$ is missing, i.e., $w(s) = 0$.

The virtual bus of the CLC by disturbance behavior is given in Fig. 4.15. The definition of the feed-forward transfer function $G_{vz}(s)$ cased from disturbance $z(s)$ is the same as the feed-forward transfer function $G_{vw}(s)$ cased from set point $w(s)$ (see Eq. 4.12). But the applications points of input signals should be changed, i.e., instead of application point of set point $w(s)$ here should be considered the application point of the disturbance $z(s)$.

⊔ **Feed-forward transfer function $G_{vz}(s)$ regarding disturbance Z(s)** The series of function blocks $G(s)$, which are in the way from disturbance $z(s)$ to the right bus wire $x(s)$, is the feed-forward transfer function $G_{vz}(s)$ of the disturbance $z(s)$.

The feed-forward transfer function $G_{vz}(s)$ of Fig. 4.14 and 4.15 will be described as follows:

$$G_{vz}(s) = G_S(s) \tag{4.15}$$

The open loop transfer function of the open loop by disturbance behavior it the same as by set point behavior, namely as in Eq. 4.13. The transfer function of the CLC by disturbance behavior is

$$G_z(s) = \frac{G_{vz}(s)}{1 + G_0(s)} = \frac{G_S(s)}{1 + G_{sens}(s)G_R(s)G_{act}(s)G_S(s)} \tag{4.16}$$

The CLC transfer function of the virtual bus are summarizing shown in Fig. 4.16.

4.2.2 Bus-Creator and Bus-Selector

In the previous section was shown, how easy it is to identify the transfer functions using bus-approach. But this is not the only advantage of the bus-approach. For the practical use is very important that the Simulink-Library of the well-known and widely implemented

Fig. 4.15 Virtual bus of the CLC by disturbance behavior of Fig. 4.14

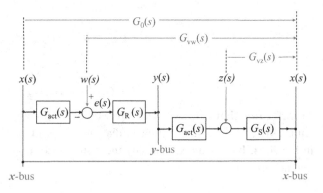

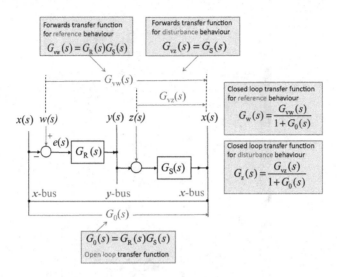

Fig. 4.16 Summary of transfer functions of a virtual bus

simulation-software MATLAB® contents tools like "Bus Creator" and "Bus-Selector", which are easy configurable and suitable for the bus-approach (Fig. 4.17).

Example

Let us simulate with the virtual bus the simple CLC consisting of the PI-controller with the transfer function $G_R(s)$ and the PT2 plant with the transfer function $G_S(s)$, which are given in [7, p. 76], with the following parameters:

$$K_{PR} = 5,0 \; T_n = 95 \text{ s}$$
$$K_{PS1} = 0,8 \; T_1 = 6 \text{ s} \; K_{PS2} = 1 \; T_2 = 95 \text{ s}$$

$$G_R(s) = \frac{K_{PR}(1 + sT_n)}{sT_n} \tag{4.17}$$

$$G_S(s) = \frac{K_{PS1}}{1 + sT_1} \cdot \frac{K_{PS2}}{1 + sT_2} \tag{4.18}$$

The virtual bus simulated with MATLAB®/Simulink tools is shown in Fig. 4.18. ◀

4.2.3   Applications of Virtual Buses for SISO- and MIMO CLC

The applications of the bus-approach for different kinds of CLC are proposed in [1–3, 7]. There are virtual buses:

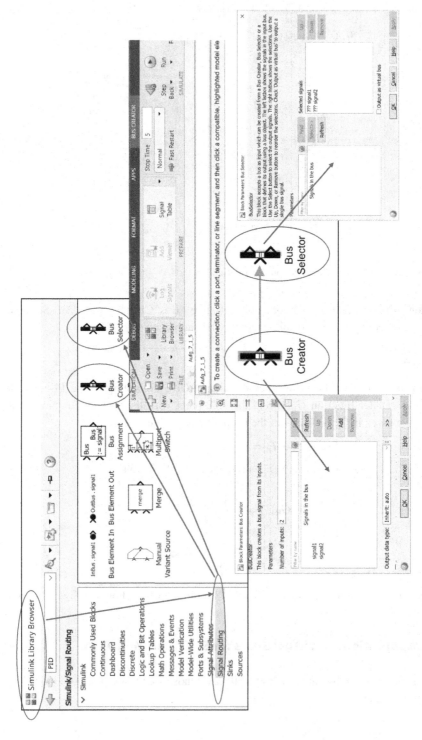

Fig. 4.17 MATLAB®/Simulink tools for simulation of virtual bus

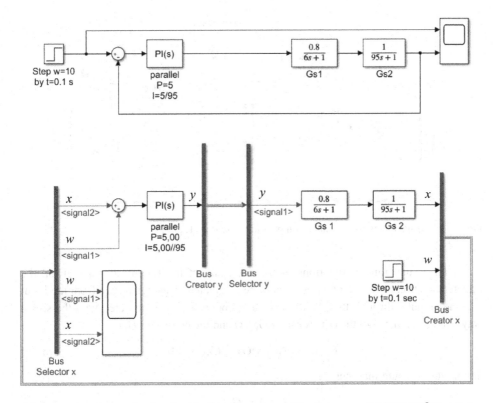

Fig. 4.18 Virtual bus simulated with Bus-Creator and Bus-Selector of MATLAB®/Simulink. *Source* [7, p. 77]

- for simple loops with only one controller and one controller variable (SISO—Single Input Single Output),
- for CLC with one controller and more as one controlled variable (SIMO—Single Input Multi-Output),
- for two controllers with two controlled values:
  - redundant control,
  - cascade control,
  - override control,
- for separated MIMO plants (Multi Input Multi Output),
- for coupled MIMO plants.

Some of the CLC listed above are given below with references to sources.

Disturbance compensation

It is supposed that the disturbance z which acts by $t = 20$ s after the step of the set point is measurable (Fig. 4.19).

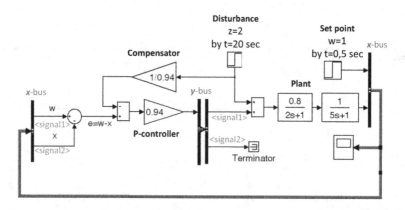

Fig. 4.19 Virtual bus for disturbance compensation. *Source* [1, p. 9]

As far as the transfer functions of the plant and of the P-controller are given, the transfer function of the compensator could be calculated. The signal way from the input z to the output x through the plant $G_S(s)$ should be exactly compensated with the signal way from z to x through the compensator $G_{Rz}(s)$ and the controller $G_R(s)$:

$$G_{vz}(s) = G_{Rz}(s)G_R(s)G_S(s) = 0 \tag{4.19}$$

Taking into account that there is

$$G_{vz}(s) = G_S(s) \tag{4.20}$$

the transfer function of the compensator will be defined as given below

$$G_{Rz}(s) = \frac{1}{G_R(s)} = \frac{1}{K_{PS}} = \frac{1}{0,94} \tag{4.21}$$

Redundant control

Supposing the PT2-plant should be controlled with the main PI-controller. It is also supposed that this controller has a failure. Its gain is suddenly changed: instead of configured value $K_{PR} = 1$ it becomes by $t = 0$ s the negative value $K_{PR} = -1$. The safety system (is not shown in Fig. 4.20) recognizes the failure and should switch off the main controller.

Further it is supposed that the switch needs $t = 10$ s to recognize the failure. That means that during the first $t = 10$ s the loop will be controlled by failed controller and will be unstable. After $t = 10$ s the redundant controller with the correct gain takes over the control.

SIMO-CLC of a plant with distributed parameters

The plant with distributed parameters, which is controlled with the single PI-controller, is simulated in Fig. 4.21 with three transfer functions:

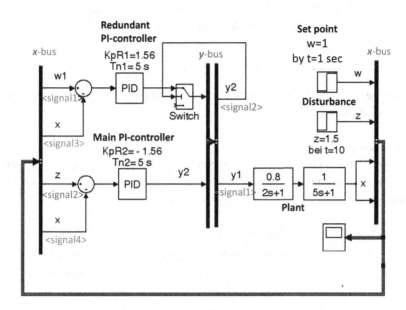

Fig. 4.20 Virtual bus an example of redundant control. *Source* [1, p. 15]

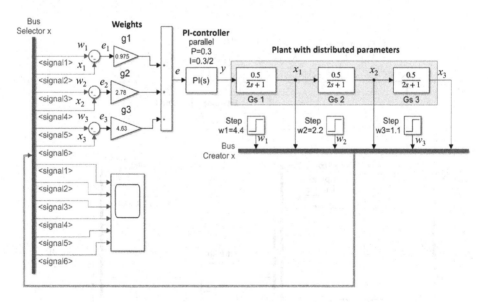

Fig. 4.21 Bus-approach for a CLC with distributed plant parameters. *Source* [7, p. 86]

$$G_{S1}(s) = G_{S2}(s) = G_{S3}(s) = \frac{0,5}{1+2s} \qquad (4.22)$$

The entire control deviation $e(t)$ is built of the partial deviations $e_1(t)$, $e_2(t)$, $e_3(t)$ as the weighted sum:

$$e(t) = g_1 e_1(t) + g_2 e_2(t) + g_3 e_3(t) \tag{4.23}$$

The weights g_1, g_2, g_3 are optimized, so that the optimal damping and overshoot for all three controlled variables are achieved.

Cascade control

An example of the well-known cascade control is shown in Fig. 4.22. The classical functional block diagram is simulated with bus-approach. Important is the corresponding configuration of the bus selector, which is given below

Signal of the loop	Signal in the bus	Selected signals
w	signal 1 (signal 1)	signal 1 (signal 1)
w1	signal 1 (signal 2)	signal 4 (signal 4)
×1	signal 3 (signal 3)	signal 1 (signal 2)
x	signal 4 (signal 4)	signal 3 (signal 3)

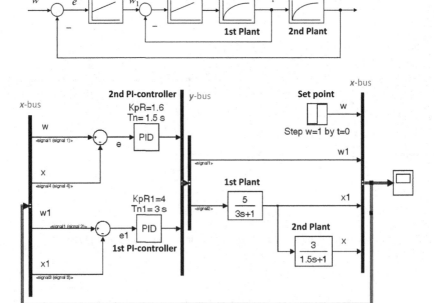

Fig. 4.22 Example of cascade control simulated with virtual bus. *Source* [1, p. 12]

Override-control

The override control consists of two closed loops, each of them is controlled with separate controllers, like shown in Fig. 4.23.

The main controller is intended for the main plant with the main controlled variable x:

$$G_S(s) = \frac{3}{(1+1,8s)} \cdot \frac{0,5}{(1+2,5s)}$$ (4.24)

The override controller is intended to control the part of the main plant with the controlled variable x_{over}:

$$G_{S\_over}(s) = \frac{3}{(1+1,8s)}$$ (4.25)

Both controllers work in parallel with different set points. The aim of the whole system is to bring each controlled variable to corresponding set point keeping on the following condition:

$$x_{over}(s) > x(s)$$ (4.26)

The controlled variables x and x_{over} will be compared via a selection box. The selection box carried out automatically the condition Eq. 4.26 switching the actuating values Y and Y_{over} according to the maximum of x_{over}.

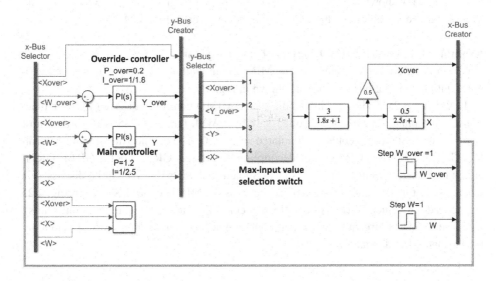

Fig. 4.23 Override-control simulated with virtual bus. *Source* [7, p. 94]

4.3   Bus-Approach for MIMO Control

Plants with n controlled variables and n actuating values are called MIMO plants of the order n (Multi-Input Multi-Output). Generally, the controlled variables of a MIMO plant could be coupled or not coupled.

4.3.1   Kinds of MIMO Plants

In this chapter are considered the following kinds of MIMO plants:

- Separated plants: every controlled variable is controlled with corresponding actuated value.
- Coupled plants: every actuated value influenced many or all controlled variables.

The separated, i.e., not coupled MIMO plants with n controlled variables $x_1, x_2, \ldots x_n$ and n actuating values $y_1, y_2, \ldots y_n$ are n separate plants. Each of them has respectively one controlled variable and one corresponding actuating value. Each separate plant will be controlled with the single controller.

Example of single-loops MIMO control of separated plants
One example of MIMO control with $n = 4$ separated controlled variables with 4 controllers was already shown in Fig. 4.4. Another example of single-loops MIMO control with $n = 6$ separated variables and the step responses are given in Fig. 4.24.

Example of decoupled MIMO control of coupled plants
An example of decoupled control of the MIMO plant with $n = 3$ coupled controlled variables with the virtual bus was already shown in Fig. 4.9.

In the Fig. 4.25 is shown an example of the MIMO control with given transfer functions of the plant with $n = 4$ controlled variables. With this Fig. 4.25 is illustrated how easy and understandable could be simulated such CLC. The engineering of main controllers GR1, GR2, GR3, GR4 and of decoupling controllers GR12, GR13, GR14, GR21, etc., with classical methods and with Data Stream Managers (DSM) will be discussed in Chap. 10. Let us here only note, that for the classical MIMO control, shown in Fig. 4.25 are needed altogether 16 controllers (4 main controllers and 12 decoupling controllers). According to Data Stream Management, there are needed only 8 controllers (4 main controllers and 4 DSM Routers).

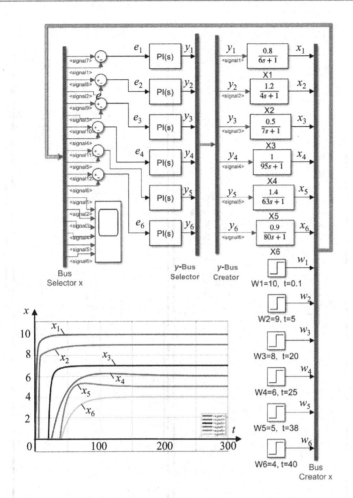

Fig. 4.24 Virtual bus for $n = 6$ single-loops MIMO control. *Source* [7, p. 90]

4.3.2   Kinds of Coupled MIMO Plants

The engineering of MIMO control for not coupled plants consists in the stability analysis and design of each single CLC, which is well known in the classical control theory and will not discussed in this chapter. However, the use of bus-approach brings not many advantages in comparison to classical functional block diagram.

There are known two kinds of coupled MIMO plants, classical functional block diagrams and corresponding virtual buses are shown in the figures below

- in P-canonical form (Fig. 4.26).
- in V-canonical form (Fig. 4.27).

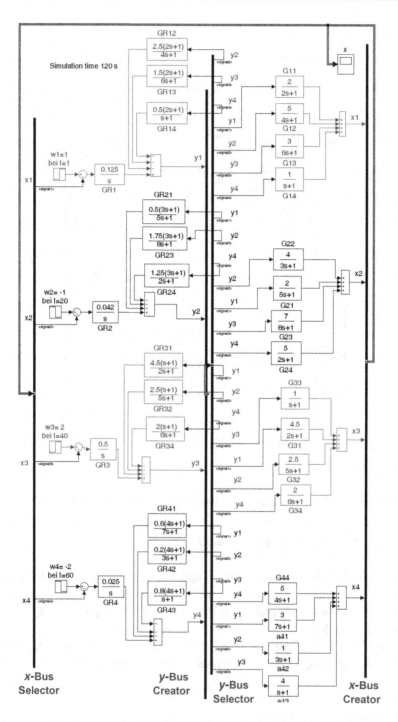

Fig. 4.25 Virtual bus for decoupled MIMO control of $n = 4$ control variables. *Source* [1, p. 90]

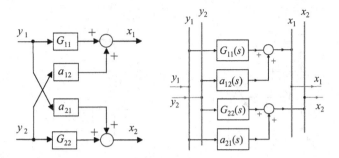

Fig. 4.26 FF-plant with $n = 2$ variables. *Source* [1, p. 40]

Obviously, there is no need to explain, what is the difference between both kinds of plants. We note only that the V-plants have feedback while the P-plants have only feed-forward signal ways. Respectively, the transfer functions of the MIMO plants of P- and V-form are different.

To avoid the exact but long academic definition, like "MIMO plant in P-canonical form", this kind of plants will be called in the following as "Feed-Forward plants" or shortly "FF-plants". Accordingly, the MIMO plants in in V-canonical form are further called "Feed-Back plants" or shortly "FB-plants".

Feed-forward plants (Fig. 4.26)
The blocks $G_{11}(s)$ and $G_{22}(s)$ which describe the relationship between main inputs y_1, y_2 and their outputs x_1, x_{12} are called *main* plants:

$$G_{11}(s) = \frac{x_1(s)}{y_1(s)} \qquad G_{22}(s) = \frac{x_2(s)}{y_2(s)}$$

The blocks $a_{12}(s)$ and $a_{21}(s)$ which transfer signals from one main input to another main output are *coupling* plants:

$$a_{12}(s) = \frac{x_1(s)}{y_2(s)} \qquad a_{21}(s) = \frac{x_2(s)}{y_1(s)}$$

The input/output relation of the FF-plants is given below

$$x_1(s) = G_{11}(s)y_1(s) + a_{12}(s)y_2(s) \tag{4.27}$$

$$x_2(s) = a_{21}(s)y_1(s) + G_{22}(s)y_2(s)$$

Feed-back plants (Fig. 4.27)

$$G_{11}(s) = \frac{x_1(s)}{y_{11}(s)} \qquad V_{12}(s) = \frac{y_{12}(s)}{x_2(s)} \qquad V_{21}(s) = \frac{y_{21}(s)}{x_1(s)} \qquad G_{22}(s) = \frac{x_2(s)}{y_{22}(s)}$$

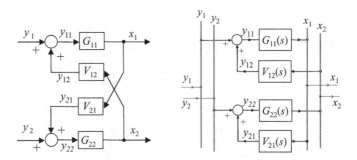

Fig. 4.27 FB-plant with n = 2 variables. *Source* [1, p. 40]

$$x_1(s) = G_{11}(s)y_{11}(s) = G_{11}(s)\,[y_1(s) + V_{12}(s)\,x_2(s)] \tag{4.28}$$

$$x_2(s) = G_{22}(s)y_{22}(s) = G_{22}(s)[y_2(s) + V_{21}(s)\,x_1(s)]$$

The transfer function of a FB-plant could be converted in the transfer function of a FF-plant. For this purpose, the signal way beginning from the input y_{11} of the FB-plant will be traced until the same variable y_{11}, which is noted as the transfer function of the open loop inside the plant:

$$G_0(s) = G_{11}(s)V_{12}(s)G_{22}(s)V_{21}(s) \tag{4.29}$$

The same transfer function of Eq. 4.29 results by the tracking the signal way from the input y_{22} to the same variable x_{22}.

The transfer function $G_{p11}(s)$ from input y_1 to the output x_1 of the FB-plant describes the disturbance behavior with the disturbance y_1, and open loop transfer function $G_0(s)$ is given in Eq. 4.29. The same relations are concerning all other transfer functions $G_{p12}(s)$, $G_{p21}(s)$ and $G_{p22}(s)$:

$$G_0(s) = G_{11}(s)V_{12}(s)G_{22}(s)V_{21}(s) \tag{4.30}$$

$$G_{p11}(s) = \frac{G_{11}(s)}{1 - G_0(s)} \qquad G_{p22}(s) = \frac{G_{22}(s)}{1 - G_0(s)} \tag{4.31}$$

$$G_{p12}(s) = \frac{G_{11}(s)V_{12}(s)G_{22}(s)}{1 - G_0(s)} \qquad G_{p21}(s) = \frac{G_{11}(s)V_{21}(s)G_{22}(s)}{1 - G_0(s)} \tag{4.32}$$

Finally, a FB-plant, converted in the FF-plant according to Eqs. 4.29–4.31, is given in Eq. 4.32.

FB-plant converted in the FF-plant

$$x_1(s) = G_{p11}(s)\, y_1(s) + G_{p12}(s)\, y_2(s) \tag{4.33}$$

$$x_2(s) = G_{p21}(s) y_1(s) + G_{p22}(s) y_2(s)$$

4.3.3 Stability of Coupled MIMO Plants

Stability of FF-plants
An FF-plant (Eq. 4.27) is stable if all partial transfer functions $G_{11}(s)$, $a_{12}(s)$, $a_{21}(s)$, $G_{22}(s)$ are stable.

Example: Stability of a feed-forward plant

The partial transfer functions are given:

$$\mathbf{G_S}(s) = \begin{pmatrix} G_{11}(s) & a_{12}(s) \\ a_{21}(s) & G_{22}(s) \end{pmatrix} = \begin{pmatrix} \frac{0,5}{1+5s} & 0,48 \\ 1,25 & \frac{0,24}{1+2,5s} \end{pmatrix}$$

The alone plant, i.e., the plant without controller, is stable because pole places s_p of characteristic equations of all partial transfer functions are negative:

$$s_{p11} = -\frac{1}{5} \text{ and } s_{p22} = -\frac{1}{2,5} \blacktriangleleft$$

Stability of FB-plants
An FB-plant of Eq. 4.28 is stable if all partial transfer functions $G_{11}(s)$, $V_{12}(s)$, $V_{21}(s)$, $G_{22}(s)$ are stable and besides of this all partial closed loops of Eq. 4.31 are stable. As far as all partial closed loops of Eq. 4.31 have the same characteristic equations

$$1 - G_0(s) = 0 \tag{4.34}$$

the stability condition will be formulated as follows:
An FB-plant of Eq. 4.28 is stable if all partial transfer functions $G_{11}(s)$, $V_{12}(s)$, $V_{21}(s)$, $G_{22}(s)$ are stable and besides of this if all roots of characteristic equation Eq. 4.34 have negative real parts.

Stability of a feed-back plant

A FB-plant is given:

$$\mathbf{G_S}(s) = \begin{pmatrix} G_{11}(s) & V_{12}(s) \\ V_{21}(s) & G_{22}(s) \end{pmatrix} = \begin{pmatrix} \frac{3}{1+s} & \frac{1}{1+3s} \\ K_{p21} & \frac{4}{1+2s} \end{pmatrix}$$

By what values of the gain K_{21} is the FB-plant stable?

To answer this question, we need to proof the stability of each partial closed loop with transfer functions according to Eq. 4.31 and the roots of characteristic equation Eq. 4.34.

The partial closed loops are stable, if K_{21} is positive, i.e., if $K_{21} > 0$:

$$s_{p11} = -\frac{1}{1} \quad s_{p12} = -\frac{1}{3} \quad s_{p22} = -\frac{1}{2}$$

The characteristic equation according to Eq. 4.30 is

$$1 - \frac{3}{1+s} \cdot \frac{1}{1+3s} \cdot K_{P21} \cdot \frac{4}{1+2s} = 0$$

$$6s^3 + 11s^2 + 6s + 1 - 12K_{P21} = 0$$

From the last equation of the 3rd order follows that according to the Hurwitz stability criterion the gain K_{P21} should fulfill the conditions given below

$$\begin{cases} 1 - 12K_{P21} > 0 \\ 11 \cdot 6 > 6(1 - 12K_{P21}) \end{cases} \rightarrow \begin{cases} 1 - 12K_{P21} > 0 \\ 11 \cdot 6 > 6(1 - 12K_{P21}) \end{cases}$$

It results in the solution that the FB-plant will be stable by following values of the gain K_{P21}:

$$0 < K_{P21} < 0,0833$$

Otherwise, is the FB-plant unstable. ◀

4.3.4   Kinds of MIMO Control

The following kinds of closed loop control (CLC) are defined for MIMO plants, shortly called MIMO CLC:

- Single-loops control for not coupled MIMO plants.
- Separated control for coupled MIMO plants.
- Decoupled control for coupled MIMO plants.

In the following are described the application of bus-approach for first two kinds of classical MIMO CLC. The application of bus-approach for decoupled control is given in Chap. 10. The bus-approach lets to simplify the structural representation of closed control loops for coupled MIMO plants and to track the signal ways, like it was described in the previous sections of this chapter. It becomes possible to define the transfer functions of the whole MIMO system and to analyze its stability upon characteristic equation of the defined transfer functions.

Before we begin to discuss the stability of MIMO CLC let us classify the MIMO control loops.

Single-loops MIMO control

If n controlled variables of a MIMO plant are not coupled with each other than the MIMO control is realized with n separated control loops independent one from another, with one single controller in each loop. It is called single-loop MIMO control. The examples of such control are given in Fig. 4.4 for $n=2$, in Fig. 4.8 for $n=4$ and in Fig. 4.24 for $n=6$ variables. The whole single-loops MIMO control is stable if each separated loop is stable. As far as the stability proof of single CLC could be successfully checked by well-known methods of the classical control theory without application of bus-approach, this topic is not considered in this chapter.

Separated MIMO control

As separated MIMO control of the coupled MIMO plant of the n-th order with n controlled variables is designated the control system realized with n separate controllers. Each controller is tuned for its single control loop independent of other loops. But the controlled variables of a coupled MIMO plant influence each other. As far as in the reality the loops are connected one to another, the controlled variable of one loop acts as the disturbance for another loop.

The example of separated MIMO control of the coupled MIMO plant of the 2nd order is given in Fig. 4.28. The actuating value y_1 of the 1st controller R_1 acts as disturbance for the 2nd loop with controller R_2. Respectively the actuating value y_2 of the 2nd controller R_2 acts as disturbance for the 1st loop with controller R_1.

Decoupled MIMO control

The examples of decoupled MIMO control were illustrated without explanation in Fig. 4.9 for MIMO plant of the 3rd order and in Fig. 4.25 for MIMO plant of the 4th order. The decoupled MIMO control of coupled plants is described in the Chap. 10.

Vector–matrix representation of MIMO control

All three kinds of MIMO control could be unified represented with input vectors $\mathbf{W}(s)$, output vectors $\mathbf{X}(s)$, with transfer matrices of controllers $\mathbf{R}(s)$ and plants $\mathbf{G}_S(s)$, building the transfer matrices of the open $\mathbf{G}_0(s)$ and closed loop $\mathbf{G}_w(s)$ for reference behavior:

$$\mathbf{W} = \begin{pmatrix} w_1 \\ w_2 \end{pmatrix} \qquad \mathbf{X}(s) = \begin{pmatrix} x_1(s) \\ x_2(s) \end{pmatrix} \tag{4.35}$$

$$\mathbf{G}_0(s) = \mathbf{R}(s)\mathbf{G}_S(s) \qquad \mathbf{G}_w(s) = \frac{\mathbf{G}_0(s)}{1 + \mathbf{G}_0(s)} \tag{4.36}$$

Fig. 4.28 Separated MIMO
control of the coupled MIMO
plant of the 2nd order.
Source [1, p. 56]

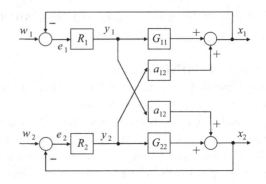

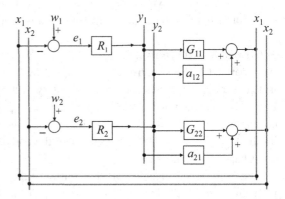

Single-loops MIMO control

$$\mathbf{R}(s) = \begin{pmatrix} R_1(s) & 0 \\ 0 & R_2(s) \end{pmatrix} \quad \mathbf{G}_S(s) = \begin{pmatrix} G_{11}(s) & 0 \\ 0 & G_{22}(s) \end{pmatrix} \tag{4.37}$$

Separated MIMO control

$$\mathbf{R}(s) = \begin{pmatrix} R_1(s) & 0 \\ 0 & R_2(s) \end{pmatrix} \quad \mathbf{G}_S(s) = \begin{pmatrix} G_{11}(s) & a_{12}(s) \\ a_{21}(s) & G_{22}(s) \end{pmatrix} \tag{4.38}$$

Decoupled MIMO control

$$\mathbf{R}(s) = \begin{pmatrix} R_1(s) & R_{12}(s) \\ R_{21}(s) & R_2(s) \end{pmatrix} \quad \mathbf{G}_S(s) = \begin{pmatrix} G_{11}(s) & a_{12}(s) \\ a_{21}(s) & G_{22}(s) \end{pmatrix} \tag{4.39}$$

4.3.5   Interaction Transfer Function and Coupling Gain

The interaction between main plants $G_{11}(s)$, $G_{22}(s)$ of the feed-forward plants, given in
Eq. 4.26, is depended on the values of the coupling transfer functions $a_{12}(s)$, $a_{21}(s)$. To
evaluate the intensity and dynamics between main plants is introduced so-called interaction transfer function $C(s)$:

$$C(s) = \frac{a_{12}(s)a_{21}(s)}{G_{11}(s)G_{22}(s)} \tag{4.40}$$

The interaction transfer function usually leads to a complicated expression, so by proportional PT1, PT2, ... PTn plants it is reasonable to simplify it to the steady state value by $t \to \infty$ or $s \to 0$, which is called static coupling factor $C(0)$ or simply coupling gain C_0:

$$C(0) = C_0 = \lim_{s \to 0} C(s) = \frac{K_{12}K_{21}}{K_{11}K_{22}} \tag{4.41}$$

If a MIMO plant has at least one separate channel, i.e., $K_{12}=0$ or $K_{21}=0$ or both, then the static interaction gain C_0 is also zero:

$$C_0 = 0 \tag{4.42}$$

4.4 Stability of MIMO Control

In the following is described how to proof the stability of two kinds of MIMO control only for coupled feed-forward plants. Should a need arise, the stability of a coupled feed-back plants to proof, such feed-back plants should first be converted into feed-forward plants like it is done in Eq. 4.33.

4.4.1 Stability of MIMO Single-Loops Control

There are no problems to check the stability of control loops with MIMO plants, which are not coupled like Fig. 4.4, Fig. 4.8, Fig. 4.24 because the methods of stability proof for separated CLC are well known.

Example

Let us check the stability of one single control loop with transfer functions of the plant $G_4(s)$ and controller $R_4(s)$, given in Fig. 4.24:

$$G_4(s) = \frac{1}{1+95s} \qquad R_4(s) = \frac{K_{PR}(1+sT_n)}{sT_n}$$

By what values of the controller gain K_{PR} is this single loop stable supposing that the reset time of controller is $T_n = 45$ s?

To answer this question, we should define the characteristic polynomial $P(s)$ of the closed loop:

$$G_0(s) = R_4(s)G_4(s) = \frac{K_{PR}(1+40s)}{40s} \cdot \frac{1}{1+95s} = \frac{N(s)}{D(s)}$$

$$P_0(s) = N(s) + D(s) = K_{PR}(1 + 40s) + 40s(1 + 95s)$$

$$P_0(s) = 3800s^2 + 40 \cdot (1 + K_{PR})s + K_{PR} \qquad (4.43)$$

The last equation is the polynomial of the 2nd order with coefficients:

$$a_2 = 3800$$
$$a_1 = 40 \cdot (1 + K_{PR})$$
$$a_0 = K_{PR}$$

According to Hurwitz stability criterion, the closed loop of the 2nd order is stable if all coefficients of its characteristic polynomial are positive. The considered CLC will be stable if the following condition is met:

$$K_{PR} > 0$$

On the same way we can check the stability all single loops of Fig. 4.24. ◀

4.4.2 Stability of Separated MIMO Control

In this section is shown, how to check the stability of the closed loops, consisting of coupled feed-forward plants, which are controlled with single controllers without decoupling. In previous section, it was called separated MIMO control. The stability analysis in this section will be done sequentially beginning with systems of 2nd order with two controlled variables x_1, x_2 and two actuating values y_1, y_2, then of 3rd and lastly of 4th order. Finally, the stability conditions will be generalized commonly for separated MIMO CLC of order n, i.e., with controlled variables x_1, x_2, ... x_n, and actuating values $y_1, y_2, \cdots y_n$.

MIMO control of 2nd order
Although such MIMO control was already given in Fig. 4.28 let us show it again in Fig. 4.29 following the signal ways like in is done in [1, p. 57].

The transfer functions of each separated closed loop are

$$G_{w1}(s) = \frac{G_{01}(s)}{1 + G_{01}(s)} = \frac{R_1(s)G_{11}(s)}{1 + R_1(s)G_{11}(s)} \qquad (4.44)$$

$$G_{w2}(s) = \frac{G_{02}(s)}{1 + G_{02}(s)} = \frac{R_2(s)G_{22}(s)}{1 + R_2(s)G_{22}(s)} \qquad (4.45)$$

Supposing that steps of set points w_1 and w_2 don't occur at the same time and assuming that actually only the step of w_1 is applied, we recognize from virtual bus of Fig. 4.29 the following two forward signal ways from set point w_1 to the output x_1 and the corresponding transfer functions $G_{v1}(s)$ and $G_{v2}(s)$.

Fig. 4.29 Separated MIMO
CLC of the order $n=2$

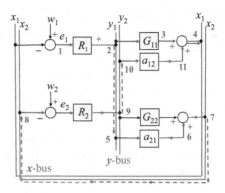

- The first signal way **1-2** goes through controller $R_1(s)$ then follows the way 3-4 through main plant $G_{11}(s)$. It results in the transfer function:

$$G_{v1}(s) = R_1(s)G_{11}(s) \qquad (4.46)$$

- The second signal way **1-2** goes also through controller $R_1(s)$ but then follows the way 5-6 through the coupling plant $a_{21}(s)$, enters the closed loop of the controller $R_2(s)$ and main plant $G_{22}(s)$, built with paths 7-8-9, and ends with the path 10-11-4 going through the coupling plant $a_{12}(s)$.

Taking into account that the transfer function of the path 7–8-9 is

$$\frac{-R_2(s)}{1 + R_2(s)G_{22}(s)}$$

the entire second signal way will be described with the transfer function

$$G_{v2}(s) = R_1(s) \cdot a_{21}(s) \cdot \frac{-R_2(s)}{1 + R_2(s)G_{22}(s)} \cdot a_{12}(s) \qquad (4.47)$$

The transfer functions $G_0(s)$ and $G_{w1}(s)$ of the entire MIMO control with input w_1 and output x_1 are

$$G_0(s) = G_{v1}(s) + G_{v2}(s) \qquad (4.48)$$

$$G_{w1}(s) = \frac{G_0(s)}{1 + G_0(s)} \qquad (4.49)$$

The characteristic equation $1+G_0(s)$ of the closed loop $G_{w1}(s)$ is

$$1 + \underbrace{R_1(s)G_{11}(s)}_{\text{First signal way } G_{v1}(s)} \underbrace{- R_1(s)a_{12}(s)a_{21}(s)\frac{R_2(s)}{1 + R_2(s)G_{22}(s)}}_{\text{Second signal way} G_{v2}(s)} = 0 \qquad (4.50)$$

Considering Eqs. 4.44 and 4.45, then referring to Eq. 4.40 and putting the interaction transfer function $C(0)$ in Eq. 4.50 we get the stability condition, as it was described in [1, p. 59]:

$$\begin{cases} 1 + G_{01}(s) = 0 & 1^{st} \text{ separated closed loop} \\ 1 + G_{02}(s) = 0 & 2^{nd} \text{ separated closed loop} \\ 1 - C(s)G_{w1}(s)G_{w2}(s) = 0 & \text{interaction between loops} \end{cases} \tag{4.51}$$

"We can check the stability of the MIMO control loop, solving these characteristic equations and defining the poles of each equation. Otherwise, we can use the Hurwitz stability criterion, like it is done below for an example." (Quote: Source [1, p. 59])

Example: Stability check of a MIMO CLC with N = 2 inputs/outputs

Given is the MIMO CLC, consisting of the FF-plant $\mathbf{G_S}(s)$ and two compensated PI-controllers shown below as diagonal matrix $\mathbf{R}(s)$:

$$\mathbf{G_S}(s) = \begin{pmatrix} \underbrace{\dfrac{1}{1+s}}_{G_{11}(s)} & \underbrace{1}_{a_{12}(s)} \\ \underbrace{0,5}_{a_{12}(s)} & \underbrace{\dfrac{2}{1+2s}}_{G_{22}(s)} \end{pmatrix} \qquad \mathbf{R}(s) = \begin{pmatrix} \underbrace{\dfrac{K_{R1}(1+sT_{n1})}{sT_{n1}}}_{R_1(s)} & 0 \\ 0 & \underbrace{\dfrac{K_{R2}(1+sT_{n2})}{sT_{n2}}}_{R_2(s)} \end{pmatrix}$$

Checking the stability conditions of Eq. 4.51, it is clearly seen that the separate closed loops are stable for all positive values of gains K_{R1} and K_{R2}.

The characteristic equation of interaction leads to the following equation:

$$s^2 \left(\frac{2}{K_{R1}K_{R2}} - 1 \right) + 2s \left(\frac{1}{K_{R1}} + \frac{1}{K_{R2}} - 0,75 \right) + 1,5 = 0 \tag{4.52}$$

The characteristic equation Eq. 4.50 has 2nd order. According to the Hurwitz stability criterion, the system is stable if the coefficients of this equation are positive. According to this criterion the system will be stable if the following conditions are met:

$$\begin{cases} K_{R1}K_{R2} < 2 \\ \frac{1}{K_{R1}} + \frac{1}{K_{R2}} > 0,75 \end{cases} \tag{4.53}$$

It leads to the condition for gains, namely, the MIMO CLC is stable if the product of gains is limited by 2:

$$K_{R1}K_{R2} < 2 \tag{4.54}$$

◀

MIMO control of 3rd order
Repeating the deviations of previous section for the MIMO CLC of the order $n=3$ (Fig. 4.30) we have to described next to the 1st and 2nd signal ways of Fig. 4.29 one signal way more, like it was done in [1, p. 63].

- The transfer function $G_{v1}(s)$ of the first signal way **1-2-3-4** is the same as Eq. 4.46.
- The second signal way **1-2-5-6-7-8-9-10-11** results in the transfer function $G_{v2}(s)$ with same equation Eq. 4.47.
- The third signal way **1-2-3-4** goes through controller $R_1(s)$ then on the way **12-13** through the coupling plant $a_{31}(s)$, enters the closed loop of the controller $R_3(s)$ and main plant $G_{33}(s)$, built with paths **14-15-16**, and ends with the path **16-17-4** going through the coupling plant $a_{13}(s)$. The third signal way results in the transfer function $G_{v3}(s)$:

$$G_{v3}(s) = R_1(s)a_{31}(s)\frac{-R_3(s)}{1+R_3(s)G_{33}(s)}a_{13}(s) \qquad (4.55)$$

Considering the transfer functions of the 1st and 2nd separated loops by Eqs. 4.44 and 4.45, then defining the 3rd separated closed loop as

$$G_{w3}(s) = \frac{R_3(s)G_{33}(s)}{1+R_3(s)G_{33}(s)} \qquad (4.56)$$

and the interaction transfer functions as

$$C_{12}(s) = \frac{a_{12}(s)a_{21}(s)}{G_{11}(s)G_{22}(s)} \qquad C_{13}(s) = \frac{a_{13}(s)a_{31}(s)}{G_{11}(s)G_{33}(s)} \qquad (4.57)$$

the stability condition for separated MIMO CLC will be formulated as follows:

Fig. 4.30 Separated MIMO CLC of the order $n=3$

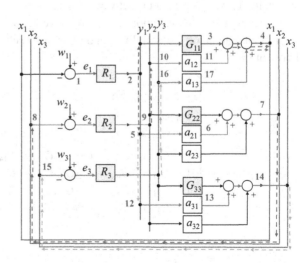

$$\begin{cases} 1 + G_{01}(s) = 0 \\ 1 + G_{02}(s) = 0 \\ 1 + G_{03}(s) = 0 \end{cases}$$

$$\begin{cases} 1 - C_{12}(s)G_{w1}(s)G_{w2}(s) - C_{13}(s)G_{w1}(s)G_{w3}(s) = 0 \\ 1 - C_{23}(s)G_{w2}(s)G_{w3}(s) - C_{12}(s)G_{w1}(s)G_{w2}(s) = 0 \\ 1 - C_{13}(s)G_{w1}(s)G_{w3}(s) - C_{23}(s)G_{w2}(s)G_{w3}(s) = 0 \end{cases} \qquad (4.58)$$

Example of stability proof of MIMO CLC of the order N = 3

The given system consisting of the FF-plant and three compensated PI-controllers, shown below

$$\mathbf{G_S}(s) = \begin{pmatrix} \frac{1}{1+s} & 1 & 0,5 \\ 0,5 & \frac{2}{1+2s} & 0,1 \\ 0,2 & 1 & \frac{3}{1+3s} \end{pmatrix} \qquad \mathbf{R}(s) = \begin{pmatrix} \frac{K_{R1}(1+s)}{s} & 0 & 0 \\ 0 & \frac{K_{R2}(1+2s)}{2s} & 0 \\ 0 & 0 & \frac{K_{R3}(1+3s)}{3s} \end{pmatrix}$$

is stable if the following conditions according to Eq. 4.58 are met:

$$\begin{cases} 2 - K_{R1}K_{R2} - 0,2K_{R1}K_{R3} > 0 \\ 2 - K_{R1}K_{R2} - 0,2K_{R2}K_{R3} > 0 \\ 1 - 0,1K_{R1}K_{R3} - 0,1K_{R2}K_{R3} > 0 \end{cases} \qquad (4.59)$$

Bei $K_{R1} = 0{,}8$; $K_{R2} = 1{,}2$ and $K_{R3} = 4$ is the MIMO CLC stable. The MATLAB®-Script for the inequalities system Eq. 4.59 and Simulink model are given in [1, p. 67]. ◀

MIMO control of 4th order

The bus-approach allows increase the number of inputs/outputs without efforts like it is shown in Fig. 4.31 for $n = 4$ output variables.

Repeating the same operations as for $n = 3$ we get the stability condition like Eq. 4.58 as a combination of the stability conditions for each separate loop and interactions between loops:

$$\begin{cases} 1 + G_{01}(s) = 0 \\ 1 + G_{02}(s) = 0 \\ 1 + G_{03}(s) = 0 \\ 1 + G_{04}(s) = 0 \end{cases}$$

$$\begin{cases} 1 - C_{12}G_{w1}G_{w2} - C_{13}G_{w1}G_{w3} - C_{14}G_{w1}G_{w4} = 0 \\ 1 - C_{21}G_{w2}G_{w1} - C_{23}G_{w2}G_{w3} - C_{24}G_{w2}G_{w4} = 0 \\ 1 - C_{31}G_{w3}G_{w1} - C_{32}G_{w3}G_{w2} - C_{34}G_{w3}G_{w4} = 0 \\ 1 - C_{41}G_{w4}G_{w1} - C_{42}G_{w4}G_{w2} - C_{43}G_{w4}G_{w3} = 0 \end{cases} \qquad (4.60)$$

Fig. 4.31 Separated MIMO
CLC of the order n=4

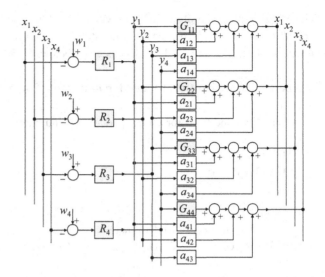

with the transfer functions of the 1st and 2nd separated loops by Eqs. 4.44 and 4.45, the
transfer function of the 3rd separated closed loop by Eq. 4.56 and of the 4th separated
closed loop given below

$$G_{w4}(s) = \frac{R_4(s)G_{44}(s)}{1 + R_4(s)G_{44}(s)} \tag{4.61}$$

The interaction transfer functions are defined like Eq. 4.57 in common form for $j=1, 2,$
3, 4 and $k=1, 2, 3, 4$:

$$C_{jk}(s) = C_{kj}(s) = \frac{a_{jk}(s)a_{kj}(s)}{G_{kk}(s)G_{jj}(s)} \tag{4.62}$$

Example of stability proof of MIMO CLC of the order N=4

The separated MIMO CLC given in Eqs. 4.63 and 4.64 is stable according to
Eq. 4.60–4.62 by $K_{R1}=0{,}8$; $K_{R2}=1{,}2$; $K_{R3}=2$ and $K_{R4}=1$.

$$\mathbf{G_S}(s) = \begin{pmatrix} \frac{1}{1+s} & 1 & 0{,}5 & 0{,}6 \\ 0{,}5 & \frac{2}{1+2s} & 0{,}1 & 0{,}8 \\ 0{,}2 & 1 & \frac{3}{1+3s} & 0{,}9 \\ 0{,}4 & 0{,}2 & 0{,}3 & \frac{4}{1+4s} \end{pmatrix} \tag{4.63}$$

$$\mathbf{R}(s) = \begin{pmatrix} \frac{K_{R1}(1+s)}{s} & 0 & 0 & 0 \\ 0 & \frac{K_{R2}(1+2s)}{2s} & 0 & 0 \\ 0 & 0 & \frac{K_{R3}(1+3s)}{3s} & 0 \\ 0 & 0 & 0 & \frac{K_{R4}(1+4s)}{4s} \end{pmatrix} \tag{4.64}$$

Generalizing all given above stability conditions for separated MIMO CLC of 2nd, 3rd, and 4th order we can derive the common stability condition for the MIMO plant of order n with $j = 1, 2, \dots n$ and $k = 1, 2, \dots n$ as given below.

u **Separated MIMO control of a feed-forward-plant of order N**

- stability conditions for each separate loop and interactions between loops

$$
\begin{cases}
1 + G_{01}(s) = 0 \\
1 + G_{02}(s) = 0 \\
\dots\dots\dots\dots\dots \\
1 + G_{0n}(s) = 0
\end{cases}
$$

$$
\begin{cases}
1 - C_{12}G_{w1}G_{w2} - C_{13}G_{w1}G_{w3} - \dots - C_{1,n-1}G_{w1}G_{w,n-1} - C_{1n}G_{w1}G_{wn} = 0 \\
1 - C_{21}G_{w2}G_{w1} - C_{23}G_{w2}G_{w3} - \dots - C_{2,n-1}G_{w2}G_{w,n-1} - C_{2n}G_{w2}G_{wn} = 0 \\
\dots\dots\dots\dots\dots\dots\dots\dots\dots\dots \\
1 - C_{n1}G_{wn}G_{w1} - C_{n2}G_{wn}G_{w2} - \dots - C_{n,n-2}G_{wn}G_{w,n-2} - C_{n,n-1}G_{wn}G_{w,n-1} = 0
\end{cases} \tag{4.65}
$$

- transfer functions of each separated loop

$$
G_{wk}(s) = \frac{R_k(s)G_{kk}(s)}{1 + R_k(s)G_{kk}(s)} \tag{4.66}
$$

- interaction transfer functions

$$
C_{jk}(s) = C_{kj}(s) = \frac{a_{jk}(s)a_{kj}(s)}{G_{kk}(s)G_{jj}(s)} \tag{4.67}
$$

4.4.3 Stability Check with MATLAB®-Script

Stability check of the 2nd order MIMO control
(See. Fig. 4.32)
The system is stable by $K_{R2} = 1,9$ because all calculated poles have negative real parts:

$-28.5000 + 0.0000i$
$-1.0000 + 0.0000i$
$-1.0000 - 0.0000i$
$-1.0000 + 0.0000i$
$-0.5000 + 0.0000i$
$-0.5000 + 0.0000i$
$0.0000 + 0.0000i$
$0.0000 + 0.0000i$

The system is unstable by $K_{R2} > 2$.

```
 1 -   s=tf('s');                          % Laplace-operator
 2 -   K11=1; K12=1;K21=0.5;K22=2;          % plants gains
 3 -   T11=1; T12=0;T21=0;  T22=2;          % plants time delay
 4 -   G11=K11/(1+s*T11); G22=K22/(1+s*T22); % main plants
 5 -   a12=K12/(1+s*T12); a21=K21/(1+s*T21); % coupling plants
 6 -   KR1=1;   KR2=1.9;                    % controller gains KR
 7 -   Tn1=T11; Tn2=T22;                    % reset times Tn
 8 -   R1=KR1*(1+s*Tn1)/(s*Tn1);     % 1th PI-controllers
 9 -   R2=KR2*(1+s*Tn2)/(s*Tn2);     % 2nd PI-controller
10 -   C=a21*a12/(G11*G22);  % Interaction between lopps 1 and 2
11 -   Gw1=R1*G11/(1+R1*G11);        % closed loop 1 is stable
12 -   Gw2=R2*G22/(1+R2*G22);        % closed loop 2 is stable
13 -   Interaction = 1 - C*Gw1*Gw2;  % stability condition
14 -   poles = tzero(Interaction)    % stable if all poles are negative
```

Fig. 4.32 MATLAB-script for separated MIMO control of the order $n = 2$

Stability check of the 3rd order MIMO control
(See. Fig. 4.33)
The system is stable by $K_{R3} = 4$ but unstable by $K_{R3} = 5$.

Stability check of the 4th order MIMO control
(See. Fig. 4.34)
The system is stable by $K_{R4} = 1$ but unstable by $K_{R4} = 2$.

4.5   Summary

Finally, let us compare the classical methods with bus-approach.

Graphic representation
- Classical approach
  - A graphic view of MIMO plants with transfer functions more than two input/output signals is voluminous, so it has no sense to draw them. The only possible way is to convert the transfer functions to the state space model.
- Bus-Approach
  - A graphic view of MIMO plants of two and higher order is simple and allows to follow the signal ways directly on the block diagram.

Identification
- Classical approach
  - As mentioned above plants, primary given with transfer functions should be derived into state space representations. But the identification of plants, given with transfer functions is easier as it is with state space models.

```
 1 -   s=tf('s');
 2 -   KR1=0.8; KR2=1.2; KR3=4; % unstable by K3=5
 3 -   Tn1=T11; Tn2=T22; Tn3=T33;
 4 -   R1=KR1*(1+s*Tn1)/(s*Tn1);   % 1th PI-controllers
 5 -   R2=KR2*(1+s*Tn2)/(s*Tn2);   % 2nd PI-controller
 6 -   R3=KR3*(1+s*Tn3)/(s*Tn3);   % 3rd PI-controller
 7 -   K11=1;   K12=1; K13=0.5;
 8 -   K21=0.5; K22=2; K23=0.1;
 9 -   K31=0.2; K32=1; K33=3;
10 -   T11=1;    T12=0; T13=0;
11 -   T21=0;    T22=3; T23=0;
12 -   T31=0;    T32=0; T33=3;
13 -   G11=K11/(1+s*T11); a12=K12/(1+s*T12); a13=K13/(1+s*T13);
14 -   a21=K21/(1+s*T21); G22=K22/(1+s*T22); a23=K23/(1+s*T23);
15 -   a31=K31/(1+s*T31); a32=K32/(1+s*T32); G33=K33/(1+s*T33);
16 -   C12=a12*a21/(G11*G22);  % Interaction between loops 1 and 2
17 -   C13=a13*a31/(G11*G33);  % Interaction between loops 1 and 2
18 -   C23=a23*a32/(G22*G33);  % Interaction between loops 1 and 2
19 -   Gw1=R1*G11/(1+R1*G11);     % closed loop 1 is stable
20 -   Gw2=R2*G22/(1+R2*G22);     % closed loop 2 is stable
21 -   Gw3=R3*G33/(1+R3*G33);     % closed loop 2 is stable
22 -   Int_1 = 1 - C12*Gw1*Gw2-C13*Gw1*Gw3; % stability condition 1
23 -   Int_2 = 1 - C23*Gw2*Gw3-C12*Gw1*Gw2; % stability condition 2
24 -   Int_3 = 1 - C13*Gw1*Gw3-C23*Gw2*Gw3; % stability condition 3
25 -   poles1 = tzero(Int_1)      % stable if all poles are negative
26 -   poles2 = tzero(Int_2)      % stable if all poles are negative
27 -   poles3 = tzero(Int_3)      % stable if all poles are negative
```

Fig. 4.33 MATLAB-script for separated MIMO control of the order n = 3

- Bus-Approach
  - Most of industrial plants are easy to identify directly with transfer functions without any conversion, analyzing their step responses or frequency responses.

Analysis and design
- Classical approach
  - The known analysis and design of MIMO control is based upon state space models in the time domain. The use of methods of closed loop control is not possible.
- Bus-Approach
  - The analysis and design of MIMO control is based upon well-known methods of closed loop control (transfer functions, frequency responses) and is easy to use.

Tuning
- Classical approach
  - The commonly used method of state space control is pole placing for state feedback. For the practical use of state feedback, the state space model should be converted back to the transfer functions model.

```
 1 -   KR1=0.8;  KR2=1.2;    KR3 = 2;   KR4 = 1;    % unstable by KR4=2
 2 -   T1 = 1; T2 = 2;    T3 = 3; T4 = 4;   % time constants
 3 -   C12 = tf([0.5*T1*T2    0.5*(T1+T2)    0.5], [2]); % interactions
 4 -   C13 = tf([0.1*T1*T3    0.1*(T1+T3)    0.1], [3]);
 5 -   C14 = tf([0.24*T1*T4    0.24*(T1+T4)    0.24], [4]);
 6 -   C23 = tf([0.1*T2*T3    0.1*(T2+T3)    0.1], [6]);
 7 -   C24 = tf([0.16*T2*T4    0.16*(T2+T4)    0.16], [8]);
 8 -   C34 = tf([0.27*T3*T4    0.27*(T3+T4)    0.27], [12]);
 9 -   Gw1= tf([K1], [1  K1]);          % closed loop 1
10 -   Gw2= tf([K2], [1  K2]);          % closed loop 2
11 -   Gw3= tf([K3], [1  K3]);          % closed loop 3
12 -   Gw4= tf([K4], [1  K4]);          % closed loop 4
13 -   Int1= 1-C12*Gw1*Gw2-C13*Gw1*Gw3-C14*Gw1*Gw4;
14 -   Int2= 1-C12*Gw2*Gw1-C23*Gw2*Gw3-C24*Gw2*Gw4;
15 -   Int3= 1-C13*Gw3*Gw1-C23*Gw3*Gw2-C34*Gw3*Gw4;
16 -   Int4= 1-C14*Gw4*Gw1-C24*Gw4*Gw2-C34*Gw4*Gw3;
17 -   poles1 = tzero(Int1)      % poles of characteristic equation Int1
18 -   poles2 = tzero(Int2)      % poles of characteristic equation Int2
19 -   poles3 = tzero(Int3)      % poles of characteristic equation Int3
20 -   poles4 = tzero(Int4)      % poles of characteristic equation Int4
```

Fig. 4.34 MATLAB-script for separated MIMO control of the order n = 4

- Bus-Approach
  - There are many different well-known tuning methods of standard controllers (P-, I-, PI-, PD-, and PID-controllers) which are easy for practical applications.

Retained static error
- Classical approach
  - The state feedback consists of proportional P-blocks, so the retained static error could not be eliminated. To eliminate it, the feedback should be completed with scaling factor, output feedback or with additional I-terms in the control loop, which complicates the practical use of the state space control.
- Bus-Approach
  - The retained static error is easy to eliminate applying standard controllers with I-term, such as PI- and PID-controllers.

References

1. Zacher, S. (2014). *Bus-approach for feedback MIMO-control*. Verlag Dr. S. Zacher.
2. Zacher, S. (2019). *Bus-approach for engineering and design of feedback control*. Denver. In *Proceedings of ICONEST*, October 7–10, 2019, published by ISTES Publishing (pp. 26–27).
3. Zacher, S. (2020). Bus-approach for engineering and design of feedback control. *International Journal of Engineering, Science and Technology, 2*(1), 16–24. https://www.ijonest.net/index.php/ijonest/article/view/9/pdf. Accessed Jan. 2022.

4. Bellman, R. E. (1961). *Adaptive control processes*. Princeton University Press.
5. Bellman, R. E. (1957). *Dynamic programming*. Princeton University Press.
6. Course of dimensionality. https://en.wikipedia.org/wiki/Curse_of_dimensionality. Accessed Jan. 2022.
7. Zacher, S. (2021). *Regelungstechnik mit Data Stream Management*. Verlag Springer Vieweg. https://link.springer.com/book/10.1007/978-3-658-30860-5. Accessed 20. Jan. 2022.

ASA: Antisystem Approach

5

5.1 What is ASA?

5.1.1 Definition of ASA

ASA (Antisystem Approach) is a method for the analysis and design of a wide range of dynamic systems. The definition of ASA was first published in [2]. This method is the interpretation of the third physical Newton's law for system theory:

> "Whenever one object exerts a force on another object, the second object exerts an equal and opposite on the first. His third law states that for every action (force) in nature there is an equal and opposite reaction. If object A exerts a force on object B, object B also exerts an equal and opposite force on object A. In other words, forces result from interactions". (Quote [3]: *Newtons Law's of Motion*)

According to ASA, for every dynamic system that transfers its inputs $\mathbf{Y}$ into outputs $\mathbf{X}$ with a matrix operator $\mathbf{A}$ in one direction, there is an equal system with the transposed matrix operator $\mathbf{A}^T$, which transfers other inputs $\mathbf{W_x}$ into outputs $\mathbf{W_y}$ in opposite direction. In other words, a single isolated dynamic system is not to be investigated alone. The abstract algebra and group theory denote systems $\mathbf{A}$ and $\mathbf{A}^T$ as antisymmetric:

> Two things are called antisymmetric, if they are symmetric, but are acting in opposite directions (Quote [4]).

The most important feature of ASA is the balance between the system $\mathbf{A}$ and its antisystem $\mathbf{A}^T$ by any values of input vectors $\mathbf{X}$ and $\mathbf{W_y}$, namely, a balance of scalar values E_x and E_y, which are called "energy" or "intensity"

$$E_x = E_y \tag{5.1}$$

In Gl. 5.1, the values of E_x and E_y are dot products of input and output vectors of the system and antisystem

$$E_x = \mathbf{X} \cdot \mathbf{W}_x^T \tag{5.2}$$

$$E_y = \mathbf{Y} \cdot \mathbf{W}_y^T \tag{5.3}$$

An example is shown in Fig. 5.1. The first system (called the original system) transfers its inputs $\mathbf{Y}$ into outputs $\mathbf{X}$ in one direction, and the second system (called the antisystem) transfers the inputs $\mathbf{W_x}$ into outputs $\mathbf{W_y}$ in the opposite direction on the original system

$$\mathbf{Y} = \begin{pmatrix} y_1 \\ y_2 \end{pmatrix} \quad \mathbf{X} = \begin{pmatrix} x_1 \\ x_2 \end{pmatrix} \quad \mathbf{W}_x = \begin{pmatrix} w_{x1} \\ w_{x2} \end{pmatrix} \quad \mathbf{W}_y = \begin{pmatrix} w_{y1} \\ w_{y2} \end{pmatrix}$$

The input vector $\mathbf{X}$ of an original system is usually given by the engineering task and could not be changed. Instead of it, the input vector $\mathbf{W_y}$ of the antisystem could be the chosen arbitrary by the user. This feature is the main advantage of ASA, namely:

> The antisystem input vector $\mathbf{W_y}$ could be arbitrary chosen in such a way that the engineering task of dynamic system $\mathbf{A}$ will be solved upon antisystem $\mathbf{A}^T$ easier and more optimal than the solution of the original system $\mathbf{A}$.

In other words, the ASA enables to analyze an antisystem $\mathbf{A}^T$ with vectors $\mathbf{W_x}$ and $\mathbf{W_y}$ instead of the original system $\mathbf{A}$ with vectors $\mathbf{X}$ **and** $\mathbf{Y}$. The antisystem does not have to be a physical system; it can also be a mathematical model of the original system. By $n=1$ or

Fig. 5.1 System and antisystem with $n=2$ variables

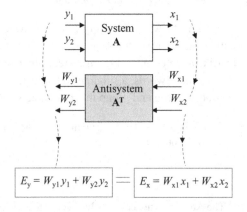

$n = 2$ ASA has no significant advantages against conventional analysis with original vectors. The bigger is n, the more reduction of calculations is expected by the use of ASA.

In this chapter, the basics of ASA are given and its features are shown. In the twenty-five years since the first publication of the ASA, there were developed different applications, which enable engineering and control of physical systems by analyzing the corresponding virtual antisystems.

It is briefly shown that how the ASA was used for electrical engineering, chemical engineering, and informatics, focusing on the ASA solutions for control and data stream management developed in this book.

5.1.2   The Long Way to ASA

The assistant researcher *Sacharjan* introduced in 1968, in [5], a new method to define the transfer functions of multivariable multi-stage control systems and called it "compression of variables". This method was developed in [2] as an option for artificial neural networks and generalized in [6] to the antisystem approach for the engineering of control systems. But the way to ASA began long ago before the "compression of variables". In Fig. 5.2, it is illustrated how the Newton's 3rd law of motion turn into an antisystem approach (ASA). A short review of methods that led to ASA is given below in this section.

5.1.2.1 Duality
The way to the antisystem approach began with *yin and yang* circle, a widespread symbol from ancient Chinese philosophy (Fig. 5.3).

> "The circle is dual as it consists of two different colors. There are two parts that complement each other. Within each part is a small circle of the opposite color, indicating the interpenetration of reciprocity. The borders and the colors seemed constantly moving. The only thing that remains constant is duality." (Quote: Source [6, p. 89])

In general, duality can be found wherever things appear in pairs, be it electrical signals (positive and negative charge) or geometric lines and figures. *J.C. Maxwell* (1873) established the duality between electric and magnetic fields. Complex conjugate numbers are an example of duality in mathematics. They always appear in pairs, and have the same real and imaginary parts with the imaginary parts having different signs. The next example of duality, which fits the way to ASA, provides the matrix calculation. A regular square matrix $\mathbf{A}$ and its inverse matrix $\mathbf{A}^{-1}$ can be viewed as dual objects since they complement each other to form the identity matrix $\mathbf{E}$

$$\mathbf{A} \cdot \mathbf{A}^{-1} = \mathbf{E} \tag{5.4}$$

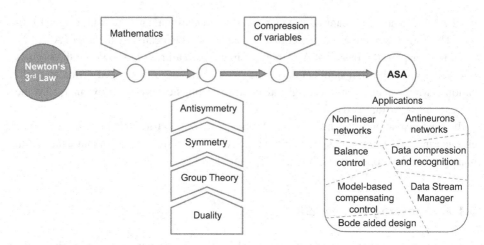

Fig. 5.2 From Newton's 3rd law to antisystem approach

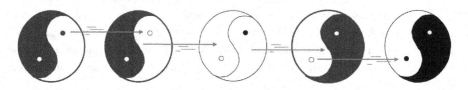

Fig. 5.3 Yin and yang symbol for polar opposites and related dual forces or principles that do not fight but complement each other. *Source* [6, p. 89]

At this place, let's finish the review of duality. More examples of duality in mathematics, electrotechnics, and system theory are given in [6] and [21]. The next stop on the way to the antisystem approach is group theory.

5.1.2.2 Group Theory

Group theory deals with groups of objects that transform themselves into themselves with identity transformations. In other words, the objects in a group have identical topological features. In abstract group theory, no reference is made to the nature of the group elements. The objects could have different physical natures, e.g., prime numbers in algebra, cells in biology, etc. There are many ways to define a group of objects according to similar structures and algebraic properties.

Otherwise, two objects that always appear in pairs form a duality. In other words, group theory is a mathematical description of duality. It is well known that objects of group theory are mostly geometric figures, chemical and biological elements, or objects of physics. The system theory and control engineering are only slightly affected by group theory. The Laplace transform as an example of group theory application in the control theory is discussed in [7]. The graphical representation of the Laplace transform is shown in Fig. 5.4.

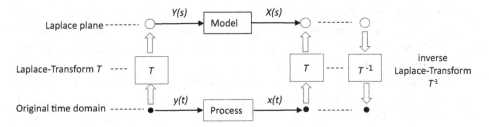

Fig. 5.4 Laplace transform and inverse transform as dual functions. *Source* [7, p. 129]

"Two levels of signal processing can be seen, namely: the real level (process) with the input *x(t)* and the output *y(t)* and the transformed level (model) with transformed inputs/outputs *X(s)* and *Y(s)*. By means of an inverse Laplace transform, it is possible to return from the Laplace plane to the original." (Quote: Source [7, p. 129])

According to group theory, the transformation T and the inverse transformation T^{-1} form a group. Mathematically, it is denoted with the unit element I of the group theory as follows:

$$T \cdot T^{-1} = I \qquad (5.5)$$

The last expression leads from group theory to the symmetry operation:

"A group that consists of only two separate objects that can be transformed into themselves or remain identical after symmetry operations is called dual." (Quote: Source [7, p. 129])

5.1.2.3 Symmetry
The symmetrical representations of nature are well known and common in natural sciences and in technology. The simplest definition of symmetry in relation to geometric figures is given by *Shafranovsky:*

"A figure is called symmetrical if it consists of the same, regularly repeating parts". (Quote: Source [8])

The definition of the symmetry operation as transformations is given by *Weyl* in [9]:

"The symmetry operation of a geometric figure corresponds to a transformation S of the point P to the point P* and an inverse transformation S^{-1} of the point P* to the point P." (Quote: Source [9])

The mathematical tools of symmetry were developed in the 1930s and since then have been massively applied in many areas of the natural sciences such as physics, classical mechanics, mathematics, and chemistry because they significantly reduce the dimension of a problem. Even slight deviations from symmetry by dual objects or processes immediately indicate an error. However, only a few symmetry properties or symmetry operations have been used specifically for the design of closed control loops although it should bring the same advantages in control engineering as in the natural sciences. In [6] is given review of symmetry operations by control engineering. In [7] is pointed on the method of symmetric optimum, described in [10], as an example of effective use of symmetry by tuning of standard controllers.

The symmetric optimum method was developed by *C. Kessler* (1958). The method is suitable for the relatively complicated control loops, which consist e.g., of a double I term (integrated block) in addition to proportional elements with a delay, controlled with a PI controller:

$$G_0(s) = \frac{K_{PR}(1 + sT_n)}{sT_n} \cdot \frac{K_{PS}}{1 + sT_1} \cdot \frac{K_{IS}}{s}$$

The PI controller is adjusted so that the crossover frequency

$$\omega_d = \omega_m$$

is set exactly in the middle of the frequency section of the two corner frequencies, as shown in Fig. 5.5.

$$\omega_{E1} = \frac{1}{T_n}$$

$$\omega_{E2} = \frac{1}{T_1}$$

Thus, the amplitude response and the phase response of the open loop in the Bode plot are symmetrical with respect to the crossover frequency ω_d. A factor k is introduced such as:

$$T_n = kT_v \tag{5.6}$$

It is recommended to choose $k = 4$ (see Chap. 3). The tuning rule for setting the controller is derived in [10] with two conjugate complex poles arise, which of course also lie symmetrically with respect to the real axis in the s-Laplace plane, like shown in Fig. 5.5 of Chap. 3. Through such a symmetrical representation or through the corresponding control setting of Eq. 5.6 is reached the maximum phase margin $\varphi_{Rd} = 37°$.

5.1.2.4 Antisymmetry

The principle of antisymmetry based upon operations of symmetry, but with the opposite direction of variables, has been known since 1929. Simple but illustrative examples are given in the fundamental work [11] about symmetry/antisymmetry in mathematics, physics, and chemistry.

Atomic physics, according to which every particle has its antiparticle, delivers convincing successes of antisymmetry. The mathematical description of a system and its

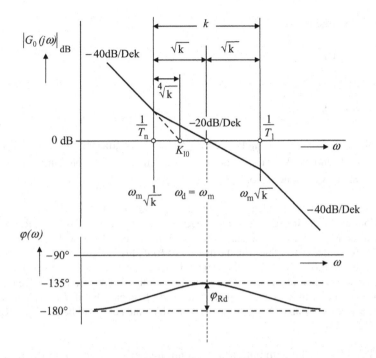

Fig. 5.5 Bode-plot of the open control loop with controller tuned upon symmetric optimum. *Source* [10, p. 244]

antisystem was proposed by *Paul Dirac* as brackets for system $\mathbf{X^i}>$ and for antisystem $<\mathbf{W^i}$. Applying this description for Gl. 5.1–5.3, the „energy" balance will be written as follows:

$$< \mathbf{X} \cdot \mathbf{W}_x^T >=< \mathbf{Y} \cdot \mathbf{W}_y^T > \qquad (5.7)$$

In crystallography, a distinction is made between the following types antireflection, antiidentity, antiinversion:

Antireflection: $L^+ \to R^+$ and $L^- \to R^-$

Antiidentity: $L^+ \to L^-$ and $R^+ \to R^-$

Antiinversion: $L^+ \to R^-$ and $L^- \to R^+$

Even more complicated operations lead to Color antisymmetry

Red $(1,+) \to$ Blue $(2,+) \to$ Green $(p,+) \to \ldots$
Red $(1,-) \to$ Green $(2,-) \to$ Blue $(p,-) \to \ldots$

and to Cross-Symmetry

$$S_1 + S_2 + \ldots \rightarrow T_1 + T_2 + \ldots$$
$$T_1 + T_2 + \ldots \rightarrow S_1 + S_2 + \ldots$$
$$S_1 + T_2 + \ldots \rightarrow T^{-1} + S^{-1} + \ldots$$
$$S_1 + T_2 + \ldots \rightarrow T_1 + S_2 + \ldots$$

However, there was no antisymmetric representation in engineering till 1933. The review of antisymmetry and duality applications in mathematics, electrotechnics, and system- and control theory was done in [6, 7] and [12].

"The periodical "VDE-Nachrichten" reported on the 23.03.2001 about project "Active-Noise-Control" of the Research Centre the German Air- and Space-Drive (DLR) co-operated with the MTU Aero-Engine and European Aeronautic Defense and Space Company Germany for the noise-damping with the method of *Paul Lueg* (patented 1933).

According to this method the noise can be damped, when two identical generators will be compared. The first one is the noise generator and the second one is the record of the first generator, i.e. with the same magnitude, the same frequency but different phase. By an observer appears the sum of both, oscillations, in which the noise will be damped." (Quote: Source [12, p. 55])

The MATLAB®/Simulink simulation for a simplified example was published in [6] and [12] and shown in Fig. 5.6. The trigonometric functions *sin(t)* and *cos(t)* are considered antisymmetric. The balancing is based on the known equation

$$\sin^2 t + \cos^2 t = 1 \tag{5.8}$$

The *cos* function is simulated using a sine wave generator with the same amplitude and frequency as the *sin* generator, but with a phase $\pi/2$, i.e., $cos(t) = sin(t + \pi/2)$ The disturbance is given by a periodical pulse generator. Without disturbance, there is a balance Gl. 5.8, i.e., the output shown by the scope of Fig. 5.6 is equal to 1. When a disturbance occurs, duality is violated, resulting in a loss of balance.

Another example of antisymmetry, given in [7], is the well-known block diagram of a closed loop (Fig. 5.7). The plant and the controller are considered functionally as antisymmetric. Through double symmetry operations, they both are represented as one "balanced" system with "balanced energy", written corresponding to Dirac's equation Gl. 5.7

$$< y_R(t) \cdot x(t) > = < w(t) \cdot z(z) > \tag{5.9}$$

5.1.2.5 Compression of Variables
The first application of antisymmetry for control, publishes in [5], indicates a balance between input and output variables of a multi-stage multivariable system. Such systems are mathematically described with linear partial differential equations with variable coefficients. It is known that the solution, optimization, and control of multi-stage multi-

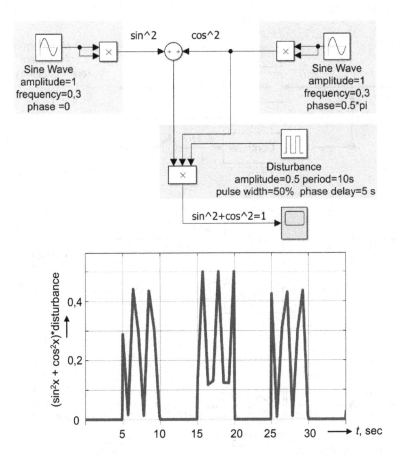

Fig. 5.6 Sine and cosine generator simulated as a system and antisystem with the periodical disturbance

variable systems are possible only by low order differential equations or with numerical methods like dynamic programming [13, 14]. To avoid this problem, the state vectors were compressed in [5] into scalar values. As a result, the computing time and memory requirements were significantly reduced and the solution to the high dimensionality of systems was possible (Fig. 5.8).

Significant is that *Richard Bellman*, the developer of the dynamic programming, called this problem as "a curse of dimensionality". With his remark, *Bellman* pointed out problems, which can be easily solved by low dimensions, but require vastly more computer time, if an extra dimension is added to the mathematical description of a system.

The method, introduced in [5], was generalized in [15] for different kinds of typical system connections like series, parallel, and feedback. The "compression's theorem" introduced in [15] is briefly represented below.

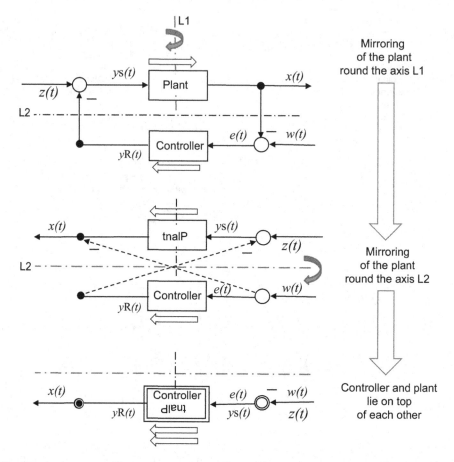

Fig. 5.7 Antisymmetry of plant and controller. *Source* [7, p. 148]

Compression's theorem according to [15], pp. 13–15.

Let us consider the main process P, the inputs X^{i-1} and the outputs X^{i-} of which are column vectors of dimension n, and a reversible process P_w, called compressing process, which transforms a set of variables W^i into W^{i-1}, whose elements are row vectors of the same dimension n (Fig. 5.9a)

$$P : X^{i-1} \rightarrow X^i$$

$$P_w : W^i \rightarrow W^{i-1}$$

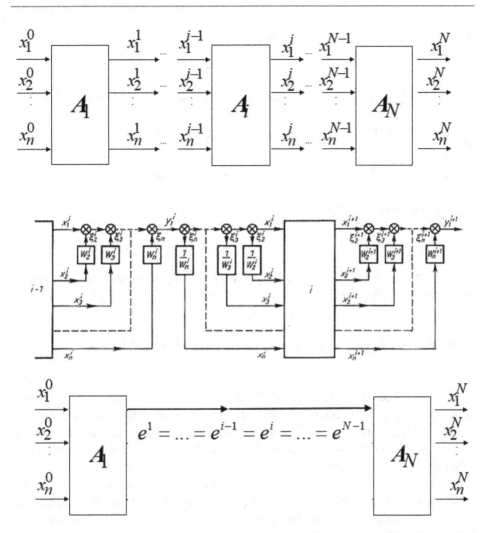

Fig. 5.8 Compression of variables for calculation of transfer functions of a multivariable multi-stage system. *Source* [5]

If we combine the main P process and the compressing P_w process in the form of a dot product, the input and output of the connection are scalars

$$e^{i-1} = W^{i-1}X^{i-1} \qquad e^i = W^iX^i,$$

and if we use the main process as a compressing process, i.e.

$$P = P_w,$$

a   **b**

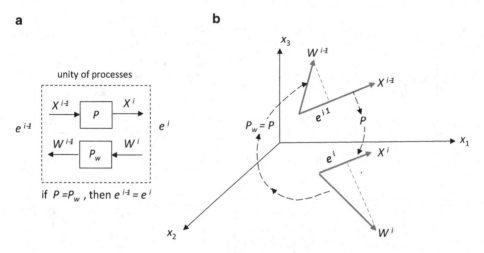

Fig. 5.9 **a** Process P combined with the compressing process P_w; **b** Invariant of projections e^{i-1}- and e^i of the vectors W onto vectors X

then the scalar variables e^{i-1} and e^i will be equal for any input variables X^{i-1} and W^i, i.e.

$$e^{i-1} = e^i \tag{5.10}$$

The transformation of the input vector X^{i-1} through process P is illustrated in [15] and shown in Fig. 5.9b in three-dimensional space as the rotation of this vector by an angle φ and a change of its amplitude. According to the compression's theorem, there is an invariant, which consists in the equality of the projections of the vectors W onto vectors X for any inputs X^{i-1} and W^i.

Finally, we arrived at the last "station" on Fig. 5.2, finished the "trip" from 3rd Newton's law of motion to ASA. Looking back and summarizing this section, we can of course notice the similarities between Gl. 5.1, 5.4, 5.5, 5.7, and 5.10, although they describe different features of a pair of systems: duality, topology, symmetry, and antisymmetry. It is a kind of indication or cue that it exists as a common feature by certain pairs of systems which could be used by engineering. This common feature is called ASA in [2, 6, 7, 10, 12].

It remains only to show examples of the practical implementation of ASA in various fields of technology before concentrating on the main goal of the book, namely, the use of ASA for control and data stream management.

Fig. 5.10 Discretized non-linear electric field (above) and its antisystem. *Source* [12, p. 63]

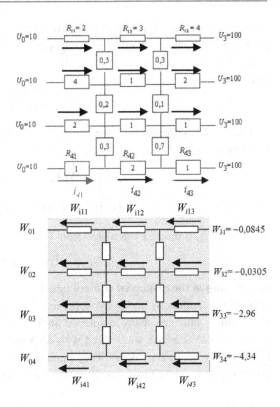

5.1.3   Examples of ASA Implementations

Non-linear electric fields

Supposing given is a non-linear electric field that is discretized and linearized as a grid of ohmic resistances (Fig. 5.10, above). The initial and final values of electrical voltages U_0 and U_3 are given as a row vector. To define is the current on only one resistance, e.g., the current i_{41}. The known solution consists of representing the given multi-stage system through matrices **A** and **B**

$$\mathbf{I} = \mathbf{B}^{-1}(\mathbf{U}_0 - \mathbf{A}\mathbf{U}_3) \tag{5.11}$$

Then the system Gl. 5.11 of $N = 18$ linear equations with $N = 18$ unknown values will be solved and after that will be the desired value of current i_{41} as the 4th component of vector **I**.

Using ASA, we can calculate only the desired value i_{41} without calculating the whole vector **I**. For this purpose, we first build the antisystem as shown in Fig. 5.10 below.

The "energy" balance

$$\mathbf{W}_3\mathbf{U}_3 = \mathbf{W}_0\mathbf{U}_0 + \mathbf{W}_i\mathbf{I} \tag{5.12}$$

Then we choose and apply to the input of antisystem such a vector $\mathbf{W}_3$ so that all components of this vector should be orthogonal to the components of the given vector $\mathbf{U}_3$ except for only one component, namely, the 4th component. It is the following for the given system

$$\mathbf{W}_3 = [-0,0845 - 0,0305 - 2,96 - 4,34]$$

As a result, all components of the "energy" vector $\mathbf{E}_0$ are equal to zero except for the 4th component

$$\mathbf{E}_0 = \mathbf{W}_0\mathbf{U}_0 = [00027,66] \tag{5.13}$$

$$\mathbf{E}_3 = \mathbf{W}_3\mathbf{U}_3 = [0,3530,2782,875 - 11,1] \tag{5.14}$$

The current i_{41} is directly calculated from Gl. 5.12

$$i_{41} = (\mathbf{W}_i)^{-1}(\mathbf{E}_3 - \mathbf{E}_0) = 24,9$$

Supervising of the displayed control panel

One of the frequently tasks by supervising of technological processes is to recognize, which value of many displayed indicators, e.g., of 300 or more sensors, has been changed. This task was solved with ASA in [2]. The sensors (indicators), shown in Fig. 5.11, are defined as "data". They build matrix $\mathbf{A}$ that is exported in SCADA (Supervisory Control and Acquisition). There are two matrices in the SCADA:

- actual matrix $\mathbf{A}^2$ (system),
- pattern matrix $\mathbf{A}^1$ (antisystem)

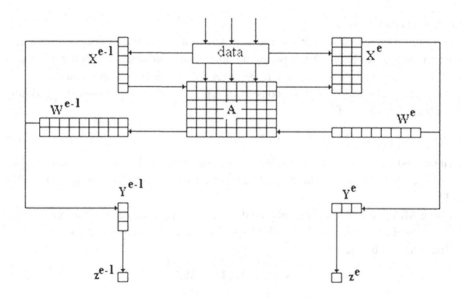

Fig. 5.11 Data compression and recognition with ASA. *Source* [12, p. 64]

The actual matrix $\mathbf{A}^2$ is compared with the pattern matrix $\mathbf{A}^1$. The sensor with the changed values should be recognized and located.

The test vector $\mathbf{X}^0$ will be applied to the input of the system. The test matrix $\mathbf{W}^1$ (not a vector!) will be applied to the inputs of the antisystem. In this case, the "energy" is not more a scalar like in Gl. 5.1 or 5.10, and not a vector, like in Gl. 5.13 and 5.14, but a matrix. The "energy" matrices, denoted in Fig. 5.11 as $\mathbf{Y}^0$ and $\mathbf{Y}^1$, will be calculated and the checksum of their components z_x and z_w will be compared to each other. If $z_x = z_w \, z_x$, then there is no change in the whole supervised panel. Otherwise, the condition $z \neq z_{new}$ means that a variable is changed. Comparing values of $\mathbf{Y}^1$, $\mathbf{X}^0$, $\mathbf{X}^1$, $\mathbf{W}^0$, and $\mathbf{W}^1$, it is possible to exactly define the number of the block with the changed component and the error.

An example is given in [12] and shown in Fig. 5.12. It results in the solution

$$\Delta a_{45} = \frac{4,5}{3 \cdot 5} = 0,3$$

The changed component a_{45} of the matrix $\mathbf{A}^2$ is calculated as

$$a_{45new} = a_{45old} + \Delta a_{45} = 4,0 + 0,3 = 4,3$$

Artificial neural networks with antineurons

Artificial neural networks are well known since 1960. The examples one can find in [10], pp. 415–421, [16], pp. 149–192, and [17], pp. 157–176. Based upon ASA, the new kind of neural network was proposed in [16] and implemented with PLC. This network consists of neuron-antineuron pairs, two of them are shown in Fig. 5.13. The white circles are neurons and the black circles are antineurons. Both have the same weight W.

"Each pair has the following features:
- there is balance of "energy" $e_{i-1} = e_i$ for each pair.
- the weights W_i will be taken on from patterns a_i by training
- the "energy" of the network

$$E = \sum_i e_i$$

If the energy is minimal, then the actual input $\mathbf{X}$ is equal to the training input $\mathbf{A}$. On this way, the network could recognize the input X as pattern A." (Quote: Source [12, p. 65])

5.2   Basics of ASA

"Symmetry can be found almost everywhere if you know how to look for it." Quote: Marjorie Senechal and George Fleck (1977) Patterns of symmetry. University of Massachusetts Press.

a) No change by parameters

Matrix A^1:

$A=$

2,4	4,4	5,2	4,8	4,8	2,0	3,6	3,6	2,4
5,6	4,8	4,0	5,6	3,2	5,6	5,6	4,0	4,0
4,4	4,8	4,8	4,4	3,6	3,2	2,4	4,8	4,0
3,6	2,0	4,8	4,0	4,0	4,8	4,4	2,8	3,6
2,4	3,6	2,0	3,2	4,0	5,6	2,8	4,4	5,6
3,2	5,2	4,8	5,6	3,2	3,6	3,6	4,8	2,8

$W^1=$ | 6 | 5 | 4 | 3 | 2 | 1 |

$X^0=$
1
2
3
4
5
6
7
8
9

$X^1=$
26,8	55,2	75,6
27,2	72,0	107,2
28,4	54,8	91,2
22,0	64,8	85,6
15,6	66,4	105,2
28,0	60,0	88,8

$W^0=$
60,0	69,6	70,4	74,4	59,2	52,8	59,2	60,8	50,4
18,8	18,4	23,2	24,0	23,2	29,2	22,4	22,0	24,8

$Y^0=$ | 2675,2 |
| 1068,4 |

$Y^1=$ | 535,6 | 1297,6 | 1910,4 |

$z=$ | 3743,6 |

b) One sensor changed it's value

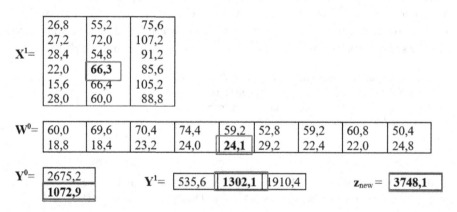

$X^1=$
26,8	55,2	75,6
27,2	72,0	107,2
28,4	54,8	91,2
22,0	**66,3**	85,6
15,6	66,4	105,2
28,0	60,0	88,8

$W^0=$
60,0	69,6	70,4	74,4	59,2	52,8	59,2	60,8	50,4
18,8	18,4	23,2	24,0	**24,1**	29,2	22,4	22,0	24,8

$Y^0=$ | 2675,2 |
| **1072,9** |

$Y^1=$ | 535,6 | **1302,1** | 1910,4 |

$z_{new}=$ | **3748,1** |

Fig. 5.12 Recognition of the changed value of sensors with ASA. *Source* [12, p. 64]

5.2.1   Introduction: ASA is Everywhere

Repeating the above given quote, there is every reason to believe the same for antisystems, namely:

Antisystems can be applied almost everywhere if you know how to build them.

Fig. 5.13 Artificial neuron and antineuron. *Source* [16, p. 184]

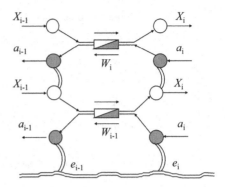

According to the ASA, an antisystem can be determined for each system, resulting in a balance of "energy" between these two systems. The prefix "anti" indicates that the two systems are identical but act against each other. There are possible different rules on how to build an antisystem to the given system and their "energy" balance:

- Dot product by Gl. 5.1,
- Regular and inverse matrices by Gl. 5.4,
- Non-linear symmetry operations by Gl. 5.5,
- The sum of two functions by Gl. 5.8.

The antisystem does not have to be a physical system; it could be the mathematical model of the original system. Through the balance between the two systems, the state of equilibrium is reached. If the identity between system and antisystem is disturbed, the balance is also disturbed, and the equilibrium is lost as shown in Fig. 5.6.

The balance is thus a feature by which the original system can be monitored and controlled. The dimension of a multivariable system is reduced because:

- the state vectors are compressed into scalars, e.g., like Gl. 5.10.
- the state matrices of dimension are compressed into vectors, e.g., Gl. 5.13, 5.14.

The main advantage of ASA is that engineering and monitoring of antisystem or of balance can be done much easier than the original system. The computing time and memory requirements of such systems are significantly reduced.

In this way, some new control engineering applications are developed. Despite the complexity of these applications, the ASA can be easily explained, which is certainly one of the advantages of the method. Some simple examples of antisystems and of the use of the balance of "energy" are given below.

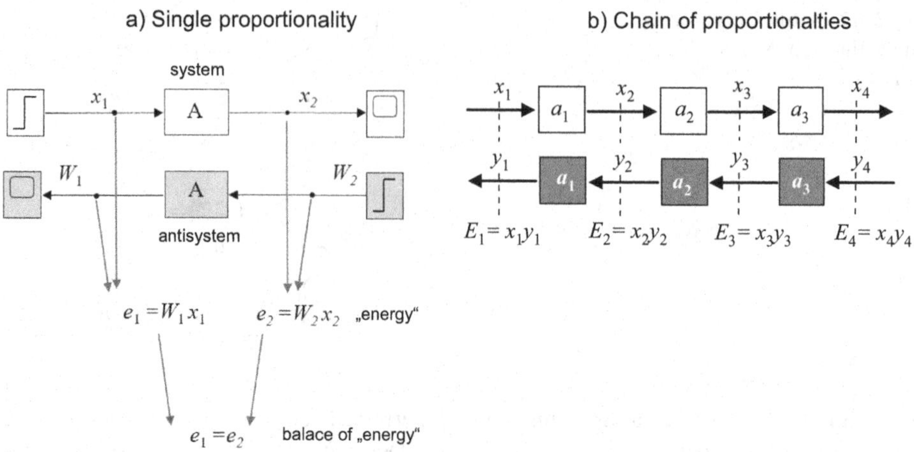

Fig. 5.14 Simple examples of system and antisystem; **a** Single proportionality; **b** Chain of proportionalities

5.2.2   Balance of Functions and Numbers

Proportionality

A simple system is shown in Fig. 5.14a with the proportionality factor A. The system transfers the input variable x_1 into x_2:

$$x_2 = A \cdot x_1$$

Another input variable W_2 is transferred into output variable W_1 with the same proportionality factor A but in the opposite direction

$$W_2 = A \cdot W_1$$

There is no doubt that by all arbitrary chosen values of input variable x_1 and W_1, there is a balance of the so-called "energies" e_1 and e_2

$$\begin{aligned} e_1 &= W_1 \cdot x_1 \\ e_2 &= W_2 \cdot x_2 \end{aligned} \qquad \rightarrow \qquad e_1 = e_2$$

Without going into the discussion about practical use of a system and antisystem, let us only notice that there are four levels for system analysis and design:

- system,
- antisystem,
- "energy",
- balance of "energy".

In other words, ASA is a kind of transformation from the original level (system) to the next level.

Chain of proportionalities
Let's consider two chains with the same proportionality factors a, b, and c, as it is shown in Fig. 5.14b, which transfer two different input values x_1 and y_4. Let's note the direction of the arrows: they are directed toward each other on the upper and lower chain. To make this difference even clearer, the bottom chain is shown shaded. The upper chain is called the system, the shaded chain is the antisystem.

The chain of proportionality factors is described below in series connection

$$
\begin{aligned}
x_2 &= a_1 x_1 & y_3 &= a_3 y_4 \\
x_3 &= a_2 x_3 & y_2 &= a_2 y_3 \\
x_4 &= a_3 x_3 & y_1 &= a_1 y_2
\end{aligned}
\tag{5.15}
$$

To each cross-section is calculated the corresponding product, which is called "energy"

$$
\begin{aligned}
E_1 &= x_1 y_1 \\
E_2 &= x_2 y_2 \\
E_3 &= x_3 y_3 \\
E_4 &= x_4 y_4
\end{aligned}
\tag{5.16}
$$

It is very easy to proof, that for arbitrary chosen input values x_1 and y_4 except of zero, the "energy" balance is valid:

$$
E = E_1 = E_2 = E_3 = E_4
\tag{5.17}
$$

The balance of energy (let us further write this word without quotation marks!) is immediately disturbed if at least one factor a, b, and c, in the system (chain above in Fig. 5.14b) differs from the corresponding factor of antisystem (shadowed chain below).

No matter how strange it may sound, the examples of Fig. 5.14 represent the same antisystem approach, which made it possible to create many applications mentioned above and which are described in the following chapters of this book. Of course, such applications as Gl. 5.15 and 5.16 don't fit engineering tasks and could be used for simple algebraic tasks. It was done in [18, 19], [20].

Compression and extraction of numbers
The system shown in Fig. 5.15 above compresses three inputs x_1, x_2, and x_3 in one output y, which is the weighted sum of inputs with factors a_1, a_2, and a_3

$$
\begin{aligned}
y_3 &= a_1 x_1 \\
y_2 &= a_2 x_2 & y &= y_1 + y_2 + y_3 \\
y_3 &= a_3 x_3
\end{aligned}
\tag{5.18}
$$

The antisystem, shown in Fig. 5.15 below, extracts three outputs p_1, p_2, and p_3 from one arbitrary chosen input q

$$q_1 = q_2 = q_3 = q \qquad \begin{aligned} p_1 &= b_1 q_1 \\ p_2 &= b_2 q_2 \\ p_3 &= b_3 q_3 \end{aligned} \tag{5.19}$$

There is the following relationship between numbers x_1, x_2, x_3 and p_1, p_2, p_3

$$\frac{a_1}{b_1} p_1 x_1 + \frac{a_2}{b_2} p_2 x_2 + \frac{a_3}{b_3} p_3 x_3 = yq \tag{5.20}$$

Let us denote the ratios between proportionality factors a_k and b_k as factors c_1, c_2, and c_3

$$\begin{aligned} c_1 &= \frac{a_1}{b_1} \\ c_2 &= \frac{a_2}{b_2} \\ c_3 &= \frac{a_3}{b_3} \end{aligned} \tag{5.21}$$

Then the balance of energy between inputs x_1, x_2, and x_3 of the system and inputs q_1, q_2, and q_3 of the antisystem is given below

$$c_1 p_1 x_1 + c_2 p_2 x_2 + c_3 p_3 x_3 = yq \tag{5.22}$$

5.2.3   Balance of Algebraic Equations

Linear algebraic equations

As an example, in [18] is described the system of two algebraic equations with two unknowns (Fig. 5.16)

Fig. 5.15 Compression and extraction of numbers

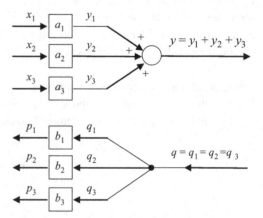

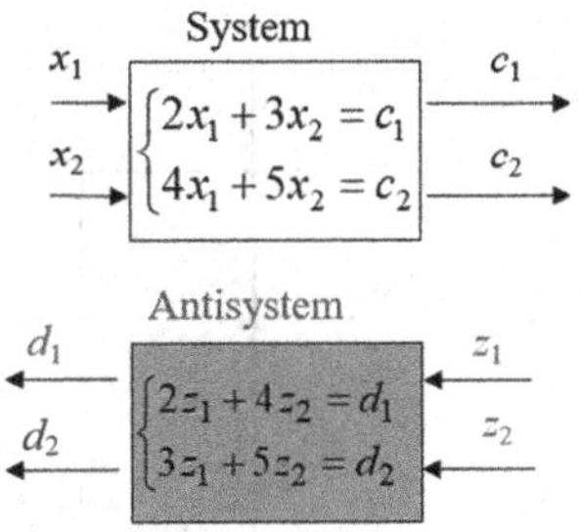

Fig. 5.16 ASA-solution of the system of two linear algebraic equations. *Source* [18, p. 3]

$$\begin{cases} 2x_1 + 3x_2 = 28 \\ 4x_1 + 5x_2 = 48 \end{cases} \tag{5.23}$$

The corresponding antisystem is

$$\begin{cases} 2z_1 + 4z_2 = d_1 \\ 3z_1 + 5z_2 = d_2 \end{cases} \tag{5.24}$$

Choosing arbitrary

$$z_1 = 1$$
$$d_2 = 0$$

we define from the antisystem values of z_2 and d_1

$$3z_1 + 5z_2 = d_2 = 0 \quad \rightarrow \quad z_2 = -0,6$$

$$2 \cdot 1 + 4 \cdot (-0,6) = d_1 \quad \rightarrow \quad d_1 = -0,4$$

Finally, from the energy balance

$$d_1 x_1 + d_2 x_2 = c_1 z_1 + c_2 z_2 \tag{5.25}$$

follows the solution

$$(-0,4)x_1 + 0 \cdot x_2 = 28 \cdot 1 + 48 \cdot (-0,6) \quad \rightarrow \quad x_1 = 2$$

$$2 \cdot 2 + 3x_2 = 28 \quad \rightarrow \quad x_2 = 28$$

Another example of the linear algebraic system, which is solved with the means of ASA is given in [18] and shown in Fig. 5.17. The energies of a system and antisystem are represented as two equal surfaces corresponding to the balance between both systems of linear algebraic equations.

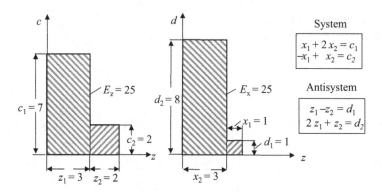

Fig. 5.17 Balance of energies is graphically illustrated by two equal surfaces. *Source* [18, p. 65]

5.3 ASA Application for Control

The balance of "energy" between a system and the corresponding antisystem by simple examples, shown in the previous section, is also valid in systems of high dimensionality with transfer functions and variables of the closed loop. It is shown in the following, which of the previously mentioned balances was implemented of this book. The details of these implementations are given in the appropriate chapters:

- "Balance control" in Chap. 11,
- "Compression and extraction of variables" in Chaps. 8, 9 and 10,
- "Multivariable control (MIMO)" in Chap. 10,
- "ASA control" in Chap. 11.

5.3.1 Balance Control

In Fig. 5.18a, it is shown that the balance of energy of Fig. 5.14 can be applied for control. The system is represented with the plant $G_s(s)$ to be controlled, the antisystem is a pattern $G_M(s)$ or the desired plant behavior. As far as both transfer functions $G_s(s)$ and $G_M(s)$ are equal, there arises no error. Otherwise, the error E will be sent further to the controller, not shown in Fig. 5.18a. Significant for this ASA option is it, that the error E is built not as classic difference between energy e_1 and e_2 but as ratio. The MATLAB® / Simulink model in Fig. 18 (b) confirms the ability to work with such controlling error. The implementation and evaluation of the quality of balance control are discussed in [23].

a

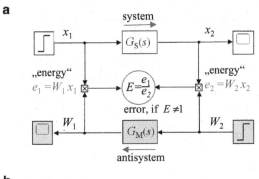

b

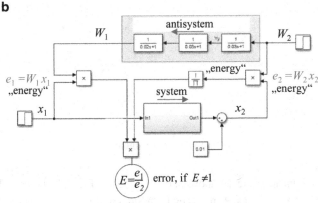

Fig. 5.18 Balance control: **a** conception; **b** simulation with MATLAB®

5.3.2   Compression and Extraction of Variables

One of the tasks by the control of multivariable systems is the decoupling of control loops from one another. Using the compression and extraction rules of Fig. 5.15, it is possible to separate certain variables from a bundle of many other variables even if instead of the factors a_1, a_2, and a_3, the transfer functions g_1, g_2, and g_3 of dynamical elements are considered (Fig. 5.19).

In Fig. 5.19 above is shown one stage of a multivariable system like given in Fig. 5.8 but with transfer functions and with three Laplace transformed inputs. To reduce the dimensionality by transfer of these three variables to the next stage is built the weighted sum

$$e_1 = g_1 y_1^{in} + g_2 y_2^{in} + g_3 y_3^{in}$$

The task of the receiver in the next stage, which gets the compressed signal $e_2 = e_1$, is to extract only one certain signal from the sum, e.g., y_3^{in}, and represent it as w_3^{out}

$$e_2 = g_{1anti} w_1^{in} + g_{2anti} w_2^{in} + g_{3anti} w_3^{in}$$

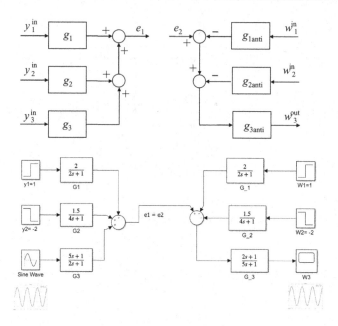

Fig. 5.19 Extraction of signal y^3_{in} from the bunch of signals e_1

The solution is shown in Fig. 5.19 above and simulated in Fig. 5.19 below. This method is successfully applicated in Chaps. 9 and 10 by Data Stream Manager *Router* and *Terminator*.

5.3.3 MIMO Control

In the following is shown, how the ASA is applied by multivariable control systems, which are also called MIMO (Multi Input Multi Output), using the same principle, as in Fig. 5.14. The classic methods of MIMO analysis and design are well developed and described in the literature, e.g., in [6, 10, 17, 21, 24–27], but the practical use is often complicated because of the high dimensionality of the system. The commonly used methods for MIMO closed loop control are state space feedback and observer design. The decoupling of MIMO subsystems brings the best results, but the realization is complicated because of derivative parts by decoupling.

The MIMO balance control based on ASA, as in Sect. 5.2.3, is shown in Fig. 5.20. Despite the apparent complexity, the MIMO control can be easily explained.

"To each system (for example, for the MIMO-plant with transfer functions G_{11}, a_{12}, a_{21}, G_{22}) could be found a proper antisystem (the MIMO-plant with transfer functions $anti\_G_{11}$, $anti\_a_{12}$, $anti\_a_{21}$, $anti\_G_{22}$) with following features:

As far as the system and antisystem are identical, the error

$$e_y - e_x$$

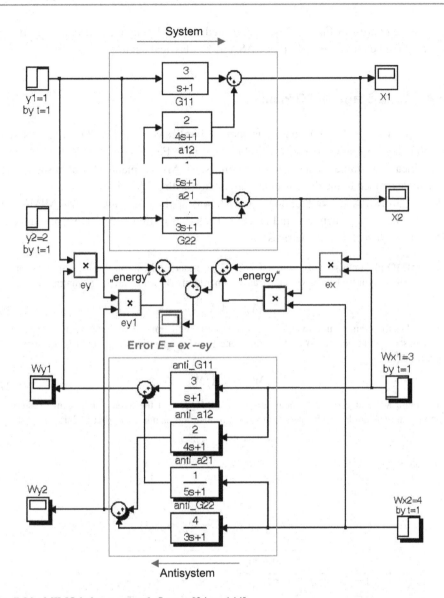

Fig. 5.20 MIMO balance control. *Source* [24, p. 144]

is zero independent of the values of input-signals X_1 and X_2 of the system and of the values of input-signals W_1 and W_2 of the antisystem.

The variables e_y and e_x are called signal-energy or signal-intensity. If input steps X_1, X_2, W_1, and W_2 are applied to the same time (e.g., $t = 1$), then the energies e_y and e_x are equal by all possible values of inputs; the error $(e_y - e_x)$ is zero.

The error will occur, if:

- the disturbance will be applied
- the parameter of the system and antisystem change
- the input steps will be applied not simultaneously." (Quote: Source [24, pp. 144, 145])

The balance control of Fig. 5.20 was applicated in [22] by the real MIMO plant. It was confirmed the better control quality by ASA as by classical control.

5.3.4   Multi-Stage MIMO Plants

The engineering of MIMO control is more complicated if a MIMO plant is spread through many stages like on the right in Fig. 5.14 or in Fig. 5.8.

The idea of variable compression for multi-stage MIMO plants was published in [5] and then developed to the dimensionality reduction with ASA in [16].

In Fig. 5.21, it is shown that how the ASA was applied by a multi-stage MIMO system using the same principle as in Fig. 5.14. The Gl. 5.15 till Gl. 5.17 are represented by MIMO through vectors and matrices.

The MIMO-plant consists of N subsystems (stages), each of them transfers the input vectors X_{i-1} to the output vectors X_i with the transfer operations A_i, which are (n, n)-matrices:

$$X_i = A_i X_{i-1} \tag{5.26}$$

To each subsystem is introduced a sub-antisystem with transposed input vectors W_i and transposed output vectors W_{i-1} with the same transfer operations A_i, which is transposed matrix A_i^T:

$$W_{i-1} = A_i^T W_i \tag{5.27}$$

The main advantage of ASA for multi-stage MIMO is that the balance between a system and the antisystem for each subsystem is dot product, i.e., is not matrix but scalar:

$$W_{i-1} X_{i-1} = W_i X_i \tag{5.28}$$

(Quote: Source [22])

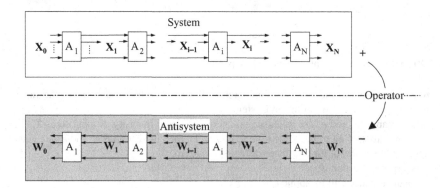

Fig. 5.21 Example of multi-stage MIMO system and antisystem. *Source* [16, p. 183]

The matrices $\mathbf{A}_1$, $\mathbf{A}_2$, ... $\mathbf{A}_N$ are supposed to be given The initial values $\mathbf{X}_0$ of the system are also given. If the task is to define the output vector $\mathbf{X}_N$, then the classic solution is the multiplication of matrices $\mathbf{A}_i$.

$$\mathbf{X}_N = \mathbf{A}_1\mathbf{A}_2...\mathbf{A}_N\mathbf{X}_0$$

If the task is to define not the vector $\mathbf{X}_N$, but only one component of it, e.g., x^N_1, then according to ASA, all components of the input vector $\mathbf{W}_N$ of the antisystem will be chosen equal to zero except for the desired component w^N_1, which will be arbitrary chosen, e.g., $w^N_1 = 1$:

$$\mathbf{W}_N = (10...0) \tag{5.29}$$

It results in the "energy" e_N

$$e_N = \mathbf{W}_N\mathbf{X}_N = x^N_1$$

The vector $\mathbf{W}_0$ is calculated from the antisystem

$$\mathbf{W}_0 = \mathbf{A}_N\mathbf{A}_{N-1}...\mathbf{A}_1\mathbf{W}_N$$

As far as the vector $\mathbf{W}_0$ is calculated above and $\mathbf{X}_0$ is given, the "energy" e_0 can be calculated directly

$$e_0 = \mathbf{W}_0\mathbf{X}_0 \tag{5.30}$$

Considering Gl. 5.29 and according to the balance of "energy" follows the solution

$$e_N = \mathbf{W}_N\mathbf{X}_N = e_0 = \mathbf{W}_0\mathbf{X}_0 = x^N_1$$

If the component x^N_2 of the output vector should be defined, the input vector $\mathbf{W}_N$ of Gl. 5.29 will be accordingly chosen

$$\mathbf{W}_N = [01...0]$$

Example: Multi-stage multivariable system

Given is a MIMO system of Fig. 5.22 with $n=3$ variables and $N=3$ stages

$$A_1 = \begin{pmatrix} 1 & 2 & 3 \\ 4 & 5 & 6 \\ 7 & 8 & 9 \end{pmatrix} \quad A_2 = \begin{pmatrix} 4 & 5 & 6 \\ 7 & 8 & 9 \\ 1 & 1 & 2 \end{pmatrix} \quad A_3 = \begin{pmatrix} 5 & 4 & 1 \\ 7 & 8 & 3 \\ 8 & 2 & 2 \end{pmatrix}$$

Also, given is the input vector $\mathbf{X}_0$. It is searched for the 2nd component x^3_2 of the output vector $\mathbf{X}_3$

$$\mathbf{X}_0 = \begin{pmatrix} 2 \\ 5 \\ 6 \end{pmatrix} \quad \mathbf{X}_3 = \begin{pmatrix} x^3_1 \\ x^3_2 \\ x^3_3 \end{pmatrix}$$

Fig. 5.22 Multi-stage
multivariable system and its
antisystem with n variables and
N stages. *Source* [22]

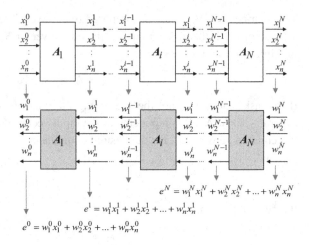

$$e^N = w_1^N x_1^N + w_2^N x_2^N + \ldots + w_n^N x_n^N$$

$$e^1 = w_1^1 x_1^1 + w_2^1 x_2^1 + \ldots + w_n^1 x_n^1$$

$$e^0 = w_1^0 x_1^0 + w_2^0 x_2^0 + \ldots + w_n^0 x_n^0$$

The solution is shown in the MATLAB® script below. ◄

```
%% -----Given ------------------------------
A1 = [1  2  3;  4  5  6;  7  8  9];
A2 = [4  5  6;  7  8  9;  1  1  2];
A3 = [5  4  1;  7  8  3;  8  2  2];
X0 = [2;  5;  6];  % Input of system
% ----Solution----------------------------
W3 = [0  1  0];          % Input of antisystem W3
W2 = A3' * W3';          % A3' is transposed matrix A3
W1 = A2' * W2;           % A2' is transposed matrix A2
W0 = A1' * W1;           % A1' is transposed matrix A1
e3 = W0' * X0;           % "Energy" balance: e0=e1=e2=e3
x32 = e3;
% Answer: the searcherd variable x32 = 22608
```

5.3.5   ASA Control

The aim of the ASA control is to build such a series connection of a plant $G(s)$ that will
compensate the plant with itself without building the inverse $1/G(s)$. For this purpose,
first is used the parallel connection, which is referred to as "system"

$$G_{\text{system}}(s) = 1 + G_S(s)$$

Then is build the feedback with the transfer function, which is viewed as an "antisystem"

$$G_{\text{antisystem}}(s) = \frac{1}{1 + G_S(s)}$$

If the system and the antisystem are series connected, as shown in Fig. 5.23, they compensate each other without forming the reciprocal transfer function of the plant as it is in
the case of the classic compensating controller (see Chap. 11)

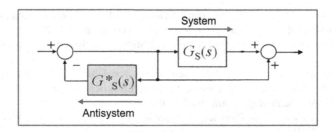

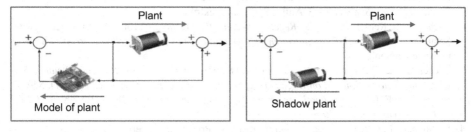

Fig. 5.23 Above: ASA controller as a series connection of the system and antisystem. Below: two options of antisystem, as a software model and as a second identical plant (shadow plant)

$$G(s) = [1 + G_S(s)] \cdot \frac{1}{1 + G_S(s)} = 1 \tag{5.31}$$

As far as the transfer function of the whole connection of Gl. 5.31 is equal to one, the plant loses its influence on the signal transfer and seems to disappear. In other words, the plant became to be "invalid". The ASA control is thus a compensation control but free from the disadvantages of conventional compensation control since an inverse or reciprocal transfer function of the controlled system is no longer necessary. As a result, the controller has no D-terms and the following is trouble-free. The antisystem of the ASA controller can be implemented in practice as a software model or as a second identical plant (Fig. 5.23, below).

The features of ASA control are successfully used in different chapters of this book for the engineering of model-based controllers.

References

1. Padmanabhan, H., Munro, J. M., Dabo, I., & Gopalan, V. (2004). *Antisymmetry: fundamentals and applications*. Department of Materials Science and Engineering, and Materials Research Institute, The Pennsylvania State University, University Park, PA, 16802, USA. https://arxiv.org/ftp/arxiv/papers/2004/2004.06055.pdf. Accessed 25. Feb. 2022.
2. Zacher, S. (1997). *Antisysteme versus Neuronale Netze: Das Prinzip und die Anwendungsbeispiele*.In: Anwendersymposium zu Fuzzy Technologien und Neuronalen Netzen, 19. bis 20. November 1997, Dortmund.

3. *Newton's Laws of Motion*. Glenn Research Center, NASA. https://www1.grc.nasa.gov/beginners-guide-to-aeronautics/newtons-laws-of-motion/. Accessed 20. Feb. 2022.
4. Zamorzaev, A. M. (1976). Theory of Simple and Multiple Antisymmetry. (in Russian). Shtiintsa.
5. Kattanek, S., & Sacharjan, S. (1968). Berechnung von Übertragungsfunktionen des Festbettreaktors durch Reduzierung des Signalflußbildes. *Chemische Technik. XX*(12), 725–728.
6. Zacher, S. (2003). *Duale Regelungstechnik*. VDE.
7. Zacher, S. (2020). *Drei-Bode-Plots Verfahren für Regelungstechnik*. Springer Vieweg.
8. Shafranowski, I. (1968). *Symmetrie in der Natur*. Nedra.
9. Weyl, H. (1968). *Symmetrie* (3. Aufl.). Springer Spektrum.
10. Zacher, S., & Reuter, M. (2022). *Regelungstechnik für Ingenieure* (16th ed.). Springer Vieweg.
11. Sivardiere, J. (1995). *La symmetrie en mathematiques, physiques et chimie*. Presse Universitaires de Grenoble.
12. Zacher, S. (2021) *Antisystem-Approach (ASA) for Engineering of Wide Range of Dynamic Systems*. International Journal on Engineering, Science and Technology (IJONEST), *3*(1), 52-66
13. Bellman, R. E. (1957). *Dynamic programming*. Princeton University Press. Republished 2003 by *Courier Dover Publications*.
14. Bellman, R. E. (1961). *Adaptive control processes*. Princeton University Press.
15. Sacharjan, S. (1983). Basics of variables compression. In: *Two methods of increased efficiency of complex technological processes*. Hayastan Publications.
16. Zacher, S. (2000). *SPS-Programmierung mit Funktionsbausteine*. VDE.
17. Zacher, S. (Ed.). (2000). *Automatisierungstechnik kompakt*. Vieweg.
18. Zacher, S. (2008). *Mobile Mathematik. Ein neues Konzept zur Lösung von linearen Gleichungssystemen*. Dr. Zacher.
19. Zacher, S. (2008). *Existentielle Mathematik*. Dr. Zacher.
20. Zacher, S. (2012). *Verbotene Mathematik.*. Dr. Zacher.
21. Zacher, S. (2021). *Regelungstechnik mit Data Stream Management*. Verlag Springer Vieweg.
22. Zacher, S., & Saeed, W. (2010). *Design of multivariable control systems using antisystem-approach*. 7th AALE Angewandte Automatisierung in der Lehre und Forschung, FH Technikum Wien, 10./11 Feb. 2010 (pp. 201–208).
23. Zacher, S. (2017). *ASA-Bilanzregelung*. https://zacher-international.com/Automation_Letters/33_ASA_Bilanzregelung.pdf. Accessed 20. Feb. 2022.
24. Zacher, S. (2014). *Bus-approach for feedback MIMO-control*. Dr. S. Zacher.
25. Zacher, S. (2019). *Bus-Approach for engineering and design of feedback control*. Denver. In: *Proceedings of ICONEST*, October 7-10, 2019, ISTES Publishing (pp. 26-27).
26. Mille, R., Wahlen, M., Wenzel, M., & Jeckel, M. (2014). *Mehrgrößenregelung eines Drei-Tank-Systems nach Bus-Approach*. Projektarbeit der Hochschule Darmstadt, FB EIT.
27. Zacher, S. (2020). Bus-approach for engineering and design of feedback control. *International Journal of Engineering, Science and Technology, 2*(1), 16-24. https://www.ijonest.net/index.php/ijonest/article/view/9/pdf. Accessed 20. May 2020.

BAD: Bode Aided Design

6

> *„Everything new is well forgotten old. "*
>
> Quote: Jacques Pesce (alias Rosa Burnet): „Memoirs " (1824)

6.1 Introduction

6.1.1 Stability

Definition

Stability is very important feature of dynamic systems. Generally, a system is stable if its output arrives at some constant value after change of its input. Regarding closed loop control systems, it means that a loop is stable if the controlled variable $x(t)$ after an input step of setpoint w or an input step of disturbance z by $t=0$ arrived at some constant value $x(\infty)=$ const by $t \to \infty$. It is illustrated with an example of Fig. 6.1 of two step responses $x(t)$ after the same step of setpoint $w=1,5$ but two different gains K_{PR} of controller $G_R(s)$. By one K_{PR}-value the loop is unstable as it is seen on the left side of Fig. 6.1, by which the step response reached in 30 s very big values of $x(t)$ without finding some steady state. By another value of the gain K_{PR} on the right of Fig. 6.1 the step response arrived at the desired steady state $x(\infty)=w$ that means the loop is stable.

The necessary and sufficient condition of stability

To derive the stability conditions let us remember, as it was shown in chapter 2, that a control system with the input w and output x could be described in three domains:

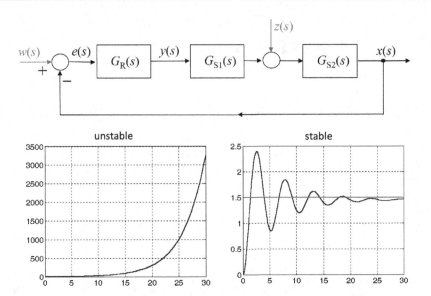

Fig. 6.1 Closed loop control and two step responses $x(t)$ after step of setpoint of $w=1,5$ by two values of controller gain

- In the time domain as differential equations, e.g., of 2nd order with constant coefficients:

$$a_2\ddot{x}(t) + a_1\dot{x}(t) + a_0 x(t) = K_{PR}K_{PS}w(t) \tag{6.1}$$

- In the Laplace domain as transfer function:

$$G_w(s) = \frac{K_{PR}K_{PS}}{a_2 s^2 + a_1 s + a_0} = \frac{x(s)}{w(s)} \tag{6.2}$$

$$(a_2 s^2 + a_1 s + a_0)x(s) = K_{PR}K_{PS}w(s) \tag{6.3}$$

- In frequency domain as frequency response:

$$G_w(j\omega) = \frac{K_{PR}K_{PS}}{a_2(j\omega)^2 + a_1(j\omega) + a_0} = \frac{x(j\omega)}{w(j\omega)} \tag{6.4}$$

According to rules known from mathematics the solution of differential equation Eq. 6.1 consists of terms

$$x_i(t) = e^{\lambda_i t} \tag{6.5}$$

with values of λ defined as roots of characteristical equation

$$a_2\lambda^2 + a_1\lambda + a_0 = 0, \tag{6.6}$$

which follows from homogeneous equation, i.e., the equation by $w(t)=0$

$$a_2\ddot{x}(t) + a_1\dot{x}(t) + a_0 x(t) = 0. \tag{6.7}$$

The system of Eq. 6.1 is stable if the real parts of λ_i values are negative because by $t \to \infty$ there is $x_i(\infty) \to 0$. Otherwise, by $t \to \infty$ there is $x_i(t) \to \infty$ and the system is unstable.

This condition of stability in time domain will be applied to Laplace domain considering that the characteristical equation Eq. 6.6 in Laplace domain corresponds to the denominator of $G_w(s)$ in Eq. 6.2. Therefore, the necessary and sufficient condition of stability is

A system is stable if all poles of its transfer function are in the left s-half plane or have negative real parts. If a pole lies on the imaginary axis or in the right s-half plane, the circle becomes critical or unstable (Fig. 6.2).

6.1.2 Hurwitz Stability Criterion

As is follows from the stability condition above, to define the stability one should solve the characteristic equation in time domain or Laplace domain. To avoid the direct solution of characteristic equations there are known methods, which are called algebraic stability criteria, like Routh criterion and Hurwitz criterion, which were introduced almost at the same time, at 1895–1897 (see e.g., [1, pp. 167–173]). The stability conditions of Hurwitz criterion for systems of the 3rd order with input $w(t)$ and output $x(t)$ upon [1, p. 20], are shown below upon [1, p. 20]. The closed control loop is described in time domain with differential equation

$$a_3\dddot{x}(t) + a_2\ddot{x}(t) + a_1\dot{x}(t) + a_0 x(t) = b_0 w(t)$$

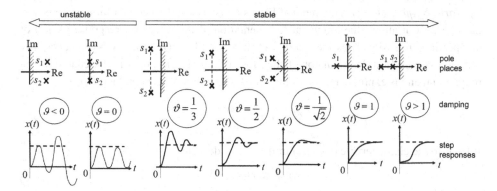

Fig. 6.2 Stability and damping of closed loop control dependent on pole places of the characteristic equation of transfer function $G_w(s)$

or accordingly in Laplace domain with the transfer function

$$G_w(s) = \frac{x(s)}{w(s)} = \frac{b_0}{a_3 s^3 + a_2 s^2 + a_1 s + a_0}.$$

According to Hurwitz stability criterion a system of the 3rd order is stable if by $a_3 > 0$ all three conditions meet:

- 1st condition:

$$a_2 \neq 0 \quad a_1 \neq 0 \quad a_0 \neq 0$$

- 2nd condition:

$$a_2 > 0 \quad a_1 > 0 \quad a_0 > 0$$

- 3rd condition:

$$a_2 \cdot a_1 > a_3 \cdot a_0$$

For systems of 1st or 2nd order only 1st and 2nd stability conditions are needed to be proofed. By systems of higher as 3rd order there are more as three conditions, because of what the use of Hurwitz stability criterion became complicated as the direct solution of characteristic equation and defining of poles places. Because of this the algebraic criteria are rarely used by control theory and are in the following not discussed.

6.2   Overview of Stability Criteria in Frequency Domain

6.2.1   Nyquist Stability Criterion

The stability criteria based upon frequency response $G(j\omega)$ were developed and used later as the algebraic criteria. First stability criterion in frequency domain was offered by *Harry Nyquist* (1932). This method has established itself since that time up to nowadays because it allows the graphical representation of the frequency response $G_0(j\omega)$ of an open loop to easily check the stability of the closed loop $G_w(j\omega)$. The Nyquist stability criterion is based on denominator of the closed loop $G_w(s)$, which is the characteristic polynomial and is expressed through the transfer function of an open loop $G_0(s)$:

$$G_w(s) = \frac{G_0(s)}{1 + G_0(s)} \tag{6.8}$$

It leads to the characteristic equation in frequency domain

$$1 + G_0(j\omega) = 0 \tag{6.9}$$

or to the equation

$$G_0(j\omega) = -1 \tag{6.10}$$

In other words, the Nyquist stability criterion is the graphic solution of Eq. 6.10 analyzes the position of the locus curve $G_0(j\omega)$ related to the critical point on the real axis $(-1, j0)$.

The Nyquist stability criterion is an important tool for illustrating the dynamic behavior of systems because of the following advantages:

- As far as the frequency response describes how the system transmits a sinusoidal input, it can be easily determined experimentally for stable controlled systems.
- The controller tuning based on Bode diagrams of an open loop is simpler than the methods in the time or Laplace domain.

"The representation of the frequency response in the Bode diagram has even more advantages:

   - Bode plots can be sketched by hand with much less effort.

   - Because of the logarithmic representation, it is possible to add the amplitude and phase responses of single loop elements connected in series to obtain the Bode diagram of the entire series connection,

   - The logarithmic scaling of the frequency axis allows easy handling of large frequency ranges, e.g., from 10^{-2} to 10^5 s^{-1}.

   - The transfer function G(s) and its parameters can easily be obtained from a Bode diagram $G(j\omega)$ or directly read out.

   - The amplitude and phase response of the inverse system $1/ G(j\omega)$ is obtained by reflecting the corresponding diagrams of the original system $G(j\omega)$ on the frequency axis. Other symmetry operations with amplitude and phase responses such as shifts up/down or right/ left are easy to use.

   The last property is very important for the symmetry operations that led to the new "Three Bode plots method" offered in this book." (Quote: Source [3, p. 41])

It should be noted that along numerous advantages, the Nyquist stability criterion needs the information about poles of an open loop because this criterion has different formulations, which depend on the number of poles with positive, negative, or zero real parts.

The Simplified Nyquist Stability Criterion in Bode Diagram

▶ **Wichtig**

The simplified criterion which applies to control loops that's transfer function of the open loop is stable or contains only one I (integrated) term.

The closed control loop is stable when by the angular frequency ω_D of the intersection point of the amplitude response $|G_0(\omega)|_{dB}$ the associated phase response is $\varphi(\omega_D) > -180°$, i.e., the associated point of the phase response is above the $(-180°)$ line. On the other hand, the closed control circuit is stable if, at the frequency ω_π, at which the phase response is $\varphi(\omega_\pi) = -180°$, the amplitude response is below the 0 dB line, i.e., $|G_0(\omega_\pi)|_{dB} < 0$.

At $|G_0(\omega_\pi)|_{dB} < 0$ the control loop becomes unstable. At $|G_0(\omega_\pi)|_{dB} < 0$, the control loop is in a critical state (Fig. 6.3).

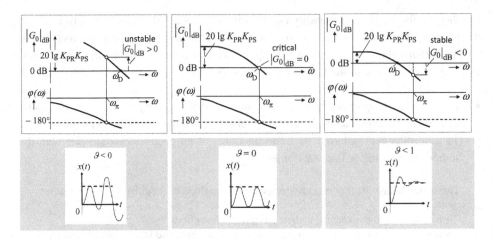

Fig. 6.3 Simplified Nyquist stability criterion in Bode diagram

The Complete nyquist stability criterion
In order to check the stability of the closed control loop are considered the transfer function $G_0(s)$ and the Bode diagram G0($j\omega$) of the open control loop. The total number of poles:

$$n = n_p + n_n + n_i$$

- *n* order of the transfer function $G_0(s)$,
- n_p number of positive poles, i.e., with positive real parts $Re>0$ on the right *s*-plane,
- n_i number of poles with $Re=0$ on the imaginary axis,
- n_n number of negative poles, i.e., with negative real parts $Re<0$ on the left *s*-plane.

A point *S* on the Bode plot, at which the phase response $\varphi_0(\omega)$ intersects the $(-180°)$ line is called simply plus S_+ point or minus S_- point as Fig. 6.4 illustrates.

The S point is positive S_+ if the $(-180°)$ line is intersected from bottom to top. Otherwise there is a negative S_- point. If the phase response only touches the $(-180°)$ line, the corresponding points are called half-intersection points. They are designated as $S_{+0.5}$ or $S_{-0.5}$ up to direction of phase response. Only the points in positive areas of the amplitude response or at $|G_0(\omega)|_{dB}>0$ dB are taken into account.

The number of positive S_+ points is denoted as a_p. The number of negative S_- intersections is a_n. The number of half-intersection points $S_{+0.5}$ or $S_{-0.5}$ is counted as half of a_p or a_n.

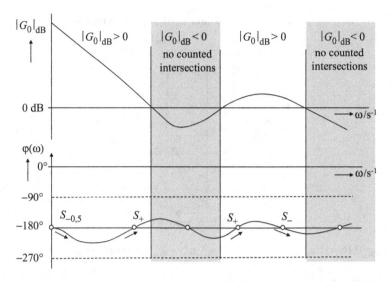

Fig. 6.4 Intersections of Bode plot

▶ **Definition of Complete Nyquist Stability Criterion in Bode Diagram**
The closed loop is stable if meets the following condition for open loop.

- without I terms or with only one I term, i.e., $n_i = 0$ or $n_i = 1$:

$$a_p - a_n = \frac{n_p}{2} \tag{6.11}$$

- with the double I term, i.e., $n_i = 2$.

$$a_p - a_n = \frac{n_p + 1}{2} \tag{6.12}$$

Example: an instable plant controlled with P controller
Given is the transfer function of an open loop

$$G_0(s) = \frac{K_{PR} K_{PS}(1 + sT_3)^3(1 + sT_5)}{(sT_1 - 1)(1 + sT_2)^3(1 + sT_4)^2} \tag{6.13}$$

with parameters: $K_{PR} = 10$; $K_{PS} = 10$; $T_1 = 100$; $T_2 = 20$; $T_3 = 1$; $T_4 = 0,1$; $T_5 = 0,01$. Therefore, the poles of Eq. 6.13 are $n = 6$; $n_p = 1$; $n_i = 0$; and $n_n = 5$.

The Bode diagram of an open loop Eq. 6.13 is shown in Fig. 6.5 (a).

As far as $a_p = 0,5$ and $a_n = 1$ the closed loop is according to Eq. 6.11 is unstable:

$$a_p - a_n = 0,5 - 1 \neq \frac{n_p}{2} \tag{6.14}$$

a a_p=0,5 and a_n=1 K_{PR}=10, unstable

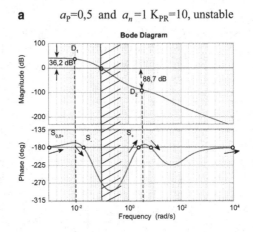

b K_{PR}=0,16 stable   **c** K_{PR}=2,7*10⁵ stable

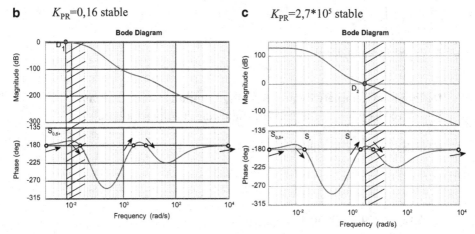

Fig. 6.5 Bode plots of given open loop Eq. 6.13 with $n_p = 1$: (a) by $a_p = 0,5$ and $a_n = 1$ the closed loop is unstable; (b) by $a_p = 0,5$ and $a_n = 0$ the closed loop is stable; (c) by $a_p = 1,5$ and $a_n = 1$ the closed loop is stable

There are two options to stabilize the closed loop:

- the 0 dB axis should be shifted upstairs of $\Delta dB = 36,3$ dB into the point D1. Then like shown in Fig. 6.5 (b) there is $a_p = 0,5$ and $a_n = 0$. The condition Eq. 6.11 is fulfilled and the loop is stable:

$$a_p - a_n = 0,5 - 0 = \frac{n_p}{2}$$

- the 0 dB axis should be shifted downstairs of $\Delta\mathrm{dB} = 88{,}7$ dB into the point D2. Then as it follows from Fig. 6.5 (c) there is $a_p = 1{,}5$ and $a_n = 1$. The stability condition Eq. 6.11 is fulfilled:

$$a_p - a_n = 1,5 - 1 = \frac{n_p}{2}$$

The calculation of needed values of KPR for axis shifting in points D1 and D2 is given in MATLAB® script below.

```
%% Given is unstable plant with P controller
clearvars
s=tf('s');
% —Plant ———————————————————————————————
Kps=10;
T1=100; T2=20; T3=1; T4=0.1; T5=0.01;
Gs=Kps*(1+s*T3)^3*(1+s*T5)/((s*T1-1)*(1+s*T2)^3*(1+s*T4)^2);
% —P controller ————————————————————————
KpR=10;  GR=KpR*s/s;
G0=Gs*GR;              % open loop
bode(G0); grid         % Bode plot of open loop
%% Step response
figure
Gw=G0/(1+G0);
step(Gw,190)
%% -- Controller tuning upon completed Nyquist criterion ——————
delta_dB=36.2;         % option D1 (0 dB axis is shifted up)
%delta_dB=88.7;        % option D2 (0 dB axis is shifted down)
dK=10^(delta_dB/20);
KpR=KpR/dK;            % option D1 (0 dB axis is shifted up)
%KpR=KpR*dK;           % option D2 (0 dB axis is shifted down)
GR=KpR*s/s;
G0=Gs*GR;
bode(G0); grid
%% —Step response of the closed loop —————————————
figure
Gw=G0/(1+G0);
step(Gw)
```

Let us notice that without information about number of positive poles n_p the stability check and the stabilizing of the loop upon Nyquist conditions Eq. 6.11 and Eq. 6.12 is not possible.

6.2.2   Mikhailov stability criterion

Four years after publishing of Nyquist stability criterion, *A.W. Mikhailov*, impressed by this criterion and by frequency methods in general, developed 1936 his own procedure, which some sources call the *Mikhailov stability criterion* and others call it the *Nyquist-Mikhailov stability criterion*.

This method is used like Nyquist criterion the graphical representation of Eq. 6.9 but with the characteristic polynomial $P(j\omega)$ of the closed loop $G_w(s)$ instead of Nyquist plot of open loop $G_0(j\omega)$:

$$P(j\omega) = 1 + G_0(j\omega) = 0 \tag{6.15}$$

or

$$P(j\omega) = N(j\omega) + D(j\omega) \tag{6.16}$$

The last expression Eq. 6.16 follows from the representation of numerator $N(s)$ and denominator $D(s)$ of the transfer function of open loop $G_0(s)$:

$$G_0(s) = \frac{N(s)}{D(s)} \tag{6.17}$$

Compering Eq. 6.15 of *Mikhailov* with Eq. 6.10 of *Nyquist* it is clear that the critical point on the real axis $(-1, j0)$ of Nyquist plot should be shifted into another critical point for Mikhailov plot, namely into $(0, j0)$. Accordingly, is formulated the Mikhailov criterion:

▶   A closed loop is stable if and only if the locus of the closed loop $P(j\omega)$ with
      increasing frequency ω from $\omega = 0$ to $\omega = \infty$ strictly sequentially circumvents the
      n quadrants of the complex plane counterclockwise, where n is the order of the
      characteristic polynomial $P(s)$.

In Fig. 6.6 is shown an example of Mikhailov plot for the transfer function of the closed control loop of the 5th order created with the MATLAB script given below. The closed control loop is stable.

$$G_w(s) = \frac{x(s)}{w(s)} = \frac{K}{s^5 + 2s^4 + 6s^3 + 5s^2 + 3s + 1} \tag{6.18}$$

```
s = tf('s');                                    % Laplace-Operator
P=s^5+2*s^4+6*s^3+5*s^2+3*s+1;                   % charateristic polynomial
xmin=-5; xmax=40; ymin=-15; ymax=10;            % limits of window
h = nyquistplot(P);                             % Mikhailov plot
setoptions(h,'ShowFullContour','off')           % disable negative frequencies – ∞ < ω < 0
axis([xmin xmax ymin ymax])
title('Mikhailov-Ortskurve')
```

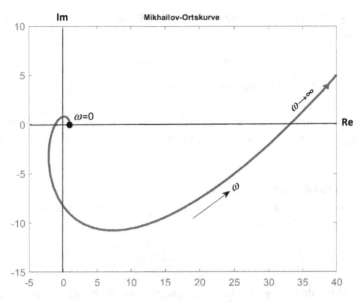

Fig. 6.6 Mikhailov plot of the stable closed loop of the 5th order: the plot goes counterclockwise the round of all $n=5$ quadrants of the complex plane. *Source* [3, p. 44]

Although no information about poles is required for Eq. 6.11, *Mikhailov* stability criterion is rarely used in practice and further is not discussed. More information about it is to be found in [3, pp. 42–53].

6.2.3   Leonhard Stability Criterion

Four years after publication of the Mikhailov stability criterion, *A. Leonhard*, who started his scientific career according to [4] with his habilitation at the technical university of Berlin and then moved to the technical university Stuttgart, published 1940 a new stability criterion [4, 5] that is now known as the two-locus curve method.

Leonhard (1940) also used as *Nyquist* and *Mikhailov* the characteristic equation Eq. 6.9 or Eq. 6.10 of the closed loop, but decomposed it in two curves, the frequency response of the controller and the negative inverse frequency response of the plant:

$$G_0(j\omega) = G_R(j\omega)G_S(j\omega) = -1$$

$$G_R(j\omega) = -\frac{1}{G_S(j\omega)} \tag{6.19}$$

The negative inverse frequency response of the plant is referred to as

$$G_{Srec}(j\omega) = -\frac{1}{G_S(j\omega)}$$

The frequency by which both modules of Eq. 6.19 are equal is called crossover frequency ω_D, i.e.,

$$|G_R(\omega_D)| = |G_{Srec}(\omega_D)| \qquad (6.20)$$

The Leonhard stability criterion is formulated very shortly:

▶   A system is stable, if at the crossover frequency ω_D the phase of the controller $G_R(j\omega_D)$ is greater than the phase of the negative inverse frequency response of the system.

Example: Leonhard stability criterion
The stability of the closed control loop consisting of the PT2 plant with the damping $\vartheta = 0{,}5$ and coefficients $b_0 = 1;\ a_0 = 0.1$

$$G_S(s) = \frac{b_0}{a_0^2 s^2 + 2a_0 d \cdot s + 1}$$

and PI controlled should be checked. The parameters of the PI controller are given: $K_{PR} = 1$ and $T_n = 0{,}08$. The solution is created upon MATLAB® script below and is shown in Fig. 6.7.

```
a0=0.1; d=0.5; b0=1;                    % given plant's parameters
KpR=1; Tn=0.08;                         % parameters of PI controller
xmin=-1.5; xmax=4.4; ymin=-3; ymax=1;   % limits of diagram
s= tf('s');                             % Laplace operator
GR=(KpR*(1+s*Tn))/(s*Tn);               % PI controller
Gs=KpS/(s^2*T2^2+s*2*d*T2+1);           % PT2 plant
Gs_neg_inv = -1/Gs;                     % negative inverse plant
h = nyquistplot(GR, Gs_neg_inv);        % two diagrams
setoptions(h,'ShowFullContour','off')   % % disable negative frequences – ∞ < ω < 0
axis([xmin xmax ymin ymax]);            % axis limits
```

From Fig. 6.7 follows that the closed loop is stable because by the crossover frequency $\omega_D = 13{,}3\ \text{s}^{-1}$ the difference $\varphi_{Rd}(\omega_D)$ between phases of the controller $\varphi_R(\omega_D)$ and of the negative inverse frequency response of the plant $\varphi_{Srec}(\omega_D)$ is positive:

$$\varphi_{Rd}(\omega_D) = \varphi_R(\omega_D) - \varphi_{Srec}(\omega_D) = 19°$$

As it is seen from Fig. 6.7, the stability check upon Leonhard criterion is a unique procedure that differs fundamentally from other stability criteria in the frequency domain. Nevertheless, the Leonhard stability criterion has not found practical application in the design of linear control loops due to its complicated and confusing definition.

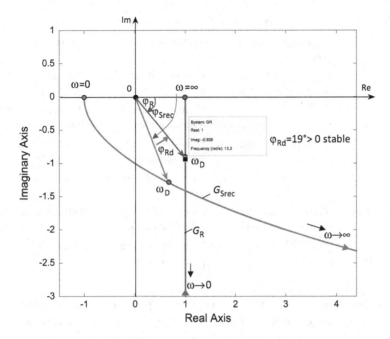

Fig. 6.7 Leonhard stability criterion. *Source* [6, p. 25]

Comparison of stability criteria in frequency domain
In Table 6.1, all three stability criteria are below and compared:

- Mikhailov criterion is based upon fixed critical point $(0, j0)$. This criterion is simple, no information of n poles is needed. Though the transfer function G_S of plant is to be given. To check the stability, it is only to check if the Mikhailov locus goes round all n quadrants (n is order of the closed loop G_w, i.e., of the characteristic polynomial P). But the use of criterion for controller design to arrive at the given phase or magnitude reserve is hardly possible. The calculation of $P(s)$ and analysis of it are not easy, but it is no problem for software programming. This criterion has not established itself for practical use.
- Nyquist criterion is based upon fixed critical point $(-1, j0)$, so the information about poles of open loop G_0 is necessary. The calculation and stability analysis of G_0 is easier as by Mikhailov criterion. Also, the practical use is easy, because of this the criterion has established itself for controller design with Bode plot.
- Leonhard criterion has no fixed critical point like $(0, j0)$ or $(-1, j0)$. This criterion is simple, although it is needed to build the negative reverse transfer function G_{Srec} of the plant G_S upon Eq. 6.19, which is no problem using software. The critical point depends on Eq. 6.20. It is advantage of criterion because no information about poles of the plant G_S or of open loop G_0 or of closed loop G_w is necessary. But it is at the same time a disadvantage because to define the crossover frequency ω_D is not easy.

Table 6.1 Comparison of stability criteria in frequency domain

Criterion of ...	Based upon transfer function of ...	Advantages	Disadvantages
Mikhailov (1936)	$P_w(s)$ characteristic polynomial of closed loop $G_w(s)$	Easy stability check going n quadrants	Difficulties by the use in Bode plot
		No information about poles is needed	The use of G_w is more difficult as of G_0
Nyquist (1932)	$G_0(s)$ of open loop	G_0 is simpler as G_w, is easier to calculate	Information about poles of open loop
		Easier use of $G_0(s)$ in Bode plot	Chang of G_R requires new calculation of G_0
Leonhard (1940)	$G_R(s)$ of controller and negative inverse $G_{Srec}(s)$ of plant $G_S(s)$	Changes of $G_R(s)$ doesn't affect $G_{Srec}(s)$	Difficulties to define crossover frequency
		No information about poles is needed	No use in Bode plot

As it is seen from Fig. 6.7, there are two different points on each curve with the same frequency ω_D. To define it the Eq. 6.20 either to be solved manually or by means of software. It is advantageous that G_R and G_{Srec} because by changing of controller parameters, only the locus of G_{Srec} needs to be drawn again. The locus of the negative inverse plant G_{Srec} remains unchanged. The transfer function G_S of plant is to be given like by all other criteria. The Leonhard criterion is not formulated for Bode plot and has found practically no use for stability check or controller design.

As it was mentioned above, the Leonhard stability criterion is rarely used by the linear control theory.

Only several researchers continued to develop this method among them W. Oppelt ([6]-[9]), L. Cremer / F. Kolberg ([10, 11]). Despite their efforts, the Leonhard stability criterion disappeared from majority of university tutorials and books and is left only in some of them like S. Zacher / M. Reuter ([1]), S. Zacher ([2, 3]), P.F. Orlowski ([12]), K. Fasol ([13]), and R. Kaerkes ([14]). On the other hand has the Leonhard criterium established itself by harmonic balance of non-linear control loops ([1] - [3, 15]).

Following Table 6.1 and comparing it with the design of controllers, an interesting trend is seen.

The design of control loops is done upon following steps going from single elements to the whole system:

• The identification of the controlled plant, from which results in the transfer function $G_S(s)$. Only after this step the adjustment of controller could begin. The transfer function $G_S(s)$ of plant is the necessary requirement for stability check and tuning of controller.

- The choice of type of standard controller $G_R(s)$.
- The defining of the open-loop transfer function $G_0(s)$.
- The calculation of the closed loop transfer function $G_w(s)$.

A chain of preparatory steps for closed loop design is thus created, namely

$$G_S(S) \rightarrow G_R(S) \rightarrow G_0(S) \rightarrow G_W(S) \rightarrow$$

The stability criteria of Table 6.1 are developed in reverse direction decomposing the system criterion for $G_w(s)$ in its terms like below

Leonhard $G_{Srec}(S)$ and $G_R(s)$ ← Nyquist $G_0(s)$ ← Mikhailov $G_w(s)$.

This trend was detected in [17], and it was set a challenging task: to change the Leonhard stability criterion and to realize it with Bode plot:

- using the negative inverse controller $G_{Rrec}(s)$ instead of negative inverse plant $G_{Srec}(s)$:

$$G_{Rrec}(j\omega) = -\frac{1}{G_R(j\omega)} \tag{6.21}$$

- decomposing GRrec(jω) further into amplitude and phase responses |GRrec(ω)|dB and φRrec(ω).
- The stability condition Eq. 6.9, 6.10, 6.19 is not affected by this modification and it is the same:

$$G_S(j\omega) = -\frac{1}{G_R(j\omega)} \tag{6.22}$$

or

$$G_S(j\omega) = G_{Rrec}(j\omega) \tag{6.23}$$

This idea was in 2017 successfully realized and published in [17] as the "Second life of the Leonhard criterion". In the following publications [1, 3, 16, 18] and projects, the new method was further developed, tested and called "Three Bode plots".

6.3 Three Bode Plots

6.3.1 Conception

The conception of the "Three Bode plots method", shortly "3BP method", as it was mentioned above is to modify the Leonhard stability criterion using its advantages and eliminating significant disadvantages shown in Table 6.1, namely

- To apply the Leonhard method in Bode diagram. The reason is to profit like established Nyquist method from all advantages of Bode plot mentioned in Sect. 6.2.1.

- To change the role distribution of controller and plant setting $G_{Rrec}(j\omega)$ according to Eq. 6.21 instead of $G_{Srec}(j\omega)$. The reason for it is clear: as far as there are only a few types of standard controllers (P, I, PI, PD, PID) it is easy also create the standard types of bode plots for negative inversed controllers.
- To define the crossover frequency ω_D as intersection point of two Bode plots, two amplitude responses:
  - $|G_S(\omega)|_{dB}$ of the plant,
  - $|G_{Rrec}(\omega)|_{dB}$ of the reciprocal (negative inverse) controller.

6.3.2   Two Bode Plots Method

The first try to implement this conception was called "2BP method" and was formulated according to Fig. 6.8 as follows:

▶ **Wichtig**

A closed control loop becomes stable when the distance $\Delta\varphi$ between phase responses φ_{Rrec} and φ_S at the intersection D of amplitude responses $|G_{Rrec}(\omega_D)|_{dB}$ and $|G_S(\omega_D)|_{dB}$ in the Bode diagram is less than $360°$:

$$\Delta\varphi = \varphi_{Rrec}(\omega) - \varphi_S(\omega_D) < 360° \qquad (6.24)$$

An example is shown in Fig. 6.9. The closed loop corresponded to the Bode diagram is unstable by $K_{PR}=20$ and $T_n=1$ s because of $\Delta\varphi=370°>360°$. Supposing the closed loop should be stabilized, and the desired phase rand (phase reserve) is given as $\varphi_{Rd}=70°$. Then the desired distance $\Delta\varphi$ should be reduced until

$$\Delta\varphi = 360° - \varphi_{Rd} = 300°$$

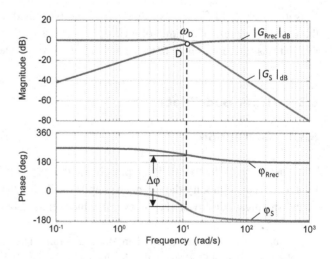

Fig. 6.8 Two bode plots criterion of *Zacher* ([1, 3, 17] - [16])

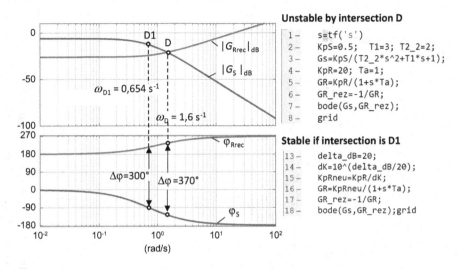

Fig. 6.9 Example of unstable control loop by intersection point D. After shifting the Bode plot $|G_{Rrec}(\omega)|_{dB}$ of reciprocal controller in the intersection point D1 the closed loop becomes to be stable with the desired phase rand of 60°

The amplitude response of $|G_{Rrec}(\omega)|_{dB}$ should be shifted upstairs to intersect the $|G_S(\omega)|_{dB}$ exactly by the point D1. The needed K_{PRnew} for the shifting is calculated in the MATLAB ® script given on the left side of Fig. 6.9.

The 2BP method was effectively used by several projects and programmed in [17] with MATLAB ® as tuning tool "Fingerprint ZBV" (Fig. 6.10). More about App Fingerprint ZBV in [3, pp. 317–326], [17] - [19].

Summarizing the results shown in Fig. 6.8, 6.9, and Fig. 6.10, we can conclude that the goal set in conception of Sect. 6.3.1 was achieved: the Leonhard stability criterion was improved, and the disadvantages were eliminated. But the use of the 2BP method is not comfortable because of necessity to define the distance $\Delta\varphi$ between two phases, $\varphi_{Rrec}(\omega_D)$ and $\varphi_S(\omega_D)$. To also improve this feature and to simplify the 2BP method, the conception of Sect. 6.3.1 was changed. Instead of stability condition Eq. 6.24 $\Delta\varphi < 360°$, it was required to create the following simple condition

$$\Delta\varphi = \varphi_{Rrec}(\omega_D) - \varphi_S(\omega_D) < 0$$

$$\varphi_S(\omega_D) > \varphi_{Rrec}(\omega_D) \tag{6.25}$$

The condition Eq. 6.25 means that a closed control loop becomes stable when the phase response of the plant $\varphi_S(\omega_D)$ is located above the phase response $\varphi_{Rrec}(\omega_D)$ of the reciprocal controller at the intersection D of amplitude responses $|G_{Rrec}(\omega_D)|_{dB}$ and $|G_S(\omega_D)|_{dB}$ in the Bode diagram.

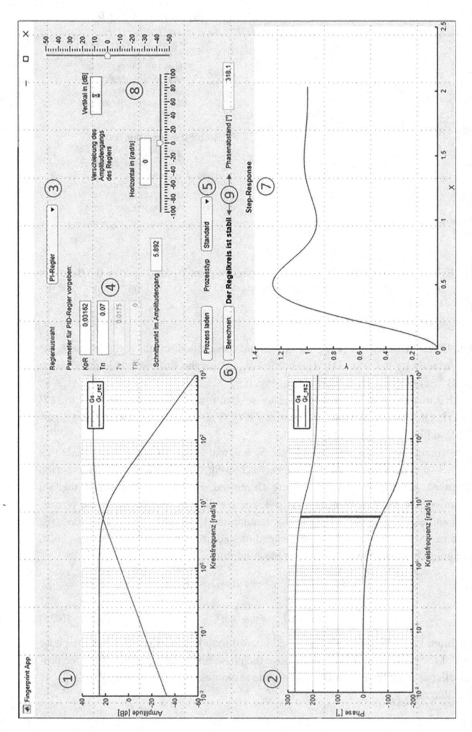

Fig. 6.10 App Fingerprint for 2BP method upon Zacher stability criterion. *Source* [3, p. 324]: 1, 2 -displays amplitude and phase responses; 3, 4-enter controller type and parameters; 5-enter plant type; 6-run simulation; 7- display step response; 8-horizontal and vertical shift of amplitude response; 9-displays "stability" and phase rand

But as far as the realization of the condition Eq. 6.25 with MATLAB® command "bode (...)" is not possible, in the following is described that despite it the condition Eq. 6.25 was implemented with means of so-called symmetrical controller.

6.3.3 Symmetry Operations With Controllers

The symmetry is a powerful tool that is massively applied in many areas of the natural sciences such as physics, classical mechanics, mathematics, and chemistry. However, only a few publications like [20] or [21] deal with symmetry operations for the analysis and design of closed control loops although it should bring the same advantages in control engineering as in the natural sciences. The first review of symmetry operations by closed loop control [22] was published in 1991 and has led to several modern methods like ASA [23] discussed in the previous Chap. 5. The use of symmetry representation for Bode plots of standard controllers is described below.

A symmetry axis is generally a curve or a line (here it is a straight line), round of which the amplitude or phase response of the controller $G_R(s)$ will be mirrored. The symmetry axes are referred to as L_G for amplitude response and L_φ for phase response. The following symmetry axes are possible as shown in Fig. 6.11 on the example of the PID-T1 controller (Source [3, pp. 214–222]):

Fig. 6.11a: Symmetrical controller $G_{Rsymm}(s)$ with $L_G = 0$ dB and $L_\varphi = 0°$

$$G_{Rsymm}(s) = \frac{1}{G_R(s)} \tag{6.26}$$

$$\begin{cases} \left|G_{Rsymm}(\omega)\right|_{dB} = -\left|G_R(\omega)\right|_{dB} \\ \varphi_{Rsymm}(\omega) = -\varphi_R(\omega) \end{cases}$$

Fig. 6.11b: Amplitude symmetrical controller $G_{Rsymm}(s)$ with $L_G = 0$ dB and $L_\varphi = -90°$

$$G_{asR}(s) = -\frac{1}{G_R(s)} \tag{6.27}$$

$$\begin{cases} \left|G_{asR}(\omega)\right|_{dB} = -\left|G_R(\omega)\right|_{dB} \\ \varphi_{asR}(\omega) = \varphi_R(\omega) + 180° \end{cases}$$

Fig. 6.11c: Phase symmetrical controller $G_{Rsymm}(s)$ with $L_\varphi = -90°$

$$G_{Rphs}(s) = -\frac{1}{s^2 G_R(s)} \tag{6.28}$$

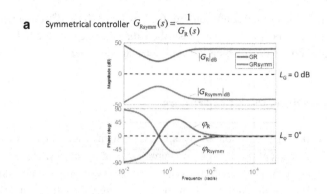

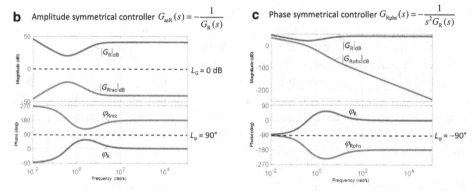

Fig. 6.11   Mirrored Bode plots of PID-T1 controller

The MATLAB script to use the symmetry operations is given below. The overview of mirrored controller is shown in Table 6.2. In this book are used the types 1 and 3.

```
s=tf('s');
wmin=10^-2;wmax=10^5;
KpR=10; Tn=5;Tv=0.25*Tn;TR=0.1*Tv;
% PID-T1 controller
GR=KpR+KpR/(s*Tn)+s*KpR*Tv/(1+s*TR);
%% Symmetrical controller with axes LG = 0 dB und Lφ = 0°
GRsymm=1/GR;
bode(GR, GRsymm,{wmin,wmax});grid
%% Amplitude symmetrical controller with axes LG = 0 dB und Lφ = +90°
GRrez =-1/GR;
bode(GR,GRrez,{wmin,wmax});grid
%% Phase symmetrical controller with axis Lφ = -90°
GRphs=1/(s^2*GR);
bode(GR,GRphs,{wmin,wmax});grid
```

Table 6.2 #L Mirrored controller

Nr	Typ of mirrored controller	Transfer function	Symmetry axis of frequency response	
			Amplitude	Phase
1	**Symmetrical** (Inverse)	$G_{\text{Rsymm}}(s) = \frac{1}{G_R(s)}$	$L_G = 0\,\text{dB}$	$L_\varphi = 0°$
2	**Amplitude symmetrical** (Reciprocal, negative inverse)	$G_{\text{asR}}(s) = -\frac{1}{G_R(s)}$	$L_G = 0\,\text{dB}$	$L_\varphi = +90°$
3	**Phase symmetrical** (Inverse with double I-term)	$G_{\text{Rphs}}(s) = \frac{1}{s^2 G_R(s)}$	no symmetry	$L_\varphi = -90°$

6.3.4 Decomposition of mirrored controllers

In order to reduce the critical phase distance $\Delta\varphi$ from 360° to 0°, the Bode diagram of negative inverse controller $G_{\text{Rrec}}(j\omega)$ of Leonhard stability criterion is taken apart in separate amplitude and phase responses, each of them was mirrored separately as follows:

- Amplitude response $|G_{\text{Rsymm}}(\omega)|_{\text{dB}}$ of the symmetrical controller $G_{\text{Rsymm}}(j\omega)$,
- Phase response $\varphi_{\text{Rphs}}(\omega)$ of the phase symmetrical controller $G_{\text{Rsymm}}(j\omega)$.

The proof and derivation of the above-given decomposition of one Bode plot of the negative inverse controllers $G_{\text{Rrec}}(j\omega)$ into two Bode plots of mirrored amplitude response $|G_{\text{Rsymm}}(\omega)|_{\text{dB}}$ and phase response $\varphi_{\text{Rphs}}(\omega)$ is to find in Chap. 8 of [3]. However, it should be clear why the method is called "Three bode plots", i.e., two above-mentioned decomposed Bode plots of controller plus Bode plot of the plant $G_S(j\omega)$.

The MATLAB® script to implement two decomposed Bode plots of controller is given below. Instead of the established commands "bode" and "margin", the amplitude response and the phase response are decoupled from each other and displayed one below the other in two separate windows "subplot (111)" and "subplot (121)", as illustrated in Fig. 6.12.

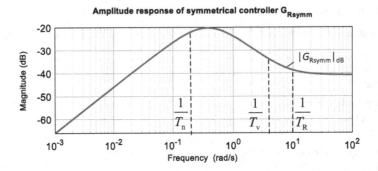

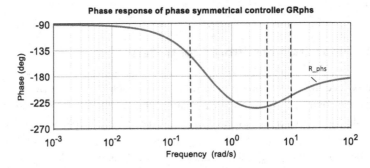

Fig. 6.12 Two decomposed Bode plots of a PID-T1 controller. *Source* [3, p. 250]

```
s=tf('s');
KpR=10; Tn=5;Tv=0.25*Tn;TR=0.1*Tv;
GR=KpR+KpR/(s*Tn)+s*KpR*Tv/(1+s*TR); % PID-T1-Regler
GRsymm_mag=1/GR;
GR_phs=1/(s^2*GR);
wmin=10^-3;wmax=10^2;
subplot(211);
opts_mag=bodeoptions('cstprefs');
opts_mag.PhaseVisible='off';
h_mag=bodeplot(GRsymm_mag,{wmin,wmax},opts_mag); hold on;grid
subplot(212);
opts_ph=bodeoptions('cstprefs');
opts_ph.MagVisible='off';
h_ph=bodeplot(GR_phs,{wmin,wmax}, opts_ph); hold on;grid
```

6.3.5   3BP Method

The term "Three Bode plots method" (3BP method) results from the following three Bode plots:

1. Amplitude and phase response of the plant $G_S(j\omega)$,
2. Amplitude response $|G_{Rsymm}(\omega)|_{dB}$ of the symmetrical controller $G_{Rsymm}(j\omega)$,
3. Phase response $\varphi_{Rphs}(\omega)$ of the phase symmetrical controller $G_{Rsymm}(j\omega)$.

The joint consideration of the above three Bode plots leads to the following stability criterion, which is given here without derivation:

▶ A closed control loop becomes stable if at the intersection D of the amplitude responses of the symmetrical controller $|G_{Rsymm}(\omega)|_{dB}$ and the plant $|G_S(\omega)|_{dB}$ in the Bode diagram the phase response of the plant $\varphi_S(\omega_D)$ is located **above** the phase response of the phase symmetrical controller $\varphi_{Rphs}(\omega_D)$.

The 3BP method is simple and universal. It applies both to stable and unstable plants. The information about poles of the plant is not required. This results in the important features of the 3BP method:

▶   **Wichtig**
   • No transfer function $G_S(s)$ or frequency response $G_S(j\omega)$ of the plant is needed, only Bode plot.
   • The controller tuning is possible upon the experimental measured Bode plot of the plant.
   • The controller tuning is possible upon only one experimental measured point of Bode plot of the plant, which is called Bode Aided Design (BAD), described in section Sect. 6.5.

Example: Tuning of PID-T1 controller

A plant, that is controlled with the PID-T1 controller $G_R(s)$, is given with the Bode plot (Fig. 6.13). The desired phase rand is

$$\varphi_{Rd} = 60°$$

To define are the parameters of the PID-T1 controller:

$$G_R(s) = \frac{K_{PR}(1 + sT_n)((1 + sT_v)}{(1 + sT_R)}$$

As far as the transfer function $G_S(s)$ of the plant in unknown the parameters of the controller are arbitrarily chosen, e.g., $K_{PR} = 10$; $T_n = 5$; $T_v = 0{,}25T_n$; and $T_R = 0{,}1T_v$.

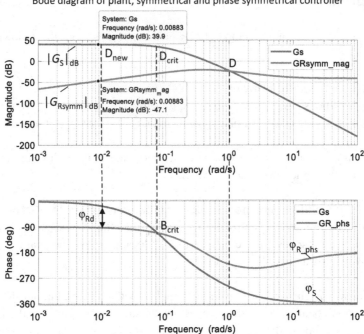

Fig. 6.13 Stability check and controller tuning upon 3BP method. *Source* [3, p. 253]

The Bode diagram of the mirrored controller with these values is drown up in Fig. 6.13. It is apparent from Fig. 6.13 that the closed control loop is unstable because at the intersection point D of amplitude responses, the phase response of the plant is not above, but is below the phase response of the phase symmetrical controller. Both phase responses intersect at the point B_{crit}, i.e., if the intersection point is D_{crit} instead of D the control loop becomes critical. On the left side of critical point B_{crit}, the phase response of the plot φ_S is located above the phase response of phase symmetrical controller φ_{Rphs}. The closed loop becomes stable. Selecting the appropriate phase reserve $\varphi_{Rd} = 60°$ in the desired point D_{new}, the amplitude response of the symmetrical controller should be shifted to this point. The required change of the gain K_{PR} is read from the Bode plot:

$$\Delta dB = 39,9 \text{ dB} - (-47,1) \text{ dB} = 87 \text{ dB}$$

$$\Delta K = 10^{\frac{\Delta dB}{20}}$$

The new value of gain needed for the upward shift of the amplitude response of a mirrored controller is calculated below

$$K_{\text{PRneu}} = K_{\text{PR}}/\Delta K$$

To test the solution below is given the MATLAB® script for the plant which was used in this example:

$$G_S(s) = \frac{100}{(10s^2 + 10s + 1)(1 + 10s)^2}$$

```
%% --- Parameters ------------------------------------
s=tf('s');
KpR=10; Tn=5;Tv=0.25*Tn;TR=0.1*Tv;
GR=KpR+KpR/(s*Tn)+s*KpR*Tv/(1+s*TR); % PID-T1 controller
Kps=100; T2=10; d=0.5;
Gs=Kps/((s^2*T2+2*d*T2*s+1)*(1+s*T2)^2);% plant
%% --- Bode diagram ---------------------------------
GRsymm_mag=1/GR;
GR_phs=1/(s^2*GR);
wmin=10^-3;wmax=10^2;
bode(Gs,{wmin,wmax}); hold on;grid
subplot(211);
bodemag(Gs,{wmin,wmax}); hold on
opts_mag=bodeoptions('cstprefs');
opts_mag.PhaseVisible='off';
h_mag=bodeplot(GRsymm_mag,{wmin,wmax},opts_mag); hold on;grid
subplot(212);
bode(Gs,{wmin,wmax}); hold on
opts_ph=bodeoptions('cstprefs');
opts_ph.MagVisible='off';
h_ph=bodeplot(GR_phs,{wmin,wmax}, opts_ph); hold on;grid
%% --- Shifting D in Dnew ----------
delta_dB=39.9+47.1;
dK=10^(delta_dB/20);
KpR=KpR/dK;
%% --- Step response --------------
figure
GR=KpR+KpR/(s*Tn)+s*KpR*Tv/(1+s*TR);
G0=Gs*GR;
Gw=G0/(1+G0);
step(Gw); grid          ◄
```

To simplify the description, the amplitude plot of symmetrical controller and the phase plot of phase symmetrical controller are both called unified as "mirrored controller".

6.3.6   Implementation of 3BP Method for Unstable Plants

The phase plots of stable plants usually intersect the phase plots of mirrored controllers, and no problems arise using 3BP method. Other is by unstable plants. The problem arises if the phase plot is located under plot of the mirrored controller and is tangent to it or doesn't intersect it. These cases are discussed below.

Unstable Controllable Closed Loop
The Bode diagram of the following unstable plant

$$G_S(s) = \frac{K_{PS}}{(1 + sT_1)(T_2^2 s^2 + 2\vartheta T_2 s - 1)} \tag{6.29}$$

with the PID-T1 controller is shown in Fig. 6.14.

The plant phase plot is located under plot of mirrored controller, so the control loop is unstable. After shifting the phase of mirrored controller from point B to point B_{new}, the

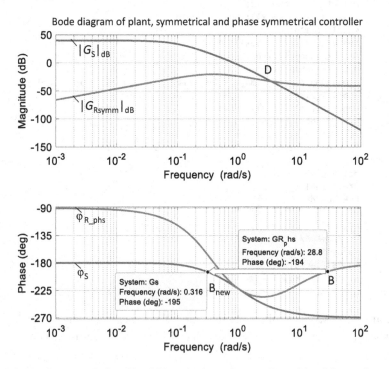

Fig. 6.14 Bode diagram of plant Eq. 6.29 and mirrored controller with arbitrary chosen initial parameters. *Source* [3, p. 256]

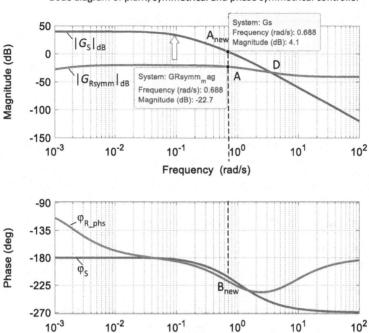

Fig. 6.15   Bode diagram of Fig. 6.14 after controller tuning upon 3BP method. *Source* [3, p. 257]

control loop becomes stable but with very small phase rand (Fig. 6.15). To increase the phase rand, the amplitude plot of mirrored controller is shifted upstairs from point A to point A_{new}.

Unstable Not Controllable Closed Loop
In the Fig. 6.16 is shown that the phase plot of the unstable plant

$$G_S(s) = \frac{K_{PS}}{(T_1 s - 1)^2 (T_2^2 s^2 + 2\vartheta T_2 s - 1)} \tag{6.30}$$

controlled with the PID-T1 controller is located under plot of mirrored controller and doesn't intersect it, so the control loop is unstable. To stabilize the control system, another controller should be used.

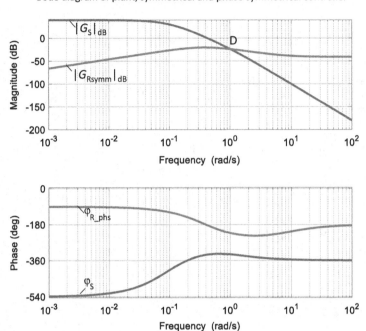

Bode diagram of plant, symmetrical and phase symmetrical controller

Fig. 6.16 Bode diagram of not intersecting phase plot of plant Eq. 6.30 and mirrored controller. The closed loop is not controllable

6.4   BAD

The methods of two Bode plots (2BP, Sect. 6.3.2) and three Bode plots (3BP, Sect. 6.3.5) differ from previously known classic methods in the frequency domain in that there is no transfer function of the plant, no transfer function of the open or closed loop control is needed for controller design. Sufficient is an experimentally determined Bode diagram of the plant alone.

In the previous section, the Bode diagram of the mirrored controller $|G_{Rsymm}(\omega)|_{dB}$ and $\varphi_{Rphs}(\omega_D)$ was first created choosing arbitrary parameters of the controller (gain K_{PR}, time constants T_n and T_v) and then changing these initial controller parameters was navigated or shifted to achieve the desired phase rand and to stabilize the control loop.

This section shows how the controller is designed directly in Bode diagram using given or experimentally defined plot of the plant. Significant is that for this method, which in [3] is called Bode-aided design, only one point of the Bode diagram of the plant is enough to determine the controller parameters. Thus, BAD (*Bode aided design*) is CAD (*Computer aided design*) of a controller using 3BP method.

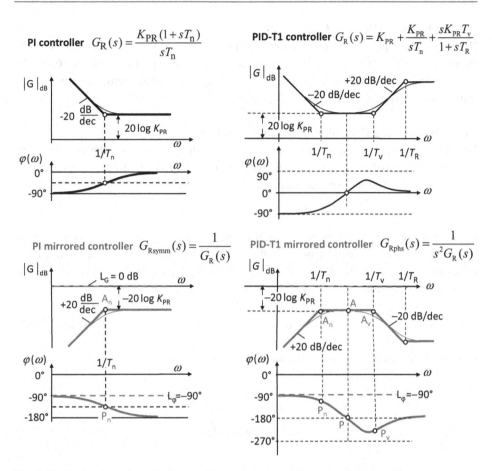

Fig. 6.17 Bode plots of standard controllers and corresponding mirrored controllers

6.4.1 Bode Plots of Mirrored Controllers

Bode Plots of PI and PID-T1 Controllers

The symmetrical controller with Bode plots amplitude frequency $G_{\text{Rsymm}}(\omega)|_{\text{dB}}$ and the phase symmetrical controller with the phase plot $\varphi_{\text{Rphs}}(\omega_{\text{D}})$, introduced in Sect. 6.3.3, are unified and denoted as mirrored controller. In Fig. 6.17 are shown Bode plots of PI and PID-T1 standard controllers and corresponding Bode plots of mirrored controllers. The amplitude plots of I term, and D term change its slope to opposite sign. The logarithmic gain of mirrored controller that corresponds to P part also changes its sign and following its value. The characteristic point of amplitude and phase plot to each break angular frequency are labeled as P_n, P_v, A_n, A_v. The middle point of the horizontal line that corresponds to P-term of PID controller is denoted by A and P.

Shifting Rules of Mirrored Controllers

The important feature of all methods of stability analysis in frequency domain is shifting of bode plots in vertical and horizontal directions (see, e.g., Fig. 6.14 and 6.15). In Fig. 6.18 and in Table 6.3 are shown rules for shifting of mirrored controllers which were realized among others by "Fingerprint" (Fig. 6.10).

Example of moving Bode plot of mirrored controller

The phase plot φ_{Rphs} of the mirrored PI controller with $T_n = 5$ s should be moved to the right from the initial point B to the desired point B_{new} (Fig. 6.19). To do it first are read out the frequencies of these points from the Bode diagram:

$$\omega_B = 0,0418 \text{ s}^{-1} \qquad \omega_{Bnew} = 0,94 \text{ s}^{-1}$$

Then the distance Δdec between points B to the desired point B_{new} is calculated upon MATLAB® script an excerpt from which is given in Fig. 6.19. Finally, are defined the values of Δdec and the time delay $T_{n\,new}$ which are necessary for the new moved Bode plot in the final position:

$$\Delta dec = |\omega_{n\,new} - \omega_n| \quad \rightarrow \quad \Delta \omega = 10^{\Delta dec} \quad \rightarrow \quad T_{n\,new} = \frac{T_n}{\Delta \omega} = 0,2223 \text{ s}^{-1}$$

The new position of phase plot is shown as dashed line in Fig. 6.19.

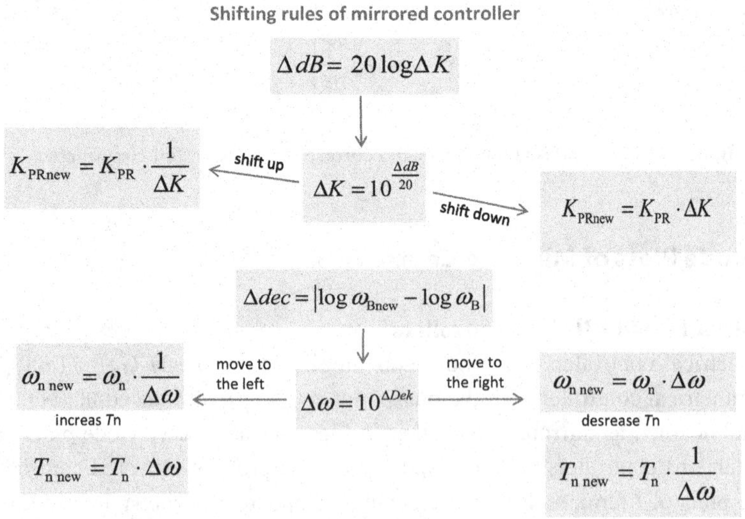

Shifting rules of mirrored controller

$$\Delta dB = 20 \log \Delta K$$

$$K_{PRnew} = K_{PR} \cdot \frac{1}{\Delta K} \quad \xleftarrow{\text{shift up}} \quad \Delta K = 10^{\frac{\Delta dB}{20}} \quad \xrightarrow{\text{shift down}} \quad K_{PRnew} = K_{PR} \cdot \Delta K$$

$$\Delta dec = |\log \omega_{Bnew} - \log \omega_B|$$

$$\omega_{n\,new} = \omega_n \cdot \frac{1}{\Delta \omega} \quad \xleftarrow[\text{the left}]{\text{move to}} \quad \Delta \omega = 10^{\Delta Dek} \quad \xrightarrow[\text{the right}]{\text{move to}} \quad \omega_{n\,new} = \omega_n \cdot \Delta \omega$$

increas Tn   desrease Tn

$$T_{n\,new} = T_n \cdot \Delta \omega \qquad\qquad\qquad\qquad\qquad T_{n\,new} = T_n \cdot \frac{1}{\Delta \omega}$$

Fig. 6.18 Shifting rules of mirrored controllers in Bode plot. *Source* [3, p. 222]

Table 6.3 #L Shifting rules of mirrored controllers in Bode diagram (*Source* [3, p. 307])

Calculation		Frequency response		Calculaltion
		Amplitude		
$K_{PRnew} = K_{PR} \cdot \frac{1}{\Delta K}$ ↑ shift up		$\Delta K = 10^{\frac{\Delta dB}{20}}$	shift down ↓	$K_{PRnew} = K_{PR} \cdot \Delta K$
		Phase		
	$\Delta dec = \|\log \omega_{Bnew} - \log \omega_B\|$			
$\omega_{n\,new} = \omega_n \cdot \frac{1}{\Delta \omega}$	move to the left	$\Delta \omega = 10^{\Delta Dek}$	move to the right	$\omega_{n\,new} = \omega_n \cdot \Delta \omega$
$T_{n\,new} = T_n \cdot \Delta \omega$	←		→	$T_{n\,new} = T_n \cdot \frac{1}{\Delta \omega}$

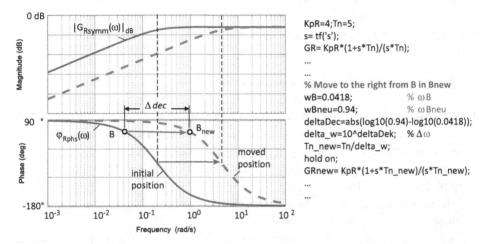

Fig. 6.19 Example of moving Bode plot of mirrored PI controller to the right. *Source* [18, p. 92]

6.4.2 BAD of P controller

P controller with unstable plant

Supposing an unstable plant should be controlled with the P controller, the phase rand should be achieved as big as possible. As far as it is not possible experimentally to measure the frequency response of unstable plant, the transfer function was defined theoretically and is given below

$$G_S(s) = \frac{s^2 + 0,5s - 0,5}{4s^4 + 4s^3 - 2s^2 - 1,5s + 0,5}$$

The Bode diagram of the plant is shown in Fig. 6.20.

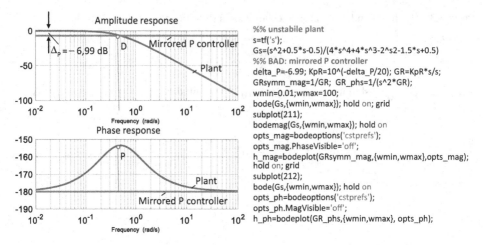

The following MATLAB script is shown to the right of the figure:

```
%% unstable plant
s=tf('s');
Gs=(s^2+0.5*s-0.5)/(4*s^4+4*s^3-2^s2-1.5*s+0.5)
%% BAD: mirrored P controller
delta_P=-6.99; KpR=10^(-delta_P/20); GR=KpR*s/s;
GRsymm_mag=1/GR; GR_phs=1/(s^2*GR);
wmin=0.01;wmax=100;
bode(Gs,{wmin,wmax}); hold on; grid
subplot(211);
bodemag(Gs,{wmin,wmax}); hold on
opts_mag=bodeoptions('cstprefs');
opts_mag.PhaseVisible='off';
h_mag=bodeplot(GRsymm_mag,{wmin,wmax},opts_mag);
hold on; grid
subplot(212);
bode(Gs,{wmin,wmax}); hold on
opts_ph=bodeoptions('cstprefs');
opts_ph.MagVisible='off';
h_ph=bodeplot(GR_phs,{wmin,wmax}, opts_ph);
```

Fig. 6.20 BAD of P controller with an unstable plant: Bode diagram and MATLAB® script

The BAD is applied doing the following steps:

1. The phase response of the mirrored P controller should be drowned in Fig. 6.20. As far as the phase response of the P controller is a straight line with a constant phase $\varphi = 0°$ and the axis of symmetry of the mirrored controller is $L\varphi = -90°$, the phase response of the mirrored controller P controller is also a straight line with a constant phase but of $\varphi = -180°$.
2. The maximum phase should be found and marked in Fig. 6.20. It is the point P with phase $\varphi = -153°$ at frequency $\omega = 0.502$ rad/s.
3. The corresponding point by the same frequency is to be denoted as D. It is the desired intersection point, that's way the amplitude plot of mirrored controller should be drowned through this point.

That's all, the BAD is finished. It is only left to read out the needed amplitude at the intersection point D from the Bode diagram

$$\Delta_P = -6,99 \text{ dB}$$

and to calculate the needed controller gain K_{PR}:

$$K_{PR} = 10^{-\frac{\Delta_P}{20}} = 2,236$$

The whole MATLAB® script of BAD procedure for P controller is shown in Fig. 6.20.

6.4.3 BAD of I controller

Controller with Stable Plant

As an example, is the frequency response of a stable plant experimentally measured and given in Fig. 6.21.

It has no sense to repeat the BAD for P controller, that's why the BAD is applied below for I controller:

$$G_R(s) = \frac{K_{iR}}{s}$$

The desired phase rand is given by $\varphi_{Rd} = 45°$.

The BAD steps of the previous case are repeated below for mirrored I controller and shown in Fig. 6.21.

1. Let us draw the phase response φ_{Rphs} of the mirrored controller. Remembering Fig. 2.16 of Kap. 2 is known that the phase plot of the I controller is a straight line located by $\varphi_R = -90°$. As far as the symmetry axis is $L_\varphi = -90°$, the phase plot of the mirrored I controller is also a straight line which is also located by $\varphi_{Rphs} = -90°$.
2. The point C is critical point which divides the Bode diagram into two frequency ranges. To the left of this point is the stability area of the closed loop because the phase response of the plant is above the phase response of the mirrored controller.
3. We mark the point P inside of stability area, at which the desired phase margin $\varphi_{Rd} = 45°$ is reached.

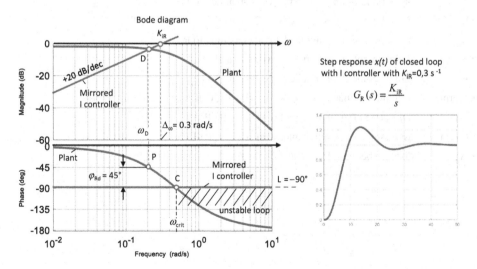

Fig. 6.21 BAD of I controller with the experimental measured Bode plot of stable plant and step response of closed loop with desired phase rand of $\varphi_{Rd} = 45°$. *Source* [3, pp. 283, 284]

4. The corresponding point D we mark in the amplitude response, at which the amplitude response of the mirrored I controller should intersect the amplitude response of the plant.

5. The amplitude response $|G_{\text{Rsymm}}(\omega)|_{\text{dB}}$ of the mirrored I controller should be drowned through point D with the slope $+20$ dB/dec.

6. We read out from Bode diagram the frequency $\Delta\omega$ of the point KiR, at which the amplitude response of the mirrored I controller intersects the 0 dB line. It is the frequency $\Delta\omega = 0.3$ rad/s.

It is the value of K_{iR}:

$$K_{\text{iR}} = \Delta_\omega = 0,3 \text{ s}^{-1}$$

6.4.4   BAD of PID controller

Characteristic Points of Mirrored PID Controller
(Siehe. (Fig. 6.22**a**))

$$G_{\text{R}}(s) = K_{\text{PR}} + \frac{K_{\text{PR}}}{sT_{\text{n}}} + sK_{\text{PR}}T_{\text{v}} \text{ (PID controller, additional form)}$$

$$G_{\text{Rsymm}}(s) = \frac{1}{G_{\text{R}}(s)} \qquad G_{\text{Rphs}}(s) = \frac{1}{s^2} \cdot \frac{1}{G_{\text{R}}(s)} \text{ (Mirrored PID)}$$

- L_{G} symmetry axis of magnitude plot $|G_{\text{Rsymm}}(\omega)|_{\text{dB}}$ with $L_{\text{G}} = 0$ dB
- L_φ symmetry axis of phase plot $\varphi_{\text{Rphs}}(\omega)$ with $L_\varphi = -90°$
- P_{n} break point of phase plot $\varphi_{\text{n}} = -135°$ by frequency $\omega_{\text{n}} = 1/T_{\text{n}}$
- A_{n} break point of magnitude plot corresponding to P_{n}
- P_{v} break point of phase plot $\varphi_{\text{v}} = -225°$ by frequency $\omega_{\text{v}} = 1/T_{\text{v}}$
- A_{v} break point of magnitude plot corresponding to P_{n}
- P_{M} middle point of phase plot between ω_{n} and ω_{v}
- A_{M} middle point of magnitude corresponding to P_{M}
- I-term   asymptote of magnitude plot with slope $+20$ dB/dec
- P-term   asymptote of magnitude plot with slope 0 dB/dec
- I-term   asymptote of magnitude plot with slope -20 dB/dec
- Δ_{p} distance between P-term and 0 dB axis: $\Delta_{\text{p}} = 20 \log(1/K_{\text{PR}})$

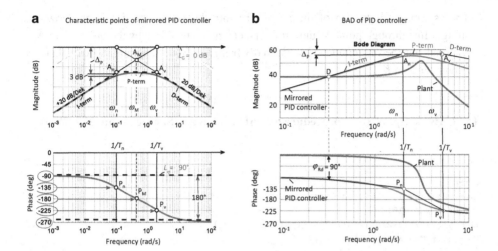

Fig. 6.22 BAD of PID controller

BAD Steps for Mirrored PID Controller

(Siehe. (Fig. **6.22b**))

The bode plot of the plant is given, the maximum possible phase reserve is desired.

1. Let us put manually points P_n and P_v by phases $\varphi = -135°$ and $\varphi = -225°$ arbitrary but under phase plot of the plant and read out from Bode diagram the frequencies $\omega_n = 2,02$ rad/s and $\omega_v = 3,31$ rad/s, from which result the controller parameters

$$T_n = \frac{1}{\omega_n} = 0,495 \text{ s}$$

$$T_v = \frac{1}{\omega_v} = 0,3 \text{ s}$$

2. We select the point D. It is the point at which the amplitude response of the mirrored controller and the amplitude response of the plant intersect by desired maximum phase margin. As it is seen from Fig. 6.22b the maximum possible phase margin is $\varphi_{Rd} = 90°$. Accordingly, we put the point D in Bode diagram.

3. The asymptote of I-term of the mirrored controller should be drowned with $+20$ dB/dec until it intersects the vertical line by the break point A_n.

4. The asymptote of P-term of the mirrored controller should be drawn as horizontal straight line through point A_n until it intersects the vertical line by the break point A_v.

5. We read out the magnitude $\Delta_P = 55{,}86$ dB from the Bode diagram and calculate the controller gain K_{PR}:

$$K_{PR} = 10^{-\frac{\Delta_P}{20}} = 0,0016$$

MATLAB® script of BAD procedure for PID controller (Source [3, pp. 279, 280])

```
%% -- Section 1: Simulation of the plant given by frequency response -----------
s=tf('s');
Kps=100; T1=0.1; T2=T1; d=0.4;   % Plant parameters are not given, only simulated
wmin=0.1;wmax=10;                % Limits
Gs=Kps/((s^2*T2+2*d*T2*s+1)*(s*T1+1));       % Plant is not given, only simulated
bode(Gs,{wmin,wmax}); hold on;grid           % Given Bode-Plot of the plant
%% --- Section 2: Calculation of controller parameters ----------------------
delta_P=55.86; w_n=2.02;  w_v=3.31;          % read out from Bode diagram
KpR=10^(-delta_P/20); Tn=1/w_n;Tv=1/w_v;TR=0.1*Tv; % Calculated parameters
%% Sektion 3: Bode-Plots
GR=KpR+KpR/(s*Tn)+s*KpR*Tv/(1+s*TR);     % PID-T1 controller
GRsymm_mag=1/GR;                         % Symmetrical controller
GR_phs=1/(s^2*GR);                       % Phase-symmetrical controller
subplot(211);                            % Amplitude plot
bodemag(Gs,{wmin,wmax}); hold on         % Window open and hold
opts_mag=bodeoptions('cstprefs');
opts_mag.PhaseVisible='off';
h_mag=bodeplot(GRsymm_mag,{wmin,wmax},opts_mag); hold on;grid
subplot(212);                            % Phase plot
bode(Gs,{wmin,wmax}); hold on            % Window open and hold
opts_ph=bodeoptions('cstprefs');
opts_ph.MagVisible='off';
h_ph=bodeplot(GR_phs,{wmin,wmax}, opts_ph); hold on;grid
%% --- Section 3: Step response -----------------------------------------------
figure
GR=KpR+KpR/(s*Tn)+s*KpR*Tv/(1+s*TR);     % PID-T1 controller
G0=Gs*GR;                                % open loop
Gw=G0/(1+G0);                            % closed loop
step(Gw,20); grid                        % step response
```

Fingerprint for BAD

The *App Fingerprint* of Fig. 6.10 for the 2BP method was further developed and adapted in [24, 25] for the 3BP method and BAD, as shown in Fig. 6.23.

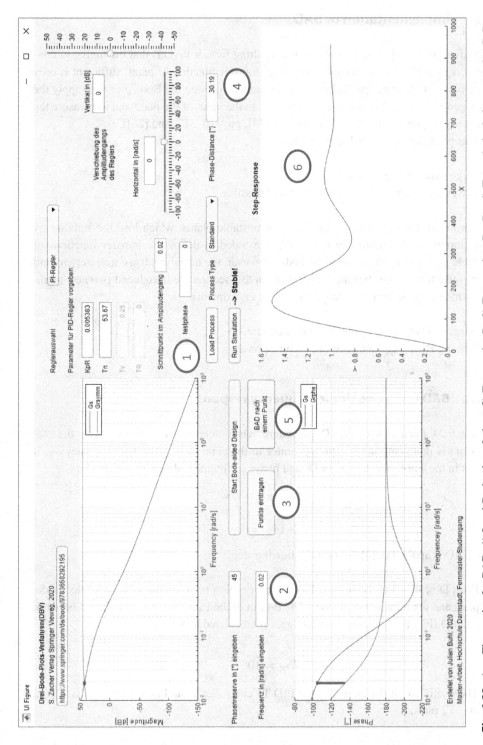

Fig. 6.23 App Fingerprint for BAD: 1) Load Bode plot of plant; 2) Enter point D; 3) Enter points Pn, Pv; 4) Enter desired phase margin; 5) Run BAD; 6) Simulated step response of closed loop. *Source* [26, p. 15]

6.5   Implementation of BAD

The advantage of 3BP method and of the resulting from it BAD is that the tuning of controller is possible without information about transfer function of plant, sufficient is only Bode plot of the plant. The well-known procedure to define the Bode plot is to apply the sinus generator with frequencies from 0 to ∞ to the input of the plant and to measure the frequency responses on the output (see, e.g., [3], pp. 310–317, and [26]).

This procedure has two disadvantages:

a. Large number of experiments,
b. The experiments are only for stable plant possible.

The problem (b) is caused by the nature of unstable plants, which lose the stability by sinus inputs. In the control theory is recommended to define the transfer functions of unstable plants using theoretical methods. However, the unstable plants are rarely to find in the industrial praxis. For the problem (a) in this section are introduced two new methods, which allow to avoid the large number of experiments:

• using BAD upon only point in Bode diagram,
• defining Bode diagram upon one step response.

6.5.1   BAD upon one single frequency response

Figure. 6.24 illustrates the BAD procedure when only one experiment with the stable plant is done entering a sinus generator to the plant input with this frequency ω_P. It results in the amplitude Δ_P (point D) and phase φ_P (point P) shown in Fig. 6.24a:

$$|G_S(\omega_P)| = \Delta_P$$

$$\varphi_S(\omega_P) = \varphi_P$$

The BAD is applied in [3] doing the following steps:

1. Point D of the plant is supposed as the intersection between the amplitude plot of the plant and the mirrored controller. The P-term of the PID-T1 controller should be set in point D (Fig. 6.24b). The controller gain is calculated as follows:

$$K_{PR} = 10^{-\frac{\Delta_P}{20}}$$

2. The width $\Delta\omega$ of the P-term of the PID T1 controller should be selected by user, e.g., $\Delta\omega = 2$ rad/s.

3. The break frequencies by both edges of $\Delta\omega$ are read out from Bode diagram. The time delay constants of the controller are calculated:

$$\omega_n = \omega_P / \Delta\omega \qquad \rightarrow \qquad T_n = \frac{1}{\omega_n}$$

$$\omega_v = \omega_P \cdot \Delta\omega \qquad \rightarrow \qquad T_v = \frac{1}{\omega_v}$$

$$T_R = 0,1T_v$$

4. The asymptotes of mirrored controller are drowned in Bode plot and shown in Fig. 6.24c.

The BAD upon only one point of Bode plot described above was tested in [3] based upon experiments of WebLab [27] and in [24, 25] through measurements on a plant for temperature control.

6.5.2   Bode plot upon single step response

The Carson-Laplace Transform is the correlation between derivative of the controlled variable $x(t)$ and Laplace transformed $x(s)$:

$$x(s) = \int_{0}^{\infty} \dot{x}(t)e^{-st}dt \tag{6.31}$$

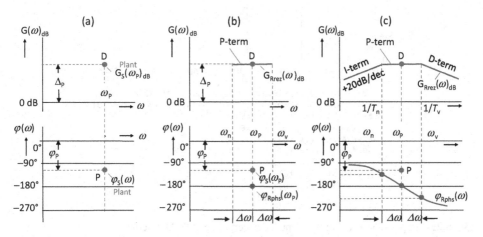

Fig. 6.24 Steps of BAD after one experiment with the plant by frequency ω_P, *Source* [3, p. 309]

If a step response $x(t)$ to divide into N time segments and to replace the curve of each segment into straight line, as illustrated in Fig. 6.25, then integral Eq. 6.31 is approximated through a sum with factors a_k, which are slopes of each secant:

$$x(s) = \sum_{k=1}^{k=N} a_k \int_{t_{k-1}}^{t_k} e^{-st} dt \qquad (6.32)$$

In other words, the step response is approximated with N secants. As an example, in Fig. 6.25 is shown the slope to point B:

$$a_B = \tan \alpha_B = \frac{x_C - x_B}{t_C - t_B}$$

Setting $s=j\omega$ in Eq. 6.32 we convert the approximated step response into frequency domain:

$$x(j\omega) = \sum_{k=1}^{k=N} a_k \int_{t_{k-1}}^{t_k} e^{-j\omega t} dt = \sum_{k=1}^{k=N} a_k \int_{t_{k-1}}^{t_k} (\cos \omega t - j \sin \omega t) dt \qquad (6.33)$$

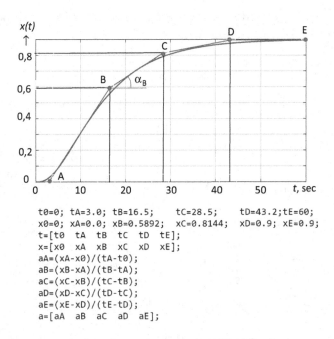

```
t0=0; tA=3.0; tB=16.5;      tC=28.5;     tD=43.2;tE=60;
x0=0; xA=0.0; xB=0.5892;    xC=0.8144;   xD=0.9; xE=0.9;
t=[t0  tA   tB   tC   tD   tE];
x=[x0  xA   xB   xC   xD   xE];
aA=(xA-x0)/(tA-t0);
aB=(xB-xA)/(tB-tA);
aC=(xC-xB)/(tC-tB);
aD=(xD-xC)/(tD-tC);
aE=(xE-xD)/(tE-tD);
a=[aA  aB   aC   aD   aE];
```

Fig. 6.25 Step response $x(t)$ of the plant and excerpt of MATLAB® script

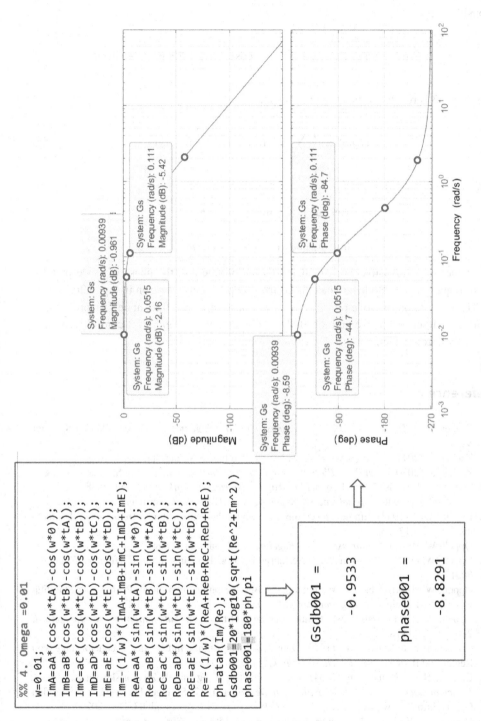

Fig. 6.26 Bode plot defined from step response and excerpt of MATLAB® script as an example of calculating the amplitude and phase of Bode plot for one value of frequency ω = 0,01 rad/s

Considering

$$\int_{t_{k-1}}^{t_k} (\cos \omega t - j \sin \omega t)dt = (\cos \omega t_k - \cos \omega t_{k-1}) - j(\sin \omega t_k - \sin \omega t_{k-1})$$

the Eq. 6.33 is results of the frequency response

$$G(j\omega) = \frac{1}{\omega} \sum_{k=1}^{k=N} \text{Re}_k(\omega) + j\frac{1}{\omega} \sum_{k=1}^{k=N} \text{Im}_k(\omega) \qquad (6.34)$$

with

$$\text{Re}_k(\omega) = a_k(\sin \omega t_{k-1} - \sin \omega t_k)$$

$$\text{Im}_k(\omega) = a_k(\cos \omega t_{k-1} - \cos \omega t_k)$$

From the last expressions are calculated the amplitude and the phase of Bode plot für each frequency. The Bode plot calculated from step response is given in Fig. 6.26.

The calculation of Bode plot from step response of a real plant upon Carson-Laplace transform described above and the BAD of the temperature and level control were successfully tested in [28].

References

1. Zacher, S., & Reuter, M. (2022). *Regelungstechnik für Ingenieure* (16th ed.). Verlag Springer Vieweg.
2. Zacher, S. (2017). *Übungsbuch Regelungstechnik* (6th ed.). Verlag Springer Vieweg.
3. Zacher, S. (2020). *Drei Bode-Plots-Verfahren für Regelungstechnik*. Verlag Springer Vieweg.
4. Sartorius, H. (1969). Adolf Leonhard *at-Automatisierungstechnik, 17*(12), 537–538.
5. Leonhard, A. (1940). *Die selbsttätige Regelung in Elektrotechnik*. Springer.
6. Leonhard, A. (1944). Ein neues Verfahren zur Stabilitätsuntersuchung. *Archiv für Elektrotechnik, 38*(1–2), 17–28.
7. Oppelt, W. (1947). *Grundgesetzte der Regelung*. Wolfenbüttel Verlag.
8. Oppelt, W. (1948). Über Ortskurvenverfahren bei Regelvorgängen mit Reibung. *VDI-Zeitung, 90*, 179–183.
9. Oppelt, W. (1960). Über die Anwendung der Beschreibungsfunktion zur Prüfung der Stabilität von Abtast-Regelungen. *Regelungstechnik*, (8), 15–18.
10. Oppelt, W. (1961). Stabilitätskriterium nach zwei Ortskurven. *Automation and remote control, 22*(9), 1175–1178.
11. Cremer, H., & Kolberg, F. (1960). Zur Stabilitätsprüfung mittels der Frequenzgänge von Regler und Regelstrecke. *Regelungstechnik*, (8), 190–194.
12. Cremer, H., & Kolberg, F. (1964). Zur Stabilitätsprüfung mittels Zweiortskurvenverfahrens. *Forschungsberichte des Landes Nordrhein-Westfalen, 1317*, Springer-Fachmedien Wiesbaden GmbH. http://www.springer.com/de/book/9783663066217. Accessed: 03. Dec. 2020.
13. Orlowski, P. F. (1999). *Praktische Regelungstechnik* (5th ed.). Springer Verlag.

14. Fasol, K. H. (1968). *Die Frequenzkennlinien.* Springer Verlag http://www.springer.com/de/book/9783709179635. Accessed: 03. Dec. 2020.
15. Kaerkes, R. (1969). Mathematische Grundlagen der Zweiortskurvenverfahren von Regelungssystemen. *Forschungsberichte des Landes Nordrhein-Westfalen*, Nr. 2033. Springer-Fachmedien Wiesbaden GmbH.
16. Gausch, F. (2009) Regelungstechnik B, Kapitel 4. *Universität Paderborn, Institut für ET und IT.* http://www-srt.upb.de/fileadmin/Lehre/Skripte/Regeltech/Regelungstechnik_B.pdf. Accessed: 28. Aug. 2017.
17. Zacher, S. (2017). *Das zweite Leben des Zweiortskurvenverfahrens.* Verlag Dr. Zacher. 978-3-937638-36-2.
18. Zacher, S. (2018). *ZBV: Zwei-Bode-Plots-Verfahren.* Verlag Dr. Zacher. 978-3-937638-37-9.
19. Zacher, S. (2018). ZBV: Zwei-Bode-Plots-Verfahren. Stuttgart: *Automation-Letters*, No. 38. https://www.zacher-international.com/Automation_Letters/38_ZBV.pdf. Accessed: 14. Apr. 2022.
20. Ellerhof, M., Haufe, A., & Krieger, T. (2019). *Fingerprint eines Regelkreises. Projektarbeit der Hochschule Darmstadt. Fernstudium M.Sc. Elektrotechnik.* https://www.zacher-international.com/C22_Team_Projekt/Fingerprint/Kurz_PraesentationFingerprint.pdf. Accessed: 16. Apr. 2022.
21. Ellerhof, M., Haufe, A., & Krieger, T. (2019). *App Fingerprint eines Regelkreises. Download.* Projektarbeit der Hochschule Darmstadt, Fernstudium M.Sc. Elektrotechnik. https://www.zacher-international.com/C22_Team_Projekt/Fingerprint/MyAppInstaller_web.zip. Accessed: 16. Apr. 2022.
22. Ellerhof, M.; Haufe, A.; Krieger, T. (2019) *Fingerprint eines Regelkreises. Tutorial.* Projektarbeit der Hochschule Darmstadt, Fernstudium M.Sc. Elektrotechnik https://www.zacher-international.com/C22_Team_Projekt/Fingerprint/Fingerprint.mp4. Accessed: 16. Apr. 2022.
23. Leonhard, W. (1965). Regelkreise mit symmetrischer Übertragungsfunktion. *Regelungstechnik, 13*, 4–12.
24. Collon, C.; Rudolph, J. (2011), Zwei Beispiele für die Berücksichtigung von Symmetrien beim Reglerentwurf. *at-Automatisierungstechnik, 59*(9), 540–551. https://www.uni-saarland.de/fileadmin/user_upload/Professoren/fr74_ProfRudolph/papersonline/collon_rudulph_2011_at.pdf . Accessed: 17. Dec. 2019.
25. Zacher, S. (1991). *Symmetrie und Antisymmetrie in der Automatisierungstechnik.* Manuskript des Kolloquiums, 16.07.1991, Fakultät Elektrotechnik, TU Dresden.
26. Zacher, S. (2003). *Duale Regelungstechnik.* VDE-Verlag.
27. Buhl, J. (2020). *Regelkreisentwurf nach dem DBV and BAD.* M.Sc. Thesis der Hochschule Darmstadt, Fernstudium Elektrotechnik.
28. Buhl, J. (2020) *App Fingerprint DBV/BAD. Download.* Projektarbeit der Hochschule Darmstadt, Fernstudium M.Sc. Elektrotechnik. https://www.zacher-international.com/Automation_Letters/FingerPrint-App.7z. Accessed: 16. Apr. 2022.
29. Zacher, S. (2020) Drei Bode Plots Verfahren und BAD, Automation Letter No. 20. https://www.zacher-international.com/Automation_Letters/11_Fingerprint_DBV.pdf accessed 16.04.2022
30. Henry, J., Zacher, S. (2010) WebLabs in Control Engineering Education: status and trends. 7. *AALE Angewandte Automatisierung in der Lehre und Forschung*, FH Technikum Wien, 10./11. Feb. 2010
31. Aigner, P., Alves Zipf, M., Schaffner, A. (2020) *Bode-Aided Design einer Temperatur- und Füllstandsregelung.* Darmstadt: Projektarbeit der Hochschule Darmstadt, Fernstudium Elektrotechnik.

Management of Identification

7

"The step response is the WANTED poster for plant"

Quote: Erwin Samal [15], p. 104

7.1 Introduction

7.1.1 Data Stream Management

An important task of supervising and control of industrial automation systems is the management of the information flow between its subsystems and inside each subsystem (Fig. 7.1).

The following types of communication are known from the basics of informatics:

- Signals
  - Signals are physical quantities that change over time and are transmitted between physical elements (sensors, actuators, transmitters, receivers). For example, the processes inside a closed control loop are described by signals.
- Messages
  - Messages are reports produced by processes of control system and transmitted using signals. One message can be conveyed with different signals.
- Data
  - Data are messages that are transmitted through networks from one physical element to another physical element by means of signals.
- Information
  - Information is the interpreted message for human operators.

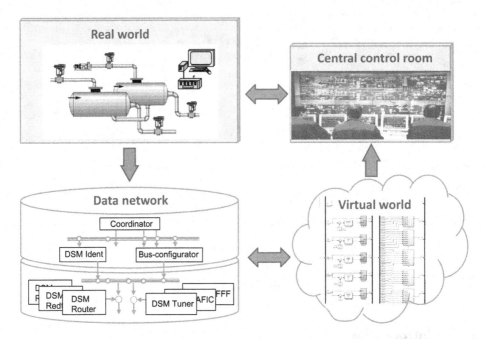

Fig. 7.1 Industrial automation system as interaction between human operators (central control room), data network, real physical control subsystems (real world), and their simulations (virtual world). *Source* [1, p. 122]

As far as the CLC is considered as a part of the entire automated system, the management of data stream between levels of CLC also becomes important.

The closed loop control during its life cycles is shown in Fig. 7.2 as the interaction between real world and virtual world supervised by human operator.

The Data Stream Managers (DSM) introduced in this book are functional blocks for supervising signals and messages. They consist of algorithms and programs of closed loop control, open loop control, and logical operations. DSM allow to manage design and control as well for the real world of physical devices as for the virtual world of mathematical models and simulations.

The DSM can be divided into two groups according to their function and location:

- DSM for control (online managers), that are a part of software of the real-world devices. They receive the input data directly from a control loop and deliver it to the controller after algorithmic revision. In other words, a DSM for control manages the control stream.
- DSM for design (offline managers), that are used by engineering of control systems which usually happens during controller setup or design phase. They interact fully or partially with human operator, with other DSM, and with the real world.

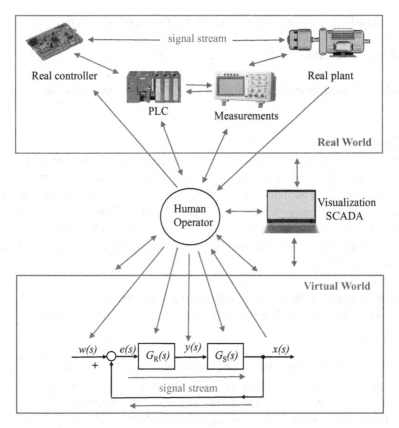

Fig. 7.2 Conception of the Data Stream Management for closed loop control as the interaction between real world and virtual world supervised by human operator (abbreviations: PLC—programmable logical control, SCADA—supervisory control and data acquisition)

7.1.2 DSM of this Chapter

In this chapter are described managers associated with the design data stream, as shown in Fig. 7.3, for two tasks of control system engineering, namely identification and controller tuning. The DSM of this chapter are developed for manual data transfer between human operator and controller. In order to convert these design managers into control managers, the communication and data transfer between individual DSMs should be additionally programmed, which is not described in this book.

DSM Ident

Two design managers Ident are programmed according to two well-known identification algorithms:

- DSM Ident-1 based upon tangent drown to the turning point of step response (see e.g., [2, 3]).
- DSM Ident-3 according to the time-percentage-values method (see, e.g., [2]. [4, 5]).

The modules formed from these algorithms are naturally offline DSM and belong to the virtual world.

DSM Tuner

The DSM tuners are derived from standardized algorithms of the controller design in the Laplace domain, such as the optimum magnitude and the symmetrical optimum described in Chap. 3. The inputs of the DSM Tuner are transfer functions of the plant determined by the above-mentioned DSM Ident. The DSM Tuner calculates the optimal parameters of standard controllers, which are its output data. The data transfer between DSM Ident, DSM Tuner, and the controller is not automated in this chapter, so that here is considered an offline DSM.

DSM AFIC (Adaptive Filter for Identification and Control)

By DSM AFIC both phases "Identification" and "Tuner" are linked in [2] as a single off-set DSM:

- Identification is done upon three consecutive following one after another input steps. From the resulting step responses are determined three coefficients K0, K1, and K2 of the plant model, built as an adaptive filter based on the conception of [6].
- With this adaptive filter is done tuning of standard controllers, such as a PI or PID controller. First is built the virtual world from filter coefficients K0, K1, and K2. Then the optimal parameters of standard controllers are determined by converting filter

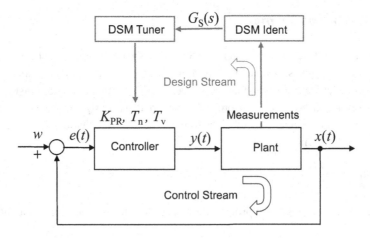

Fig. 7.3 Two kinds of data stream by engineering of closed loop control. *Source* [1, p. 131]

coefficients K0, K1, K2 into controller parameters KPR, Tn, and Tv. Finally, the controller parameters are transferred to the real controller in the real world.

7.2 DSM Ident

7.2.1 Structure

The aim of modules, which in [1] are called *Ident*, is the identification of the proportional PT1, PT2 … PTn plants or plants with transport delay building. Two DSM Ident types are considered:

- Ident 1 according to the turning-point-tangent method described in Chap. 3 and shown in Fig. 3.2.
- Ident 3 according to the time-percentage-values method.

Both modules used the experimental defined step responses and have the same structure consisting of similar blocks, as explained in Fig. 7.4. The difference between the two modules is the identification algorithm and the number of required points of step response. The DSM Ident 1 is simple and needs one point, while the DSM Ident 3 is more complicated and identifies the plant with three different points of the step response. Accordingly, the identification results by DSM Ident 3 are more precise than by DSM Ident 1.

Switch

With this block the examined control circuit is switched between two operating modes:

- Identification mode, in which the plant is separated from the controller so that the step of set point w is sent directly to the input of the plant as actuating variable y.
- Control mode, by which the set point w is disconnected from the plant and connected to controller.

Block 1

This block is used to determine the proportional coefficient K_{PS} of the controller according to the relationship between the controlled variable in the steady state $x(\infty)$ and the size of the input step y_0:

$$K_{PS} = \frac{x(\infty)}{y_0} \tag{7.1}$$

Block 2

This block is used to calculate the values of the controlled variable $x(t)$ as a percentage of the end value $x(\infty)$ of the steady state:

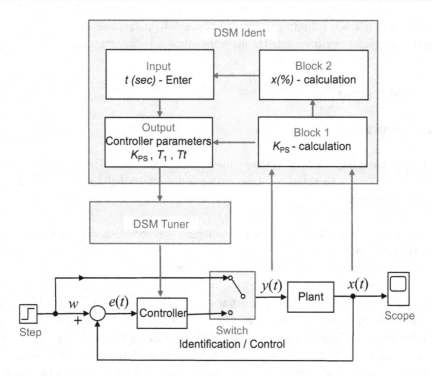

Fig. 7.4 Structure of the DSM Ident. *Source* [1, p. 135]

- For DSM Ident 1 it is $x(63)$, which means 63% of $x(\infty)$,
- For DSM Ident 3 there are $x(10)$, $x(50)$, and $x(90)$, which are accordingly 10%, 50%, and 90% of $x(\infty)$.

Input
This block summarizes the input parameters that are to be supplied to the DSM Ident by external sources. In this chapter, the input parameters are entered manually.

Output
As a result of an identification, the calculated parameters of the plant are manually transferred to the DSM tuner.

7.2.2   DSM Ident-1

The DSM Ident-1 is shown in Fig. 7.5. It is based upon turning-point-tangent method ([1]–[3]).

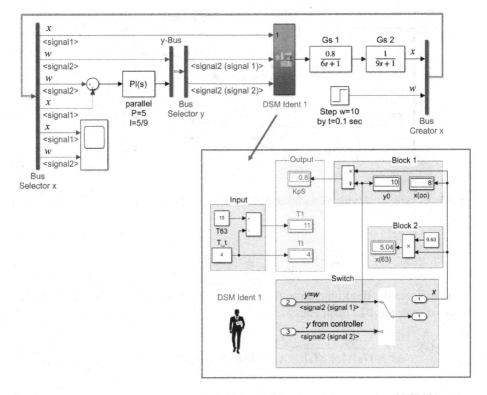

Fig. 7.5 Simulation of closed loop with DSM Ident 1 (above) and the structure of DSM Ident 1

Example of DSM Ident-1

- Input. Step and step response (Fig. 7.6):

$$y_0 = 10 \quad x(\infty) = 8$$

- Block 1. Calculation of plot's gain upon Eq. 7.1:

$$K_{PS} = \frac{x(\infty)}{y_0} = 0.8$$

- Block 2. Read out from Fig. 7.6:

$$Tt = 4 \, \text{sec} \quad T_{63} = 15 \, \text{sec} \quad T_1 = T_{63} - Tt = 11 \, \text{sec}$$

$$x(63) = 0.63 \quad x(\infty) = 5.04$$

- Output: The identified plant model according to approximation of turning-point-tangent method:

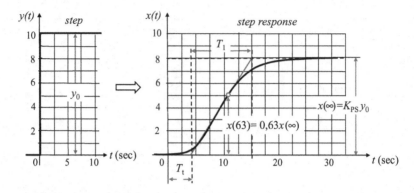

Fig. 7.6 Example of identification with DSM Ident 1. *Source* [1, p. 136]

$$G_S(s) = \frac{K_{PS}}{1 + sT_1} e^{-sT_t} = \frac{0,8}{1 + 11s} e^{-4s} \tag{7.2}$$

To eliminate the dead time term T_t in Eq. 7.2, the plant model of Eq. 7.2 is further approximated as PT2 upon *Pade* [2, 3]:

$$G_S(s) = \frac{K_{PS}}{(1 + sT_1)(1 + sT_t)} = \frac{0,8}{(1 + 11s)(1 + 4s)} \blacktriangleleft \tag{7.3}$$

7.2.3 DSM Ident 3

The Input module and Block 1 are the same as by DSM Ident 1 (Eq. 7.1).

The Block 2 (Fig. 7.4) of DSM Ident 3 is programmed upon method of the time-percentage values, which was proposed by *Schwarze* (see [1, pp. 229, 230], and [4]). According to this method, a PTn plant is identified as series connection of n PT1 terms with only one for all terms identical delay time constant T_1:

$$G_S(s) = \frac{K_{PS}}{(1 + sT_1)^n}$$

To define the time constant T_1 first are calculated values $x(10)$, $x(50)$ and $x(90)$, by which the controlled variable achieves 10%, 50%, and 90% of its steady state value $x(\infty)$. Then are read out from step response the time values t_{10}, t_{50}, and t_{90} according to ordinates $x(10)$, $x(10)$, and $x(10)$ as shown in Fig. 7.7. The characteristic parameter μ is calculated:

$$\mu = \frac{t_{10}}{t_{90}} \tag{7.4}$$

According to the value of μ is read out from Table 7.1 the number n of series connected PT1 terms and corresponding coefficients α of the linearization α_{10}, α_{50}, and α_{90},

Finally, is calculated the delay time constant T_1:

$$T_1 = \frac{\alpha_{10}t_{10} + \alpha_{50}t_{50} + \alpha_{90}t_{90}}{3} \tag{7.5}$$

The structure of DSM Ident 3 is shown in Fig. 7.8.

Example of DSM Ident 3

- Input. Step and step response (Fig. 7.7):

$$y_0 = 10 \quad x(\infty) = 8$$

- Block 1. Calculation of plot's gain upon Eq. 7.1:

$$K_{PS} = \frac{x(\infty)}{y_0} = 0,8$$

- Block 2. Read out from Fig. 7.6 and calculated:

$$x(10) = 0.8 \quad x(50) = 4 \quad x(90) = 7.2$$

$$t_{10} = 4.03 \text{ sec} \quad t_{50} = 12.6 \text{ sec} \quad t_{90} = 29.5 \text{ sec}$$

$$\mu = \frac{t_{10}}{t_{90}} = \frac{4.03}{29.5} = 0.1366$$

The order of the PTn plant and coefficients α of the linearization read out from Table 7.1:

$$n = 2$$

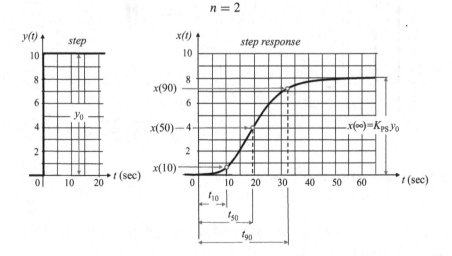

Fig. 7.7 Example of step and step response sent from Input module to blocks 1 and 2 of DSM Ident 3

Table 7.1 Table of *Schwarze* for method of the time-percentage values. (Quelle: [4, p. 176])

Characteristic value μ	Order n of PTn plant	Coefficients α of the linearization		
		α_{10}	α_{50}	α_{90}
0.137	2	1.880	0.596	0.257
0.174	2,5	1.245	0.460	0.216
0.207	3	0.907	0.374	0.188
0.261	4	0.573	0.272	0.150
0.304	5	0.411	0.214	0.125
0.340	6	0.317	0.176	0.108
0.370	7	0.257	0.150	0.095
0.396	8	0.215	0.130	0.085
0.418	9	0.184	0.115	0.077
0.438	10	0.161	0.103	0.070

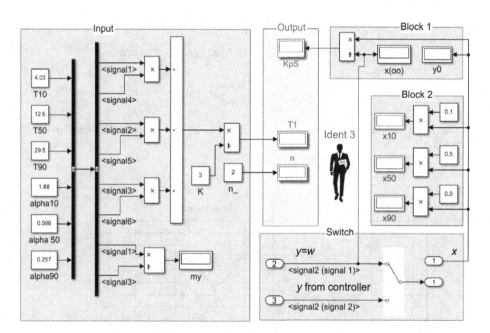

Fig. 7.8 Structure of DSM Ident 3. *Source* [1, p. 142]

$$\alpha_{10} = 1.88 \quad \alpha_{50} = 0.596 \quad \alpha_{90} = 0.257$$

Finally is calculated the delay time constant of the PTn plant:

$$T_1 = \frac{\alpha_{10}t_{10} + \alpha_{50}t_{50} + \alpha_{90}t_{90}}{3} = 7.556$$

- Output: The identified plant model according to the method of the time-percentage values:

$$G_S(s) = \frac{K_{PS}}{(1 + sT_1)^n} = \frac{0.8}{(1 + 7.556s)^2} \blacktriangleleft \qquad (7.6)$$

The identified plants of DSM Ident 1 (Eq. 7.3) and DSM Ident 3 (Eq. 7.6) of the same real physical plant are compared with measurements data and shown in Fig. 7.9. From this comparison is immediately clear the advantage of the DSM Ident 3, therefore, in the next section is applied only this DSM.

7.2.4 MATLAB® Script of DSM Ident 3

The MATLAB® script of DSM Ident 3 consists of 5 sections and is shown in Fig. 7.10 till 7.15 for manual use in Table 7.1.

- Section 1: The file 33,247.txt with measurements results of a step and step response of a real physical plant, received by online experiment from WebLab of University of Tennessee Chattanooga [9] is imported in MATLAB® and saved as *Numeric Matrix* under the name "data" in *Workspace* (Fig. 7.10).
- Section 2: The imported step and step response are graphically displayed or plotted (Fig. 7.11).
- Section 3: The step responses of Fig. 7.11 are evaluated. The values read out from step response of Fig. 7.11 should be typed after prompts into the *Command Window* to fulfill the lines 21 to 28 of Fig. 7.12. According to this information, the step response x(t) is shifted from MATLAB® to the coordinate's origin, as shown in the

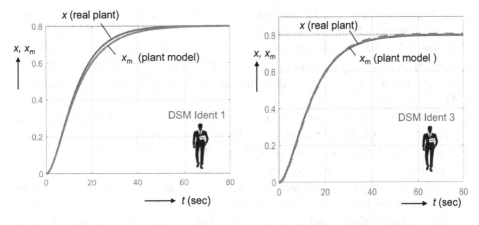

Fig. 7.9 Comparison of step responses of plant models with real plant (DSM Ident 1 on the left side, DSM ident 3 on right side)

Fig. 7.10 Sections 1 and 2 of MATLAB script of DSM Ident 3. *Source* [2, p. 428]

```
3     %% 1. Import measurement's data
4 –   clear all;
5 –   data =readmatrix('33247.txt');
6
7     %% 2. Plot measurement's data
8 –   load('32587.mat')
9 –   t = data(:,1);
10 –  y = data(:,2);
11 –  x = data(:,3);
12 –  s = tf('s');
13 –  figure
14 –  plot(t,y,t,x, 'Linewidth',3)
15 –  grid on;
```

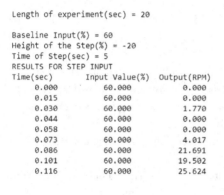
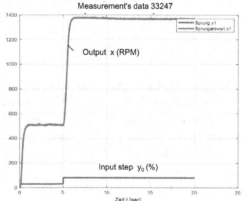

```
Length of experiment(sec) = 20

Baseline Input(%) = 60
Height of the Step(%) = -20
Time of Step(sec) = 5
RESULTS FOR STEP INPUT
Time(sec)        Input Value(%)   Output(RPM)
     0.000           60.000            0.000
     0.015           60.000            0.000
     0.030           60.000            1.770
     0.044           60.000            0.000
     0.058           60.000            0.000
     0.073           60.000            4.017
     0.086           60.000           21.691
     0.101           60.000           19.502
     0.116           60.000           25.624
```

Fig. 7.11 Step responses of measurement of the real plant. *Source* [2, p. 428]

left diagram of Fig. 7.13. The appropriate area of the step response should be zoomed as shown in the right diagram of Fig. 7.13.

- Section 4: The steady state value $x(\infty)$ of the controlled variable is determined (line 38 in Fig. 7.14). The gain Kps of the plant is calculated in the line 40. In lines 41–43 are determined the percentage values that are necessary according to the method of the time-percentage values.
- Section 5: The time-percentage values $t10$, $t50$, and $t90$ are read out from right diagram of Fig. 7.13. These values should be typed after prompts into the *Command Window* to fulfill the lines 45–47 to 28 of Fig. 7.14.

According to Eq. 7.4, the required characteristic value μ is calculated by MATLAB® in line 48 and immediately displayed so that the corresponding value n for the order of the plant can be found in Tab. 7.1. The values $\alpha10$, $\alpha50$, and $\alpha90$ are read out from Table 7.1 and entered in lines 50-52 of Fig. 7.14. The Eq. 7.5 is exe-

Fig. 7.12 Section 3 of
MATLAB® script to the DSM
Ident 3. *Source* [2, p. 429]

```
19    %% 3. Shift axes to the coordinate's origin
20
21 -  Xanf=input('Xanf=');'
22 -  Xend=input('Xend=');
23 -  Yanf=input('Yanf=');
24 -  Yend=input('Yend=');
25 -  t0=input('t0='); % i
26 -  t_ax=input('t_ax=');
27 -  x_ax=input('x_ax=');
28 -  tm=t-t0;
29 -  xm=x-Xanf;
30 -  figure
31 -  plot(tm,xm,'r','Linewidth',3)
32 -  hold on
33 -  G=s/s;
34 -  step(x_ax*G,t_ax);
35 -  hold on
36 -  grid
```

cuted in line 53, and the plant is identified (line 54) and compared with the measured values in Fig. 7.15.

7.3   DSM Tuner

The Data Stream Manager *Tuner* is used to determine the parameters of standard controllers P, I, PI, PD, or PID according to a mathematical model of the controlled system that has already been determined by the DSM Ident.

Standard Types of Control Loops
A large set of closed control loops consisted of standard controller and different types of plants like PT1, PT2...PTn, I, IT1, IT2, Tt, is divided in [1, 2, 7, 8, 10] into 6 basic types according to the transfer function $G_0(s)$ of the open loop:

Typ A: optimum magnitude, $\vartheta = \frac{1}{\sqrt{2}}$ $G_0(s) = \frac{K_{PR}K_{PS}K_{IS}}{sT_n(1+sT_1)}$

Typ B: optimum magnitude, $\vartheta = \frac{1}{\sqrt{2}}$ $G_0(s) = \frac{K_{PR}K_{PS}}{(1+sT_1)(1+sT_2)}$

Typ C: aperiodically, $\vartheta = 1$ $G_0(s) = \frac{K_{PR}K_{PS}K_{IS}}{sT_n}$

Typ D: aperiodically, $\vartheta = 1$ $G_0(s) = \frac{K_{PR}K_{PS}}{1+sT_1}$

Typ E: unstable, $\vartheta = 0$ $G_0(s) = \frac{K_{PR}K_{PS}K_{IS}}{s^2T_n}$

Typ SO: symmetrical optimum, $\vartheta = \frac{1}{\sqrt{2}}$ $G_0(s) = \frac{K_{PR}K_{PS}K_{IS}(1+sT_n)}{s^2T_n(1+sT_1)}$

The coefficients in the above formulas are denoted as follows:

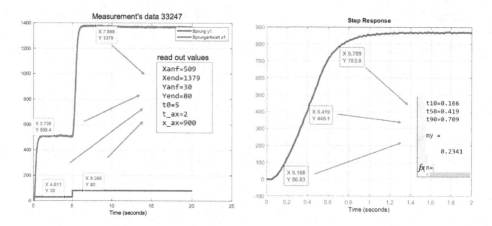

Fig. 7.13 Step responses for zooming (on the left) and after zooming. *Source* [2, p. 429]

```
37    %% 4. Define gain of the plant Kₚₛ and calculate x(10), x(50), x(90)
38 -  x_oo=Xend-Xanf;
39 -  y0=Yend-Yanf;
40 -  KpS=x_oo/y0;
41 -  x10=0.1*x_oo
42 -  x50=0.5*x_oo
43 -  x90=0.9*x_oo
44    %% 5. Algorithm of time-percentage-values and comparison of step responses
45 -  t10=input('t10=');
46 -  t50=input('t50=');
47 -  t90=input('t90=');
48 -  my=t10/t90
49 -  n=input('n=');
50 -  alpha10=input('alpha10=');
51 -  alpha50=input('alpha50=');
52 -  alpha90=input('alpha90=');
53 -  T=(alpha10*t10+alpha50*t50+alpha90*t90)/3;
54 -  Gs=KpS/(1+s*T)^n;
55 -  step(y0*Gs,t_ax,'b*')
```

Fig. 7.14 Sections 4 and 5 of MATLAB script to the DSM Ident 3. *Source* [2, p. 430]

- K_{PR} and K_{PS} are gains of the controller and plant,
- K_{IS} is integration constant of the plant,
- T_n is reset time of the controller,
- T_1 and T_2 are time constants of the plant.

Algorithms of controllers tuning

To each standard type of closed loops are derived in [1] and [2] the algorithms of the optimal tuning:

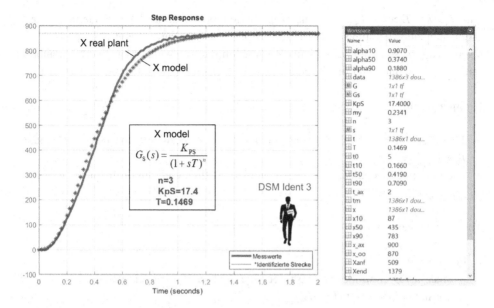

Fig. 7.15 Step response of the plant after identification compared with measurements of the step response of the real plant. *Source* [2, p. 430]

Typ A: optimum magnitude, $\vartheta = \frac{1}{\sqrt{2}}$ $\quad\quad K_{PR} = \frac{T_n}{2 \cdot K_{PS} K_{IS} \cdot T_1}$

Typ B: optimum magnitude, $\vartheta = \frac{1}{\sqrt{2}}$ $\quad\quad K_{PR} = \frac{(T_1 + T_2)^2}{2 \cdot K_{PS} \cdot T_1 \cdot T_2} - \frac{1}{K_{PS}}$

Typ C: aperiodic, $\vartheta = 1$ $\quad\quad K_{PR} = \frac{3,9 \cdot T_n}{K_{PS} K_{IS} T_{aus}}$

Typ D: aperiodic, $\vartheta = 1$ $\quad\quad K_{PR} = \frac{1}{K_{PS}} \left(\frac{3,9 \cdot T_1}{T_{aus}} - 1 \right)$

Typ E: unstable, $\vartheta = 0$ $\quad\quad$ unstable independent of gain K_{PR}

Typ SO: symmetrical optimum, $\vartheta = 0,5$ $\quad K_{PR} = \frac{1}{2 \cdot K_{PS} K_{IS} \cdot T_1}$

The step responses of closed control loops of each standard type expected after tuning upon expressions shown above are given in Fig. 7.16.

The following are given general recommendations on how to convert some current closed loop like

$$G_0(s) = \frac{K_{PR}(1 + sT_n)(1 + sT_v)K_{PS}}{sT_n(1 + sT_1)(1 + sT_2)(1 + sT_3)(1 + sT_4)(1 + sT_5)(1 + sT_6)} \quad (7.7)$$

with $T_1 > T_2 > T_3 > T_4 > T_5 > T_6$ in one of the basic types mentioned above due to compensation and equivalent time constant. The adaptation of the current control loop Eq. 7.7 to a standard type should always be implemented in the following order: first the compensation is made; only after that if required can be formed the substitute time constant T_E.

Compensation

- The reset time T_n should be chosen equal to the largest time constant of the plant: $T_n = T_{largest} = T_1$.
- The rate time T_v compensates the second largest time constant of the plant: $T_v = T_{second\ largest} = T_2$.

There are two exceptions:

- PD controller:
  - since a PD controller has no reset time T_n, then T_v compensates the largest time constant of the plant: $T_v = T_{largest}$.
- symmetric optimum (SO):
  - T_n is not compensated but set proportional to largest time constant of the plant: $T_n = k \cdot T_{largest}$ with $k = 4$.
  - T_v compensates the second largest time constant of the plant: $T_v = T_{second\ largest}$.

Aa far as compensation is done the Eq. 7.7 is converted as shown below

$$G_0(s) = \frac{K_{PR} K_{PS}}{sT_n(1 + sT_3)(1 + sT_4)(1 + sT_5)(1 + sT_6)} \tag{7.8}$$

It is clearly seen that Eq. 7.8 is of the Typ A, but instead of one term $(1 + sT)$ there are here 4 such terms. The next step to adapt the current loop to the type A is called substitute or equivalent time constant T_E.

Equivalent time constant

The sum of time constants T_3, T_4, T_5, T_6 of Eq. 7.8 is designated as the equivalent time constant T_E:

$$T_E = T_3 + T_4 + T_5 + T_6 \tag{7.9}$$

However, to build such a sum like given in Eq. 7.9 is mathematically not correct. But the error is minimal if one time constant, e.g., T_3, is 5 times larger than the sum of the remaining time constants or equal to it:

$$T_3 \geq 5 \cdot (T_4 + T_5 + T_6) \tag{7.10}$$

Supposing that condition Eq. 7.10 is met, the transfer function Eq. 7.8 is converted to the type A:

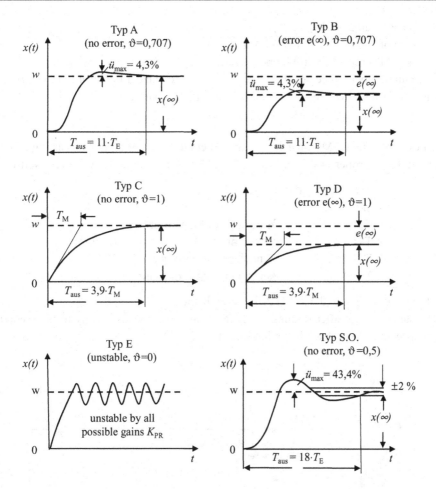

Fig. 7.16 Step responses of closed loops according to optimal tuning standard types. *Source* [10, p. 21]

$$G_0(s) = \frac{K_{PR}K_{PS}}{sT_n(1 + sT_E)} \qquad (7.11)$$

According to it follows the optimal tuning:

$$K_{PR} = \frac{T_n}{2 \cdot K_{PS} \cdot T_E}$$

Structure of DSM Tuner

The DSM Tuner consists of two blocks, input and output.

- The input block receives the following data from the DSM Ident:

– - plant parameters K_{ps}, T_1, T_2, …
– - type of controller P, I, PD, PI, PID.
– - type of standard control loop A, B, C, …
• The output block processes the inputs according to standard control loop algorithms and returns the calculated parameters directly to the controller.

As an example, the DSM Tuner for the closed control loop of type A consisting of PT3 plant and PI controller is shown in Fig. 7.17. The plant parameters were received from

```
%% Input-block of DSM Tuner
clear all;
Kps=20;          % sent from DSM Ident
T1=300;          % the largest time constant of plant
T2=1;            % time constant of plant
T3=4;            % time constant of plant
```

DSM Ident according to the MATLAB®-script below

The step response after running the DSM Tuner corresponds exactly to the theoretically expected for type A as shown in Fig. 7.17.

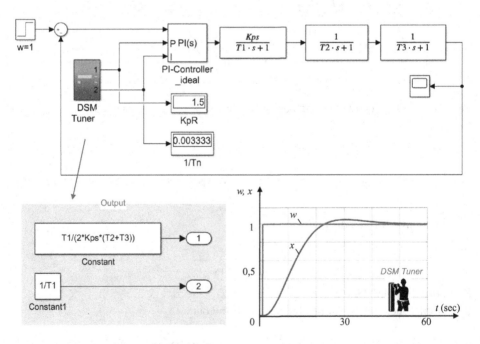

Fig. 7.17 DSM Tuner and step response of the control loop of type A. *Source* [1, p. 149]

7.4   AFIC

7.4.1   Conception of Adaptive Filter for Identification and Control

The aid of AFIC is the identification and control of an unknown plant using an adaptive filter (Fig. 7.18) according to the LMS algorithm (*Least Mean Squares*).

An adaptive filter is a transport delay with time constant T, which is connected in a closed loop with positive feedback. The method applies to P-plants with delay (PT1, PT2, … PTn) or to PT1 plants with transport delay time Tt also known as dead time. The measurement data [t, x] are the input step $y0$ and the output step response X_plant recorded by the experiment with the real plant. The variable *Error* is built as a difference between measurement data [t, x] and the output X_model of the filter after the same input step $y0$. The variable *Error* is then minimized with LMS algorithm, as a result are

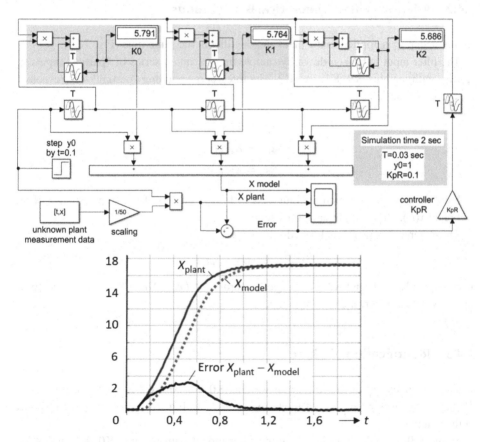

Fig. 7.18 First stage of AFIC: The identification of the real plant with three adaptive filters upon measurements data minimizing the difference between output of real X_plant and of the filter-model X_model

defined the filter factors K0, K1, and K2. The real plant is approximated and simulated with these factors, as shown in Fig. 7.19.

After this the filter model is converted into the transfer function of PT1-term with parameters K_{PS}, T_1, and transport delay T_t. In Fig. 7.19 is shown an example that the filter factors K0 = 5.791, K1 = 5.764, and K2 = 5.686 are converted to the following transfer function with $K_{PS} = 17.2$; $T_1 = 0.3$ s, and transport delay $T_t = 0.2$ s:

$$G_S(s) = \frac{K_{PS}}{1 + sT_1}e^{-sT_t} \tag{7.12}$$

Finally, the standard controller PI or PID is tuned upon Eq. 7.12 and implemented. In the following is given the detailed description of the AFIC after short introduction to the theory of adaptive filters.

7.4.2 Adaptive Filter: Theoretical Backgrounds

An input step y_0 is applied to the plant and the filter at the same time (Fig. 7.20).

The filter input is broken down by sample times T into a series of "partial" inputs y_0, y_{F1}, y_{F2} with weights K0, K1, and K2. The three corresponding "partial" step responses x_0, x_{F1}, x_{F2} of the filter

$$x_{F0} = K_0 y_0$$
$$x_{F1} = K_1 y_{F1}$$
$$x_{F2} = K_2 y_{F2}$$

build the sum $x_F(t)$, which is compared with the sum $x(t)$ of the step response of the real plant $x(t_0)$, $x(t_1)$, $x(t_2)$ at each time segment $t_0 = 0$, $t_1 = T$, $t_2 = 2\,T$.

The difference (error) e is formed:

$$e(t) = x(t) - x_F(t)$$

The weights K0, K1, and K2 are changed according to *Least Mean Squares* (LMS) [6] in such a way that the error $e(t)$ is minimized.

7.4.3 Identification of Plant

The step response x_F of the filter with weights K0, K1, and K2 is shown in Fig. 7.21a. The step responses x_0, x_{F1}, x_{F2} of filter are compared in Fig. 7.21b with the step response of the real plant.

In the following is described how the filter model with weights K0, K1, and K2 is converted into the transfer function of the plant Eq. 7.12 as a PT1 term with the dead time.

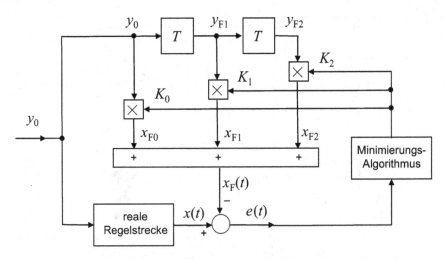

Fig. 7.19 Second stage of AFIC: the simulation of the real plant as adaptive filter with three factors K0, K1, and K2; on the left side is shown the step response *X_model* simulated with filter compared with the step response *X_plant* of the real plant

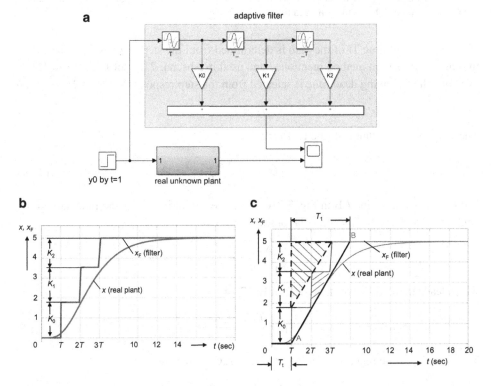

Fig. 7.20 Structure of an adaptive filter. *Source* [1, p. 151]

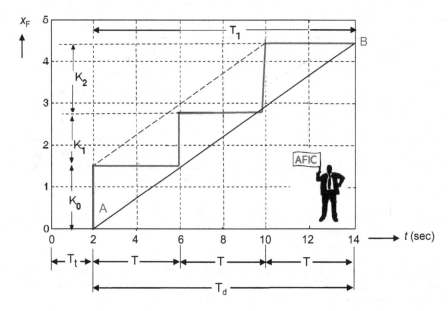

Fig. 7.21 **a** Filter with weights K0, K1, K2; **b** step response of the filter xF with sample time T and step response of the real plant; **c** conversion the filter model with weights K0, K1, K2 intoc-transfer function of PT1-term with dead time

First, the dead time Tt of the path is determined from the step response, similar to the turning-point-tangent method mentioned and used in Abschn. 7.2.2 for DSM Ident 1. For example, the following dead time is selected from the step response of Fig. 7.21c:

$$T_t = T \tag{7.13}$$

The tangent to the point A of the PT1 term

$$G_S(s) = \frac{K_{PS}}{1 + sT_1} \tag{7.14}$$

is drown as straight-line AB in Fig. 7.21c. From the similarity of red and blue dashed triangles in Fig. 7.21c follows the relationship

$$\frac{K_1}{K_1 + K_2} = \frac{T}{T_1}$$

which leads to the time constant of T_1 of Eq. 7.14:

$$T_1 = \frac{T(K_1 + K_2)}{K_1} \tag{7.15}$$

The gain K_{PS} of the plant is defined from Eq. 7.1. Considering

$$x(\infty) = K_0 + K_1 + K_2 \tag{7.16}$$

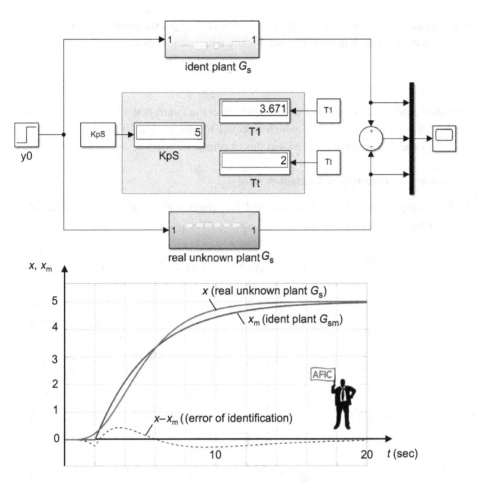

Fig. 7.22 Interrelation between parameters of filter K0, K1, K2 and of plant model K_{ps}, T_1, and T_t. The line AB is tangent to the initial point of the step response

from Eq. 7.1 follows the value of the gain of the plant:

$$K_{PS} = \frac{K_0 + K_1 + K_2}{y_0} \tag{7.17}$$

The relationship Eq. 7.13, 7.15–7.17 between filter weights and approximated parameters of the transfer function of as a PT1 term with the dead time (Eq. 7.1) is illustrated in Fig. 7.22.

Selection of the Sampling Time T

As far as the dead time T_t is selected from the step response according to Eq. 7.13, the selection of the sample time T is important for precision of simulation. The adaptive filter with not correct selected sampling time can easily become unstable. In Fig. 7.22, the selection of the sampling time T is graphically explained, namely

$$T = \frac{1}{3}T_d \tag{7.18}$$

The time interval T_d is defined as the time when the controlled variable achieved about 98% of the steady state value, which is equal to $x(\infty)=5$ in Fig. 7.22. Then follows $T_d=12$ s and T is selected as $T=4$ s in this example.

Example: Identification with adaptive filter

Let us suppose that a real plant has the PT6 behavior and is simulated with the following transfer function:

$$G_S(s) = \frac{5}{(1 + 2s)(1 + s)^2(1 + 0.8s)(1 + 0.5s)(1 + 0.1s)} \tag{7.19}$$

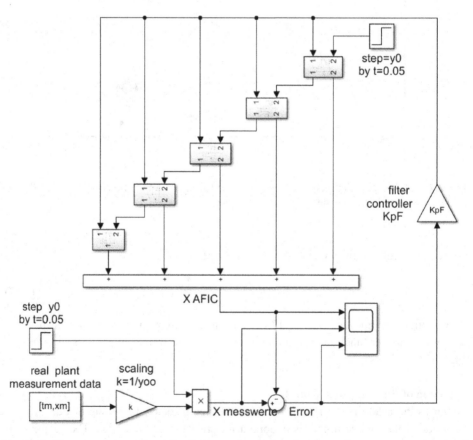

Fig. 7.23 The given plant G_s of Eq. 7.18 and its model G_{sm} as well as corresponding step responses. *Source* [1, p. 155]

The step response of Eq. 7.19 after input step of $y_0 = 1$ is given in Fig. 7.23. The time interval is $T_d = 10$ s, the sampling time is selected as $T = 2$ s. After identification with the adaptive filter shown in Fig. 7.18 with the same input step of $y_0 = 1$ are defined factors $K0 = 1.7839$; $K1 = 1.7521$; $K2 = 1.464$ upon Eq. 7.13, 7.15– 7.17. The conversion of the adaptive filter with these factors leads to the transfer function of the model:

$$G_{Sm}(s) = \frac{5}{1 + 3.67s} e^{-2s} \tag{7.20}$$

In Fig. 7.23 are shown step responses of the given plant G_s and its model G_{sm}. The error of identification is minimal. ◄

7.4.4 App DSM AFIC with MATLAB®

The DSM AFIC consists of three blocks A, B, C, and of a coordinator to execute these blocks one after another with the following MATLAB® script. The user (operator) should enter some parameters in advance as shown below.

```
% Coordinator
KpS=input ('Enter KpS=')        % initial value, e.g., KpS=1
T = input('Enter T=')           % sampling time T=2
y0=input ('Enter y0 =');        % input step
run('ident_3steps_A')           % run block A
sim('ident_3steps_A')           % the filter weights will be defined and loaded into workspace
Tt=input('Enter Tt=')           % selected dead time upon Gl. 7.18 or set Tt=0
                                % or Tt = 0,5*T or T=1,5*T upon of current step response
T1=T*(K1+K2)/K1;                % delay time constant upon Gl. 7.15
run('ident_3steps_B')           % run block B
sim('ident_3steps_B')
run('ident_3steps_C')           % run block C
sim('ident_3steps_C')
```

In Fig. 7.24 and Fig. 7.25 are shown MATLAB/Simulink programs of blocks with 5 adaptive filters. The step responses of each block are summarized in Fig. 7.26.

More details about App AFIC are to find in [11]. The DSM AFIC were successfully implemented since 2011 by several industrial projects like [12]. [13, 14].

All Data Stream Managers described in this chapter are shown in Fig. 7.27.

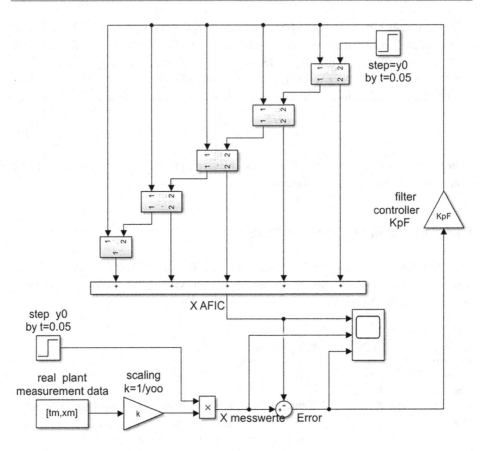

Fig. 7.24 App DSM AFIC: block A

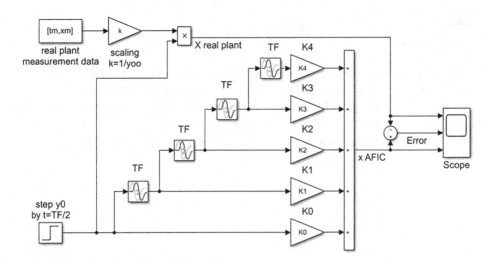

Fig. 7.25 App DSM AFIC: block B

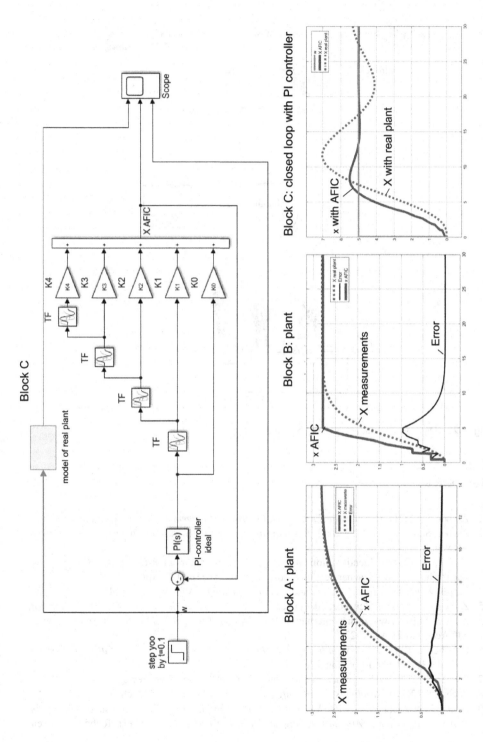

Fig. 7.26 App DSM AFIC: block C and step responses of all blocks A, B, C

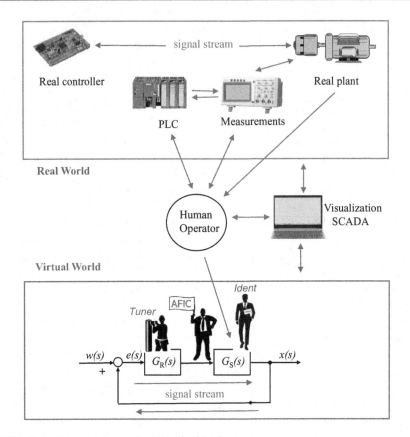

Fig. 7.27 Data stream managers described in this chapter

References

1. Zacher, S. (2021). *Regelungstechnik mit Data Stream Management.* Verlag Springer Vieweg.
2. Zacher, S., & Reuter, M. (2022). *Regelungstechnik für Ingenieure* (16th ed.). Verlag Springer Vieweg.
3. Zacher, S. (2011). Identifikation von Regelstrecken. *Automation-Letter Nr. 3*, Verlag Dr. Zacher. https://www.szacher.de/Automation-Letters/. Accessed 21 Apr 2022.
4. Latzel, W. (1995). *Einführung in die digitalen Regelungen.* VDI-Verlag GmbH.
5. Zacher, S. (2018). Zeitprozentkennwert-Verfahren. *Automation-Letter Nr. 24.* Verlag Dr. Zacher. https://www.zacher-international.com/Automation_Letters/24_Zeit_Prozentkennwert. pdf. Accessed 21 Apr 2022.
6. Unbehauen, R. (1998). *Systemtheorie 2. Mehrdimensionale, adaptive und nichtlineare Systeme* (7th ed.). R.Oldenbourg Verlag.
7. Zacher, S. (2017). *Übungsbuch Reglungstechnik, 6.* Verlag Springer Vieweg.
8. Zacher, S. (2016). *Regelungstechnik Aufgaben* (4th ed.). Verlag Dr. S. Zacher.
9. Henry, J., & Zacher, S. (2010). WebLabs in Control Engineering Education: status and trends. *7. AALE Angewandte Automatisierung in der Lehre und Forschung,* FH Technikum Wien,

10./11. Feb. 2010 http://www.rev-conference.org/REV2010/Henry_Zacher.pdf. Accessed 23 Apr 2022.

10. Zacher, S. (2020). *Drei-Bode-Plots-Verfahren für Regelungstechnik*. Verlag Springer Vieweg.

11. Zacher, S. (2021). AFIC. *Automation-Letter Nr. 19*. Verlag Dr. Zacher. https://zacher-international.com/Automation_Letters/19_AFIC.pdf. Accessed 21 Apr 2022.

12. Theiss, J. (2011). Entwurf und Realisierung einer adaptiven Regelung an einer ausgesuchten Strecke der Gebäudeautomatisierung. *Bachelorarbeit, Hochschule Darmstadt*, FB EIT.

13. Mrugalla, M. (2011). Simulation und Visualisierung eines Herstellungsprozesses mit Regelkreisen unter Anwendung des adaptiven Filters. *Diplomarbeit, Hochschule RheinMain*, FB Ingenieurwissenschaften, Studienbereich Informationstechnologie und Elektrotechnik.

14. Leichsenring, P. (2012). Regelungstechnische Analyse und Entwurf einer adaptiven Regelung für eine Industrieanlage. *Diplomarbeit, Hochschule RheinMain*, FB Ingenieurwissenschaften, Studienbereich Informationstechnologie und Elektrotechnik.

15. Samal, E., & Becker, W. (2000). *Grundriß der praktischen Regelungstechnik* (20th ed.). Oldenbourg Wissenschaftsverlag GmbH.

Management of Setpoint Behavior

8

"It is one thing to decide to climb a mountain. It is quite another to be on top of it." (Quote: Herbert A. Simon. https://www.brainyquote.com/quotes/herbert_a_simon_729501 *accessed May 06.2022)*

8.1 Control and Management

8.1.1 Setpoint behavior and disturbance behavior

The engineering and the implementation of closed loop control (CLC) are carried out in the book in two domains:

- Real world, which is the world of real devices and physical signals.
- Virtual world, which is the world of mathematical descriptions and simulations.

The feedback control is realized in the real world with hardware and software modules as functions of time t or real angular frequency ω. The systems of the real world are described in three domains of the virtual world as functions of time t or Laplace operator s or complex angular frequency $j\omega$. In between these two worlds, there is a human operator as shown in Fig. 8.1.

The first task of the operator is the engineering of the feedback control, and the first task of the engineering is the identification of a plant. The tools for this task, proposed in [1] and described in the previous chapter, are called Data Stream Managers (DSM). The next task of engineering is to define the transfer functions of the closed loop control (CLC) in the virtual world using mathematical methods, simulation, and visualization.

Fig. 8.1 Real world of feedback control and virtual world for its engineering. A human operator as supervisor and interface between both worlds

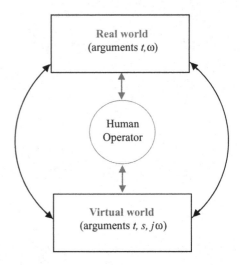

It is supposed that a CLC has two inputs:

- Set point w.
- Disturbance z.

Further, it is supposed that these inputs are applied to the CLC one after another, not simultaneously. To each of these inputs is defined in the control theory the corresponding working mode or behavior:

- Setpoint behavior, when only set point w is applied at the known place of the loop and with the known form (usually as step function):

$$G_w(s) = \frac{x(s)}{w(s)} = \frac{G_{vw}(s)}{1 + G_0(s)} \tag{8.1}$$

- Disturbance behavior, when only disturbance z is applied, which's form and place in the loop, to which it is applied, are usually unknown:

$$G_z(s) = \frac{x(s)}{z(s)} = \frac{G_{vz}(s)}{1 + G_0(s)} \tag{8.2}$$

Example: Level control of a tank

The classic CLC shown in Fig. 8.2 has set point U_w and disturbance Q_{out}. The overshoot and the settling time of the step response by setpoint behavior are smaller than those by disturbance behavior. ◄

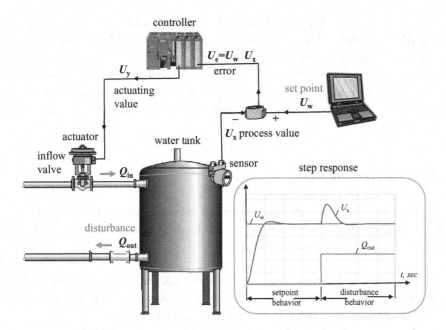

Fig. 8.2 Example of classic feedback level control with set point U_w and disturbance Q_{out}

The classical control theory threats both kinds of behaviors in the same way using the same controller. The disadvantages of such an approach are listed as follows:

- The transfer functions of setpoint behavior $G_w(s)$ and of disturbance behavior $G_z(s)$ have the same denominator but different numerators $G_{vw}(s)$ and $G_{vz}(s)$.
- The majority of the controller are usually optimally designed for setpoint behavior, but not for disturbance behavior, that's why the disturbance behavior is usually not optimal.
- As far as place of disturbance entered in the loop is not known, the classical methods of disturbance compensation are of no practical use.

The classical representation of a feedback control as a signal stream of an isolated closed loop fits no longer the current state of the control engineering. Today's automation control is based on the processing of several data streams as signals, messages, and data. To improve it, the conception of dual control was introduced in 2003 in [2]. The data streams such as material, energy, and information were analyzed. It was suggested the second closed loop be next to the main control loop. The task of this second loop is the tuning of the controller adjusting its parameters to current plant parameters. The conception of dual control was the first step in the development of Data Stream Managers.

The next two steps in this direction were the development of the bus-approach ([3–6]) and the ASA approach ([7–9]) summarized in [10] and described in previous chapters. Upon these conceptions, a single closed control loop was represented like the modern automation system with several data streams. As a result, a plant in this book is not only controlled upon transfer functions of Equ. 8.1 and 8.2, but managed for more data streams. In this and the next chapters is explained how it is realized for setpoint and disturbance behaviors.

8.1.2   CLIMB and HOLD

As far as the set point behavior occurs without disturbance and the disturbance behavior in its turn occurs without setpoint changes, two kinds of behavior are suggested in this book:

- CLIMB is the setpoint behavior realized with the classical controller together with the Data Stream Manager (DSM). The task of the CLIMB-mode is to bring the controlled variable x possibly rush and without error to the setpoint value w.
- HOLD corresponds to the disturbance behavior. But in opposite to the classical conception, as far as the set point w is arrived, i.e., as far as it is $x=w$, the disturbance behavior is supervised with the controller, but with the Data Stream Managers (DSM). This case is discussed in Chap. 9.

The DSM treated in the current chapter is shown in Fig. 8.3. The short overview follows in the next section.

8.1.3   Overview of DSM of this chapter

SPFC: Simplified Predictive Functional Control
This method is based upon a known algorithm called PFC (Predictive Functional Control). By the PFC the settling time is decomposed into small time intervals. During each time interval, the behavior of the plant is compared with the behavior of the reference model and corrected if needed. The simplification of PFC parameters, which is done by SPFC, leads to dual control. The plants and controllers are handled as LTV (linear time-varying) terms.

SFC: Surf-Feedback Controller
SFC is the further development of SPFC. Instead of one single set point, the SFC used a series of set points. It allows to adapt the controlled variable x to the reference model x_{M} during each step response.

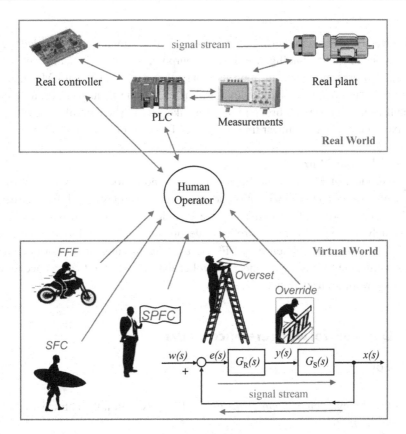

Fig. 8.3 Controllers and DSM developed described in this chapter for setpoint behavior

Override

The concept of the override control is that next to the main controller for controlled variable x is used for technological reasons another controller for the other controlled variable x_{over} like cascade control. The difference between cascade control and override control is that the override controller is supervised by the Data Stream Manager, which switches it on or off dependent on the control tasks.

Overset

The DSM Overset brings the current controlled variable $_x$ to the desired set point w applying a much larger series of set points to the desired one. The overset algorithm arises from the dead-beat control algorithm, known also as bang-bang control, and is thought for use in the open or closed loop. In both cases, the series of setpoint steps is calculated with DSM and leads the control variable to the desired steady state value without oscillations and static error.

Axon

Well-known artificial neural networks are mainly used to simulate and generate the logical operations of human intelligence. For this purpose, the single logical terms are connected in networks and adjusted upon learning rules. The Data Stream Manager Axon is built upon the model of only one biological neuron, using its mechanism and producing impulses that are spread through interfaces like axons. The control closed loop with DSM Axon functions like a linear time-varying (LTV) controller with a LTV plant.

FFF (Feed-Forward Fuzzy)

Fuzzy controllers of *Mamdani* or *Sugeno* are well known today. They consist of fuzzification of sensors and defuzzification actors, of control rules, and inference between rules. The conception of FFF is that not the fuzzification is realized but only the idea of it. Namely, the input of the controller is decomposed into parallel connected terms. Instead of rule base and inference by FFF are used tuning rules of standard controllers. In other words, the FFF is a non-linear controller that adopted the fuzzy conception to a control loop without fuzzy logic.

8.2   Dual control with reference model

8.2.1   SPFC

SPFC is the abbreviation of *Simplified PFC*, and PFC is in turn the abbreviation of *Predictive Functional Control.*

Predictive Functional Control

The idea of PFC was proposed by *Jacques Richalet* [11] for proportional plants with delays like PT1:

$$G_S(s) = \frac{x(s)}{y(s)} = \frac{K_{PS}}{1 + sT_1} \tag{8.3}$$

The current step response of the controlled system $x(t)$ is matched to the step response of a previously given dynamic reference model $x_M(t)$:

$$G_M(s) = \frac{x_M(s)}{y_M(s)} = \frac{K_{PM}}{1 + sT_M} \tag{8.4}$$

The solution

$$x_M(t) = x_M(0)e^{-\frac{t}{T_M}} + K_{PM}(1 - e^{-\frac{t}{T_M}})\hat{y} \tag{8.5}$$

of the differential equation that corresponds to the reference model Equ. 8.4

$$T_M \frac{dx_M(t)}{dt} + x_M(t) = K_{PM}\hat{y} \tag{8.6}$$

is discretized with the sampling time T_h as follows:

$$x_{Mk+h} = \alpha \cdot x_{Mk} + (1 - \alpha) \cdot K_{PM} \cdot y_k \qquad (8.7)$$

with the factor α:

$$\alpha = e^{-\frac{T_h}{T_M}} \qquad (8.8)$$

After the same initial input step y_k is applied to the plant and model, the error between outputs x_{Mk} and x_{Mk} to each time interval k is calculated:

$$e_k = w - x_k$$
$$e_{Mk} = w - x_{Mk} \qquad (8.9)$$

According to the PFC algorithm, the actuating value y_{k+h} next to y_k should minimize the error Equ. 8.9

$$x_{\text{ref}} = x_{k+h} = (1 - \lambda)e_k + x_k \qquad (8.10)$$

with factor λ as a function of the desired settling time T_{aus}:

$$\lambda = e^{-\frac{T_\lambda}{T_{\text{aus}}}} \qquad (8.11)$$

The resulting algorithm of *Richalet* for actuating value y_k is given below:

$$y_k = \frac{1}{K_{PM}} \left[x_{Mk} + \frac{1 - \lambda}{1 - \alpha} (w - x_k) \right] \qquad (8.12)$$

However, the relation between factors α and λ upon Equ. 8.8 and 8.11 depends on the relation between delay time constants T_1, T_M, the settling time T_{aus}, and sampling times T_h, T_λ. Recommended are

$$T_\lambda = \frac{1}{5}T_h \text{ or } T_\lambda = \frac{1}{3}T_h \qquad (8.13)$$

Simplified Predictive Functional Control
Instead of recommended values of Equ. 8.13 in [2] was chosen

$$T_\lambda = T_h \text{ and } T_{\text{aus}} = T_M \qquad (8.14)$$

It leads to the following algorithm instead of Equ. 8.12, which was called in [2] the simplified PFC:

$$y_k = \frac{1}{K_{PM}} [x_{Mk} + (w - x_k)] = \frac{1}{K_{PM}} [w - (x_k - x_{Mk})] \qquad (8.15)$$

The test of both algorithms had shown that although the SPFC algorithm Equ. 8.15 losses compared to the PFC of Equ. 8.12 some control quality, the implementation of the SPFC algorithm is simpler, as shown in Fig. 8.4.

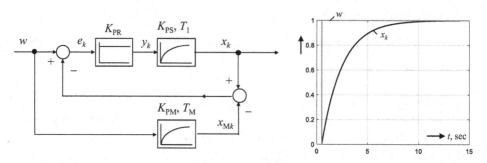

Fig. 8.4 Simplified PFC and desired step response

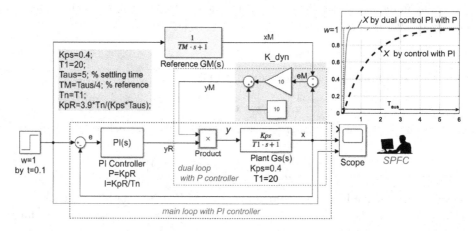

Fig. 8.5 SPFC with main PI controller and accelerating P controller and its step response compared with classic CLC with PI controller

SPFC (Simplified Predictive Functional Controller)

The dual control according to the SPFC algorithm of Equ. 8.15 with a varying gain of the PI controller is shown in Fig. 8.5.

In Fig. 8.5 is seen the block "Product" that is not usual for the classic CLC functional diagrams. The multiplication as an element of the closed loop was introduced for the first time in [2]. With this block is built from the conventional LTI (linear time-invariant) PI controller with constant parameters

$$G_R(s) = \frac{K_{PR}(1 + sT_n)}{sT_n} \tag{8.16}$$

the LTV (linear time-varying) PI controller:

$$G_R(s) = \frac{K_{LTV}(s)(1 + sT_n)}{sT_n} \tag{8.17}$$

The varying gain K_{LTV} of PI controller of Equ. 8.17 consists of two terms, initial gain K_{PR} and gain K_{dyn}.

The CLC shown in Fig. 8.5 consists in its turn of two loops:

- The loop with the PI controller of Equ. 8.16 with the gain K_{PR} to control the main error $e(s)$ between set point w and current value x of the plant.
- The loop with the P controller with the gain K_{dyn} to control the difference between the output x_{M} of the reference model and the output x of the plant:

$$e_{\text{M}}(s) = x_{\text{M}}(s) - x(s) \tag{8.18}$$

The initial gain K_{PR} is defined upon the classic standard rule, given in chapter 3, with $T_{\text{n}} = T_1 = 20$ s for the desired settling time $T_{\text{aus}} = 5$ s:

$$K_{\text{PR}} = \frac{3{,}9T_{\text{n}}}{K_{\text{PS}}T_{\text{aus}}} = \frac{3{,}9 \cdot 20}{0{,}4 \cdot 5} = 39 \tag{8.19}$$

The initial gain is not kept constant but is changed as shown below:

$$K_{\text{LZV}} = K_{\text{PR}} \cdot K_{\text{dyn}} \cdot [e_{\text{M}}(s) + 1] = 390 \cdot [e_{\text{M}}(s) + 1] \tag{8.20}$$

Such control loops with variable controller's gain are called [2] *dual control loops*.

The entire actuating variable $y(s)$ of the LTV controller of Equ. 8.17 is formed from the actuating variable $y_{\text{R}}(s)$ of the standard PI controller of Equ. 8.16 with gain K_{PR} and the actuating variable $y_{\text{M}}(s)$ of the standard P controller with the empirically chosen gain $K_{\text{dyn}} = 10$:

$$y(s) = y_{\text{R}} \cdot y_{\text{M}} = G_{\text{R}}(s)e(s) \cdot K_{\text{dyn}}[e_{\text{M}}(s) + 1] \tag{8.21}$$

Advantages of the dual control with LTV PI controller consisting of two LTI (linear time-invariant) Pi and P controller are seen in Fig. 8.5 by comparing step responses of the SPFC and of the classic control loop with the one standard PI controller that is set for the same settling time $T_{\text{aus}} = 5$ s. The dual SPFC control is accelerated without increasing overshoot or decreasing the damping, as is always the case with classic control loops.

8.2.2 SFC

Conception

Axbxbreviation SFC means *Surf-Feedback Control*. It was introduced in 2021 in [12] for closed loop control of linear plants with varying parameters. The idea of SFC is clearly and simply explained in [12] on an example of a surfer, who is standing at point A and wants to ride the wave to point B imagining his trajectory. The distance is large, and if along this way the parameters of this single wave change, the surfer's trajectory

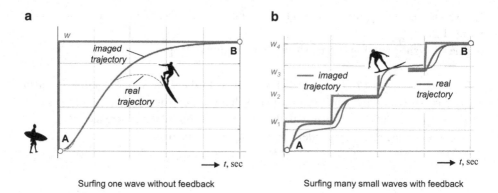

Fig. 8.6 Conception of SFC illustrated on the example of surfing: **a**) one single wave with changing features (one setpoint step w without feedback from reference model); **b**) many small waves
(many setpoints steps with feedback)

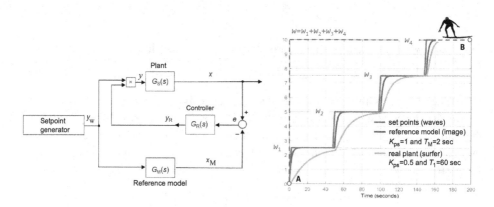

Fig. 8.7   SFC as dual control with reference model. *Source* [10, p. 398]

will change drastically, as shown in Fig. 8.6a, deviating from imagining trajectory. It is
completely different with many small waves. The surfer can quickly adjust itself to each
wave and exactly reach the end point (Fig. 8.6b) with small deviations from the imaged
trajectory. So, what is missing for the surfer to succeed? The answer is Lots of little
waves instead of the big one and the feedback after each wave!

This is exactly what the concept of the SFC is based on: the generator of waves is
represented with the generator of set points w_1, w_2, w_3, and w_4 as shown in Fig. 8.7. It is
again the dual control like in Fig. 8.4 but with one difference, the actuating variable $y(s)$
is not affected by the error $e_M(s)$ as in Equ. 8.21. It is given by the setpoint generator as
actuating variable y_w:

$$y(s) = y_R(s) \cdot y_W(s)$$

More about SFC is to find in [1, 10, 12, 13]. The SFC was tested and implemented in [14].

Implementation of SFC (Fig. 8.8)
Let's assume that the controlled variable $x(t)$ of a PT2 plant

$$G_S(s) = \frac{K_{PS}}{(1 + sT_1)(1 + sT_2)}$$

with parameters

$$K_{PS} = 2$$
$$T_1 = 15 \text{ sec}$$
$$T_2 = 5 \text{ sec}$$

is to be transmitted after a step of set point

$$w = 10$$

from the starting point A at $t = 0$ to the end point B during desired settling time

$$T_{aus} = 10$$

without overshoot or with the damping $\vartheta \geq 1$:

$$\begin{cases} x(0) = 0 \\ x(T_{aus}) = w \end{cases}$$

According to classic methods, the first is identified the transfer function $G_S(s)$ of the plant. Then is formed the open loop with a standard controller $G_R(s)$ and the transfer function $G_w(s)$ of the closed loop is determined. However, if the parameters of the real plant deviate from the parameters of the identified plant $G_S(s)$, the control will not run according to the desired $G_w(s)$ or no longer optimally, perhaps even unstably. In this case, let us use instead of classic methods the SFC approach.

According to the SFC method, the desired settling time T_{aus} is broken down into N time intervals with a respective duration T_w of each interval as in the case with the PFC Sect. 8.2.1:

$$T_w = \frac{T_{aus}}{N} \tag{8.22}$$

The step w of set point is also divided into N steps of height w_w. To adapt the step response $x(t)$ to the desired one during control, a reference model $G_M(s)$ is introduced:

$$G_M(s) = \frac{K_{PM}}{(1 + sT_M)^2} \tag{8.23}$$

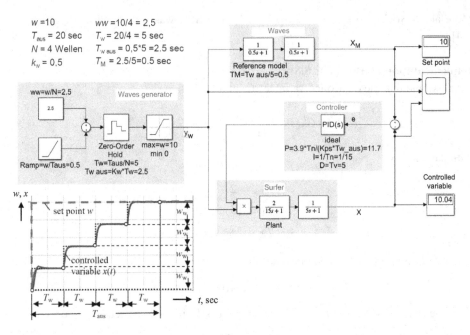

Fig. 8.8   SFC implementation with MATLAB®/Simulink

For the example shown in Fig. 8.8 the time constant T_M of the reference model is calculated according to the ratio

$$T_M = \frac{T_{aus}}{N} = \frac{20 \text{ sec}}{4} = 5 \text{ sec} \qquad (8.24)$$

The gain K_{PM} of the reference model Equ. 8.23 is always set to

$$K_{PM} = 1 \qquad (8.25)$$

in order to carry out the control without static error by $t \rightarrow \infty$, i.e., by steady state:

$$e(\infty) = 1 \qquad (8.26)$$

The step response $x_M(t)$ of the reference model $G_M(s)$ after the input step y_w will exactly reach the desired value w in the steady state at $t \rightarrow \infty$ or at $t = T_{aus}$:

$$x(\infty) = \lim_{s \to 0} G_M(\infty) \cdot y_w = \lim_{s \to 0} \frac{1}{1 + s T_M} \cdot y_w = w$$

The input step y_w applied to the input of the reference model $G_M(s)$ and to the input of the plant $G_M(s)$ eliminates the steady state error:

$$e(\infty) = x_M(\infty) - x(\infty) = 0$$

The step responses of the SFC with the LTV PID controller

$$G_R(s) = \frac{K_{PR}(1 + sT_n)(1 + sT_v)}{sT_n} \tag{8.27}$$

with parameters

$$T_n = T_1 = 15 \text{ s}$$

$$T_v = T_2 = 5 \text{ s}$$

are shown in Fig. 8.9. The gain K_{PR} is calculated like in Equ. 8.19 according to the basic type C described in Chap. 3:

$$K_{PR} = \frac{3,9 T_n}{K_{PS} T_{w\ aus}} = 11,7$$

T_{w_aus} is the settling time for each time interval

$$T_{w\ aus} = k_w \cdot T_{aus}$$

with the empirically chosen factor

$$k_w = 0,5.$$

The step responses of Fig. 8.9 confirm that the SFC is robust, i.e., the control remains optimal, if the parameter T_1 of the plant deviates fivefold from the initial value $T_1 = 15$ s of the identified plant till the shifted value $T_1 = 75$ s. In Fig. 8.10 is shown that even a tenfold shift of the plant's parameters is possible for optimal surf-feedback control.

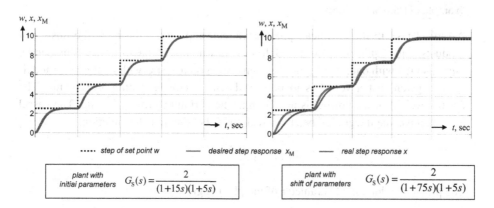

····· step of set point w — desired step response x_M — real step response x

| plant with initial parameters | $G_s(s) = \dfrac{2}{(1+15s)(1+5s)}$ |

| plant with shift of parameters | $G_s(s) = \dfrac{2}{(1+75s)(1+5s)}$ |

Fig. 8.9 Step responses of the SFC of Fig. 8.8 with the PID controller of Equ. 8.27

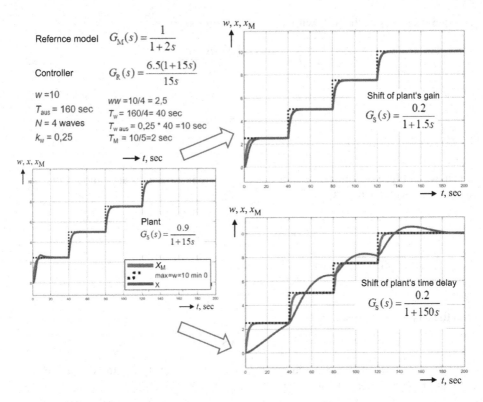

Fig. 8.10 SFC by tenfold change of plant's parameters. *Source* [13]

Example of DSM SFC: Dosage

A typical task in the chemical, pharmaceutical, and food industry is the dosing of products, e.g., liquids or powders. Below is shown an example of how dosing can be implemented with data stream management based on the SFC with a reference model.

It is given that a vessel is to be dosed with a powder within $T = 200$ ms. The desired weight of the powder is $w = 10$ mg. The permitted error is 0,1 mg. The vessel as a plant has the integrating behavior with the following transfer function:

$$G_S(s) = \frac{K_{IS}}{s}$$

The previously identified parameter of the plant is given as

$$K_{IS} = 0,2 \text{ sec}^{-1} \tag{8.28}$$

To realize the SFC, the total filling time T is divided into k time intervals of length T_h, e.g.,

$$k = 5 \quad T_h = \frac{T}{k} = \frac{200}{5} = 40 \text{ ms}$$

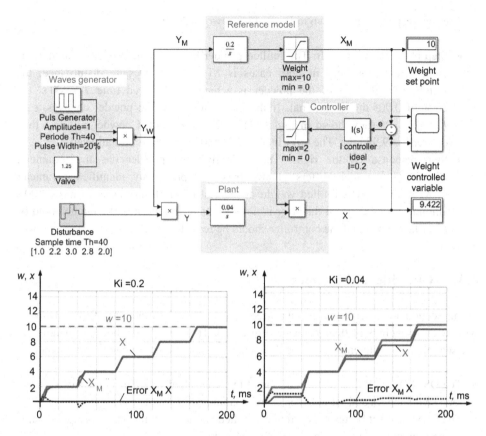

Fig. 8.11 SFC implementation for dosage of the identified plant with $K_i = 0,2$ s^{-1} and disturbed plant with $K_i = 0,04$ s^{-1}. *Source* [1, p. 182]

The dosage SFC is shown in Fig. 8.11 consisting of three modules:

- The reference model $G_M(s)$ as I term with saturation that is not affected by disturbances. The integrating factor of the reference model is Equ. 8.28, i.e., the same as
  by the plant.
- The actuator (valve).
- The I controller with saturation for adjusting the difference between powder
  weights of the plant and the reference model for each time interval. The plant is
  adjusted with the much smaller integrating factor as those of the plant:

$$K_{IR} = 0,04 \text{ sec}^{-1}$$

The step responses are shown in Fig. 8.11 for two cases:

- Step response on the left: The refilled powder has the same properties as the reference model, namely in both cases is $Ki=0{,}2\ s^{-1}$. Despite disturbance, the weight $x=9{,}991$ mg of the powder is achieved to desired time $T=200$ s, the error of 0,009 mg is minimal. If the parameter Ki of the powder changes, e.g., $Ki=0{,}05\ s^{-1}$ (not shown in Fig. 8.11), then the weight of powder x is less than desired $x=9{,}983$ mg. The error increases to 0,017 mg but is still allowed.
- Step response on the right: The currently used powder has the parameter $Ki=0{,}04\ s^{-1}$, which differs greatly from the previously identified parameter of $Ki=0{,}2\ s^{-1}$. The refilled weight at the end of the filling process at $T=200$ s amounts to 9.422 g and the error is not permitted. In this case, the plant should be identified again, and the controller should be reconfigured. ◀

8.3   Override and Overset

In this section are described two Data Stream Managers. The functions of these DSM as well as the meanings of their names are different. That's why they both are handled in one section, namely to emphasize the difference and not to confuse one with the other:

- The DSM Override is used to limit one of two controlled variables in the same control loop switching on and off between corresponding controllers.
- The DSM Overset brings the current controlled variable x to the desired setpoint value w applying much larger set points as the desired one.

8.3.1   DSM Override

Conception
If it is required that during the control of the main variable x some another variable, called x_{over}, should be kept within certain limits g_{max} and/or g_{min}, then it is a matter of the override control. For example, the temperature x of an oven should be kept constant and at the same time the pressure x_{over} should not exceed the maximum permissible value:

$$x < x_{over} \tag{8.29}$$

In order to fulfill the condition Equ. 8.29, a loop with the G_{Rover} controller is formed next to the main control loop (Fig. 8.12). The task of the DSM Override is switching between the two controllers to fulfill the condition Equ. 8.29.

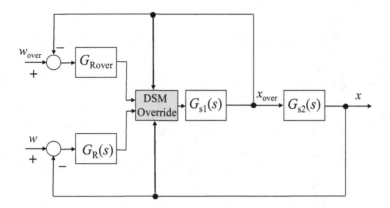

Fig. 8.12 Functional block diagram of override control. *Source* [1, p. 178]

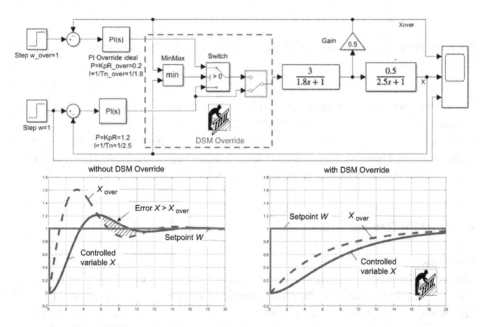

Fig. 8.13 MATLAB®/Simulink model of the override control and corresponding step responses. *Source* [10, p. 259]

Implementation

The MATLAB®/Simulink model of an override control is given in Fig. 8.13. The main control variable X is compared with the limited variable X_{over} with the *MinMax* block from the Simulink Library *MathOperations*. If the condition Equ. 8.29 is met, the output of *MinMax* is set on "1". Depending on this, the *Switch* block switches the actuating variable between the PI main controller and the PI override controller. The *ManualSwich*

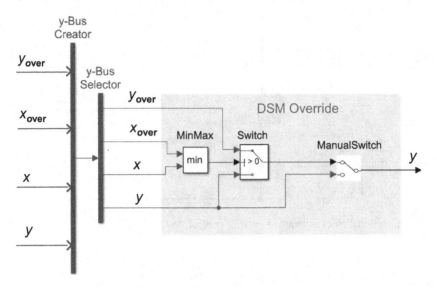

Fig. 8.14 DSM Override with MATLAB®/Simulink according to bus-approach. *Source* [1, p. 179]

block is used to manually switch between two operating modes (with or without override control).

The step responses of x and x_{over}, shown in Fig. 8.13, confirm the control quality with DSM Override. Using the bus-approach described in Chap. 4, the DSM Override is realized with MATLAB®/Simulink as shown in Fig. 8.14.

8.3.2   DSM Overset

Conception

The idea of Data Stream Manager Overset is that the controlled variable x should reach the given setpoint value w in the given (usually in the shortest possible) time T_{aus}. For this purpose, it is known in the control theory so-called *bang-bang control*, when the actuating variable y_R is switching between limits y_{Rmax} and y_{Rmin}. If such control occurs without overshooting, it is called *dead-beat control*.

The DSM Overset is built upon the same idea with the following difference: it takes place in an open loop as a series of set points w_1, w_2, and w_3, and each of them is over the needed set point w. To easily distinguish the adjusting steps w_1, w_2, and w_3 from the needed target set point w, they are called in following simply *oversets*.

In Fig. 8.15 are shown two oversets w_0 and w_1 which are larger than the needed set point $w = 5$. Nevertheless, the step response $x(t)$ reached the needed set point w without overshoot.

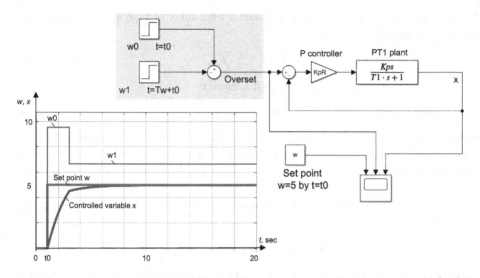

Fig. 8.15 DSM overset with a closed loop of PT1 plant with P controller and step response after two overset steps w_0 and w_1. *Source* [1, p. 175]

Overset algorithm

To derive the algorithm of DSM Overset, let us consider the simple control loop of Fig. 8.15. The transfer function $G_0(s)$ of the open loop consisting of the P controller $G_R(s)$ and the PT1 plant $G_S(s)$ is written as follows:

$$G_0(s) = G_R(s)G_S(s) = K_{PR} \cdot \frac{K_{PS}}{1 + sT_1} \tag{8.30}$$

The transfer function $G_w(s)$ of the closed loop by CLIMB-mode (setpoint behavior) follows from Equ. 8.30

$$G_w(s) = \frac{G_0(s)}{1 + G_0(s)} = \frac{K_{PR}K_{PS}}{1 + K_{PR}K_{PS} + sT_1} = \frac{K_{Pw}}{1 + sT_w} \tag{8.31}$$

The closed loop has according to Equ. 8.31 the PT1 behavior. The parameters of the closed loop of Equ. 8.31 are denoted as gain K_{Pw}

$$K_{Pw} = \frac{K_{PR}K_{PS}}{1 + K_{PR}K_{PS}} \tag{8.32}$$

and the time delay T_w

$$T_w = \frac{T_1}{1 + K_{PR}K_{PS}} \tag{8.33}$$

The settling time T_{aus} of the closed loop with a time delay of Equ. 8.33 mathematically exactly is equal to

$$T_w = 3,9 \cdot T_{aus} \tag{8.34}$$

Roughly approximated for practical cases it is equal to

$$T_w = 5 \cdot T_{aus} \tag{8.35}$$

However, the set point w could not be reached exactly because in this example a P controller is used. A static or retained error $e(\infty)$ is calculated from the known condition for the steady state:

$$x(\infty) = \lim_{t \to \infty} x(t) = \lim_{s \to 0} G_w(s)w = K_{Pw}w \tag{8.36}$$

$$e(\infty) = w - x(\infty) = (1 - K_{Pw})w \tag{8.37}$$

In order to minimize the static error $e(\infty)$ and the settling time T_{aus}, the gain K_{PR} of the P controller should be set to the maximum possible value. However, because of a limitation y_{max} and y_{min} of the actuating variable, the real control is usually done with a smaller K_{PR} value as needed. As a result, the settling time T_{aus} decreases against expected from Equ. 8.35.

To improve it with the help of DSM Overset, the set point w should be reached after a sequence of oversets. In this example is assumed to finish control after $N=2$ oversets w_0 and w_1. The period T of the pulse depends on the time constant T_w of Equ. 8.34 or 8.35:

$$T = \frac{T_w}{N}$$

According to the overset algorithm, the controlled variable $x(t)$ should reach 90% of the set point w at the end of overset w_0, which should correspond to 63% of the final value $x(\infty)$ of the controlled variable $x(\infty)$ at the steady state (see Equ. 8.36):

$$0,9w = 0,63x(\infty) = 0,63K_{Pw}w_0 \tag{8.38}$$

This condition lets to define the value of the overset w_0:

$$w_0 = \frac{0,9w}{0,63K_{Pw}} = \frac{1,4286}{K_{Pw}}w \tag{8.39}$$

With the next overset w_1, the controlled variable x should be brought by $t=T_w$ from point of Equ. 8.38

$$x(T) = 0,63x(\infty) \text{ or } x(T) = 0,9w$$

exactly to the set point w. If we let to do it with the overset w_0, i.e., $w_1=w_0$, the controlled variable will be brought above the set point w up to

$$x(T_w) = K_{Pw}w_0 \tag{8.40}$$

In order to avoid the error

$$e(T_w) = K_{Pw}w_0 - w \tag{8.41}$$

the overset w_1 should be calculated from the condition:

$$K_{PW}w_1 = K_{PW}w_0 - w \tag{8.42}$$

The following overset w_1 is therefore necessary:

$$w_1 = w_0 - \frac{1}{K_{PW}}w \tag{8.43}$$

The resulting step response is shown in Fig. 8.15.

Implementation of DSM Overset
The DSM Overset shown in Fig. 8.16 is implemented for control of the PT2 plant in an open loop without the controller. The controlled variable $x(t)$ arrived at exactly the given set point $w = 5$ after given settling time $T_{aus} = 15$ s with $N = 3$ oversets:

$$w_0 = 300$$
$$w_1 = 30$$
$$w_2 = 20$$

The advantage of the DSM controller by the CLIMB-mode (setpoint behavior) against the PI controller, which is adjusted for the same settling time $T_{aus} = 15$ s, is seen from step responses of Fig. 8.16. The PI controller causes an overshoot of about 10% while the control with DSM Overset has no overshoot. The disadvantage of DSM Overset by the HOLD-mode (disturbance behavior) is missing feedback. This disadvantage can be eliminated with the help of the DSM Terminator proposed for HOLD-mode in Chap. 9.

8.4   DSM with Neuro-Fuzzy elements

8.4.1   DSM Axon

Introduction
A biological neuron generates a series of short electrical spikes with a speed of 50 to 60 pulses/s if the sum of the electrical input signals exceeds a certain value called threshold value, which is approximately 70 mV.

"It is said that the neuron becomes active and "fires" impulses via the axon to the other neurons" (Quote: Source [2, p. 111]).

The first artificial neuron as a model of the biological neuron was introduced in 1943 by *McCulloch* and *Pitts* in [15]. In 1949, *Hebb* proposed the concept of the learning

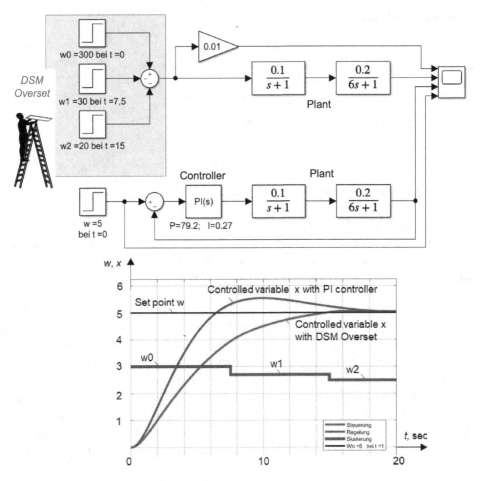

Fig. 8.16 MATLAB®/Simulink implementation for CLIMB-mode (setpoint behavior) and corresponding step responses of DSM Overset without feedback (above) and closed loop control of the same plant with PI controller. *Source* [1, p. 176]

mechanism, which states that the learning process occurs in the brain by changing synapse strengths (weights). The learning rules introduced by *Hebb* [16] motivated many researchers to design artificial neural networks as adaptive systems mostly for information technology. The single biological neuron as the model for closed loop control was neglected. 1997 in [17] was offered the closed loop with positive feedback as the model of a biological neuron and then further developed in [2, 19, 20] for control tasks. It was called "*Regelungstechnisches Neuron*" (control-technical neuron) or short *RT Neuron* (CT-neuron). The structure and the function of this CT-neuron differ fundamentally from all known iterative neuronal models of *F. Rosenblatt* (1957), *B. Widrow* (1960), *S.-J. Amari* (1972), *Ch. von der Malsburg* (1973*)*, *K. Fukushima* (1975), *S. Grossberg*

(1976), *J. Anderson* (1977), *J. Albus* (1979), *T. Kohonen* (1980), *I. Aleksander* (1981), *J. J. Hopfield* (1982), *L. Cooper* (1982), *T. Sejnowski / G. Hilton* (1984), *B. Kosko* (1985), *D. Rumelhart / G. Hilton / R. Williams* (1986), *R. Hecht-Nielsen* (1986), *R. Lippmann* (1987), *C. Mead /F. Fagin* (1987)*B. Bavarian* (1988), *R. Murphy* (1990), and *W.T. Miller / F.H. Glanz / L.G. Kraft* (1990).

Conception

In [10] is explained one example of how to simulate the periodic discharges (spikes) of biological neurons, link them to the control loop, and display them as a dual control loop without iterations. This example is given below.

Example: Closed control loop with positive feedback as the model of biological neuron (Source [1, pp. 421, 422]

The PT2 plant shown in Fig. 8.17 consists of the PT1 element with constant parameters and the linear time-varying (LTV) PT1 element with a time constant $T=0,1$ s and a changeable gain K_{PS}. In Fig. 8.17 is explained that the changeable gain K_{PS} consists of two terms:

- the neural network;
- the dynamical gain K.

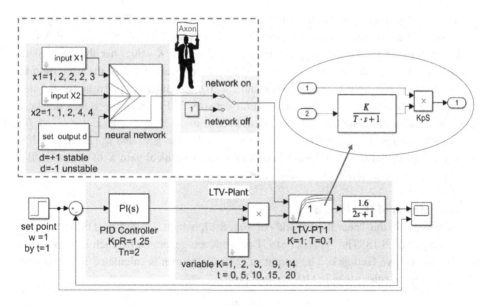

Fig. 8.17 Dual control of LTV plant with the network of two CT-neurons

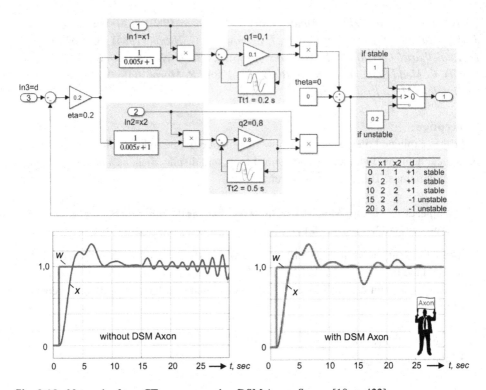

Fig. 8.18 Network of two CT-neurons used as DSM Axon. *Source* [10, p. 422]

The neural network consisting of two CT-neurons is shown in Fig. 8.18. The gain K is assumed as an internal disturbance. It changes from 1 to 14 in a cycle of 5 s, as shown in Fig. 8.17. Further, it is assumed that neither the K values nor the mathematical description of the plant are previously known or available. But the user (human operator) to each time between $t=0$ s and $t=20$ s can enter set point w, record step response $x(t)$, and read out from the neural network of Fig. 8.18 the values d. These values evaluate the stability of the control loop: $d=+1$ for stable and $d=-1$ for unstable control. The algorithm of the stability evaluation is to find in [2, 10, 20]. If it is recognized that the closed loop is unstable, the dynamical gain K of the controller will be changed to stabilize the system. ◄

Implementation of DSM Axon

The structure of the "neural network" of Fig. 8.17, which is used as DSM Axon, is explained in Fig. 8.18. The "spikes" of CT-neurons are generated by each of two closed loops with positive feedback. The output of each CT-neuron is calculated in an internal closed loop upon rules given in [10, 15] and [16]:

$$\alpha = W_1 x_1 + W_2 x_2 - \theta$$

- If $\alpha < 0$ then $y = -1$ and the system is unstable.
- If $\alpha \geq 0$ then $y = +1$ and the system is stable.

$$W_1 = W_1 + \eta \cdot (d - y) \cdot x_1$$
$$W_2 = W_2 + \eta \cdot (d - y) \cdot x_2$$

- If $\alpha < 0$ then $d = -1$ and the system is unstable.
- If $\alpha \geq 0$ then $d = +1$ and the system is stable.

The learning increment η is 0.2. The threshold θ is $\theta = 0$. The output switch has a binary transfer function:

- $K = 1$ by $d = +1$.
- $K = 0.2$ by $d = -1$.

The CLC without DSM Axon is by $K = 1$ unstable, as it is seen from the step response on the left in Fig. 8.18.

On the right of Fig. 8.18 is shown that the CLC with DSM Axon becomes unstable at $t = 15$ s and $t = 20$ s. The DSM Axon recognizes it and immediately stabilizes the system switching from $K = 1$ to $K = 0.2$.

8.4.2   FFF

Introduction

Fuzzy logic was introduced in 1965 by *Zadeh*, further developed by *Kosko* [21], and implemented by *Mamdani* and *Sugeno* for control purposes. The design and use of fuzzy controllers have been described by *Frank, Kahlert, Kindl*, and *Tilli*. The fuzzy controllers are robust and they retain stable behavior even if the parameters of the controlled system are not constant. The time and the costs required for the development of fuzzy controllers are lower than those of classic controllers. But as far as fuzzy controllers have no strict mathematical backgrounds, they are mostly used for systems from which simple and robust behavior is expected, e.g., in motor vehicles, households, and medical devices.

Conception of FFF

The feed-forward fuzzy controller called FFF was developed in [20] and treated in [2] as a "fuzzy controller without fuzzy logic but with fuzzy trial elements". The term "trial" usually means the slimmed down demo version of a software product that is offered for testing in the market. The trial elements of the FFF controller are also the stripped-down blocks of a fuzzy controller according to *Mamdani* (see [10]). They correspond to the "fuzzification" operation of conventional fuzzy logic. The rule base and the inference of

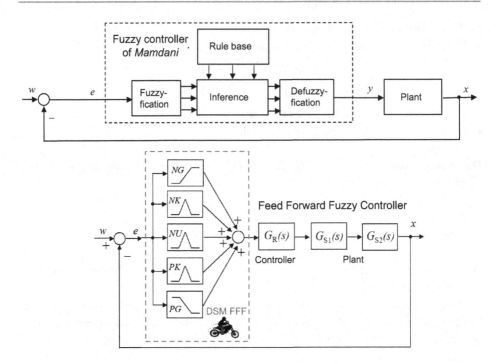

Fig. 8.19 Fuzzy controller of *Mamdani* (above) and Feed-Forward Fuzzy controller. *Source* [1, p. 179]

a classic fuzzy controller (Fig. 8.19, above) are replaced in the FFF controller by simple addition and by the algorithm of the classic standard controller (Fig. 8.19, below).

> "The rule base is already present in every control loop due to the general feedback principle. The desired non-linearity of the static characteristic of the controller can also be achieved without a rule base, inference and defuzzification by setting input membership functions. (Quote: Source [2, p. 129])

Implementation of FFF

The implementation of the FFF is given in Fig. 8.20.

The PI controller is optimally set according to the damping $\vartheta = 1$ (aperiodic behavior, no overshoot) regardless of fuzzy blocks *big, middle,* and *small*. The tuning of the PI controller is done according to the rules of chapter 3. First is defined the transfer function of the open loop:

$$G_0(s) = G_R(s)G_S(s) = \frac{K_{PR}(1 + sT_n)}{sT_n} \cdot \frac{0,8}{(1 + 2s)(1 + 20s)} \tag{8.44}$$

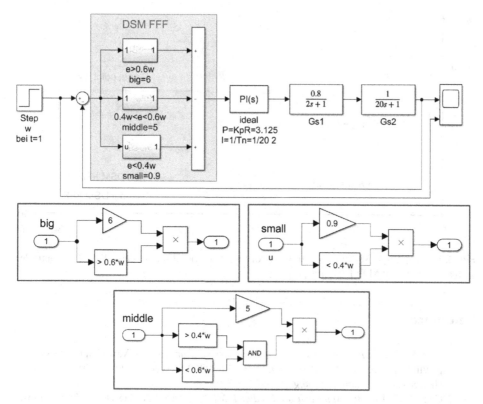

Fig. 8.20 Closed control loop with PI controller and DSM FFF (above) consisting of fuzzy trial elements big, middle, and small. *Source* [1, p. 180]

After compensation with $T_n = 20$ s follows:

$$G_0(s) = \frac{0,8K_{PR}}{sT_n(1 + 2s)}$$

The gain of the PI controller is calculated upon the optimum magnitude method, given in Chap. 3:

$$K_{PR} = \frac{T_n}{4\vartheta^2 K_{PS}T_2} = \frac{20}{4 \cdot 1 \cdot 0,8 \cdot 2} = 3,125$$

The step response of the closed control loop with DSM FFF and PI controller tuned with $T_n = 20$ s and $K_{PR} = 3.125$ is given in Fig. 8.21. For comparison, the step responses of the same control loop with the PI controller without DSM FFF for two K_{PR} values are also shown in Fig. 8.21. In both cases, the control quality without DSM FFF is worse than with it: larger overshoots and the settling time.

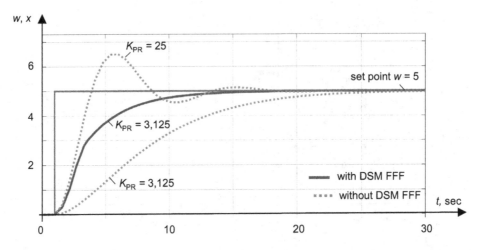

Fig. 8.21 Step responses of the control loop of Fig. 8.20 with different gains K_{PR} of PI controller, with and without DSM FFF. *Source* [1, p. 181]

References

1. Zacher, S. (2021). *Regelungstechnik mit Data Stream Management*. Verlag Springer Vieweg. https://link.springer.com/book/10.1007/978-3-658-30860-5. Accessed 20. Jan.2022.
2. Zacher, S. (2003). *Duale Regelungstechnik*. VDE Verlag.
3. Zacher, S. (2014). *Bus-Approach for Feedback MIMO-Control*. Verlag Dr. S. Zacher https://www.szacher.de/my-Books/Bus/. Accessed 20. Jan. 2022.
4. Zacher, S. (2019). Bus-Approach for Engineering and Design of Feedback Control.: *Proceedings of ICONEST*, Denver, CO, USA, October 7–10, 2019, published by ISTES Publishing, pp. 26–27. https://ijonest.net/index.php/ijonest/issue/view/4. Accessed 20. Jan. 2022.
5. Zacher, S. (2020). Bus-Approach for Engineering and Design of Feedback Control. *International Journal of Engineering, Science and Technology, 1*(2), 16–24. https://www.ijonest.net/index.php/ijonest/article/view/9/pdf. Accessed 20. Jan. 2022.
6. Zacher, S. (2021). *Bus-Approach for Feedback Control*. Verlag Dr. S. Zacher https://www.youtube.com/watch?v=dXMXKQJtuIQ. Accessed 20. Jan. 2022.
7. Zacher, S. (2020). *Antisystem-Approach (ASA)*. Chicago, USA: IConEST, October 15–18, 2020 https://www.zacher-international.com/IJONEST_Journal/Zacher_Antisystem_Presentation.pdf. Accessed 20. Jan. 2022.
8. Zacher, S. (2020). *Antisystem-Approach (ASA) for Feedback Control*. https://www.youtube.com/watch?v=UhUrWrx24Ag. Accessed 20. Jan. 2022.
9. Zacher, S. (2021). Antisystem-Approach (ASA) for Engineering of Wide Range of Dynamic Systems, *Iowa State University, USA: International Journal on Engineering, Science and Technology (IJonEST), 3*(1), 52–66 https://ijonest.net/index.php/ijonest/issue/view/4. Accessed 20. Jan. 2022.
10. Zacher, S., Reuter, M. (2022). *Regelungstechnik für Ingenieure*. 16th edition, Wiesbaden: Verlag Springer Vieweg. https://link.springer.com/book/10.1007/978-3-658-36407-6. Accessed 20. Jan. 2022.
11. Richalet, J., & O'Donovan, D. (2009). *Predictive Functional Control*. Springer-Verlag.

12. Zacher, S. (2021). *Surf Feedback Control.* Automation-Letter, No 42. Verlag Dr. S. Zacher https://www.zacher-international.com/Automation_Letters/42_Surf_Control.pdf. Accessed 20. Jan. 2022.
13. Zacher, S. (2021). Surf Control for industrial plants. Stuttgart: Verlag Dr. Zacher. https://youtu.be/_Fe5Had3sjo accessed 21.04.2022.
14. Freitag, E. (2021). *Integration einer Ultramikrowaage in einen Laborfüller.* Darmstadt: M.sc. Thesis der Hochschule Darmstadt, FB EIT.
15. McCulloch, W. S., & Pitts, W. (1943). (1943) A logical calculus of the ideas immanent in nervous activity. *Bulletin of Mathematical Biophysics, 5,* 115–133.
16. Hebb, D. (1949). *The organization of behavior.* Wiley.
17. Zakharian, S., & Ladewig-Riebler, P. (1997). Reglereinstellung mit Künstlichen Neuronalen Netzen, *Fachtagung "Moderne Methoden des Regelungs- und Steuerungsentwurfes",* Otto-von-Guericke-Universität Magdeburg, pp. 95–100.
18. Zakharian, S., Ladewig-Riebler, P., & Thoer, S. (1998). *Neuronale Netze für Ingenieure, Arbeitsbuch für regelungstechnische Anwendungen.* Verlag Vieweg.
19. Zacher, S. (2000) Neuro-Regelung. In S. Zacher (Hrsg.), *Automatisierungstechnik kompakt.* Vieweg Verlag.
20. Zacher, S. (2000). *SPS-Programmierung mit Funktionsbausteinsprache.* VDE Verlag.
21. Kosko, B. (1989). Bidirectional Associative Memories. Fuzzyness vs. Probability. *Int. Journal of General Systems, 11,* 1–45.

Management of Disturbance Behavior

9

The aim of the Data Stream Manager by disturbance behavior is to recognize the deviation of the actuating variable earlier as the controller recognizes the deviation of the controlled variable.

9.1 Introduction

9.1.1 Operating Modes of Closed Loop Control

Two operating modes of a closed loop control (CLC) are well known: setpoint behavior and disturbance behavior. They correspond to the following practical tasks:

- The setpoint behavior means that the controlled variable $x(t)$ should be transferred from an initial state x_0 at time $t=0$ (usually is assumed to be $x_0=0$) to a desired setpoint w. By $t \to \infty$ is expected the steady state $x(\infty) = w$. Depending on the physical controlled variable (e.g., level, temperature) the setpoint behavior is called "refilling" or "emptying" for level or "heating up" or "cooling down" for temperature.
- The disturbance behavior means that some external disturbance z effects the controlled variable $x(t)$, which is currently at the operating point $w = x(\infty)$ and this value should be kept constant despite the disturbance z.

Both operating modes are treated similarly in any control engineering textbook because the denominator polynomials $1 + G_0(s)$ of both transfer functions $G_w(s)$ for set point behavior and $G_z(s)$ for disturbance behavior are the same:

$$G_w(s) = \frac{x(s)}{w(s)} = \frac{G_{vw}(s)}{1 + G_0(s)} \tag{9.1}$$

© The Author(s), under exclusive license to Springer Nature Switzerland AG 2022
S. Zacher, *Closed Loop Control and Management*,
https://doi.org/10.1007/978-3-031-13483-8_9

$$G_z(s) = \frac{x(s)}{z(s)} = \frac{G_{vz}(s)}{1 + G_0(s)} \tag{9.2}$$

The only difference is, what input signal acts, setpoint w or disturbance z (Fig. 9.1).

Almost all design methods for CLC described in the literature refer to the setpoint behavior. The same results of design are applied to disturbance behavior because of the same denominator polynomials mentioned above in Eq. 9.1 and 9.2. It is expected that the controllers, which are optimally set to the command behavior, are also optimal tuned for disturbance behavior. But in the reality, it is not the case. Although the CLC by disturbance behavior remains stable with the same damping ϑ as by the setpoint behavior, the disturbance behavior differs from setpoint behavior as well regarding overshoot, settling time T_{aus}, as regarding static error $e(\infty)$, if the P or PD controller is used. There are two causes for it:

- The transfer functions $G_{vw}(s)$ and $G_{vz}(s)$ of forward signal stream are different by $G_w(s)$ for setpoint behavior and for $G_z(s)$ for disturbance behavior,
- The setpoint behavior is considered and designed for w as a step while the type of disturbance step z is usually unknown.

The classic methods of disturbance suppression are effective only if the disturbance is known and measurable, which is rather rare.

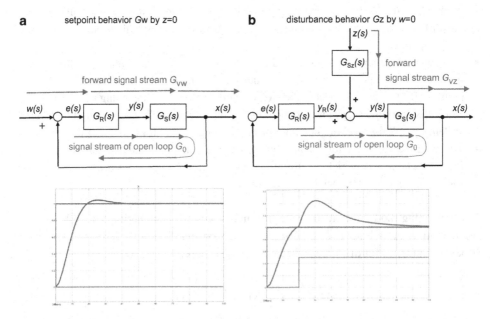

Fig. 9.1 Closed control loop by: **a)** setpoint behavior; **b)** disturbance behavior

9.1.2 Conceptions

In this chapter the disturbance behavior is treated based upon following new conceptions [1] – [11].

DSM

The Data Stream Manager (DSM) is a functional block for supervising of signals and messages. It consists of algorithms and programs of closed loop control, open loop control and logical operations. DSM allows to manage design and control as well for the real world of physical devices as for the virtual world of mathematical models and simulations. The prerequisite for DSM is the precise mathematical model of the controlled plant. More about DSM in Chaps. 1 and 7.

CLIMB and HOLD

The setpoint behavior is called CLIMB mode and is controlled with classic standard controller. The disturbance behavior is referred to as HOLD mode. The HOLD mode is supervised as well by controller, which's usual task is to keep constant the controlled values x, as by Data Stream Manager (DSM), which's task is to keep constant or adapt the actuating value y. More about CLIMB mode in Chap. 8.

ASA

According to the ASA (Antisystem Approach [1]) for any system with the transfer functions $G_1(s)$ and $G_2(s)$ an antisystem with the same transfer function $G_1(s)$ and $G_2(s)$, but with signal transmission in the opposite direction, can be formed in such a way that both systems build a balance $E_1 = E_2$ for any input signals y_1, y_2 and X (Fig. 9.2a). The ASA is detailed discussed in Chap. 5 and applicated in Chaps. 10, 11. In this chapter the ASA concept is used to extract one unknown variable of the antisystem (the variable y_1 in Fig. 9.2) from the balance variable E_2. The variables y_1 and y_2 of the system are

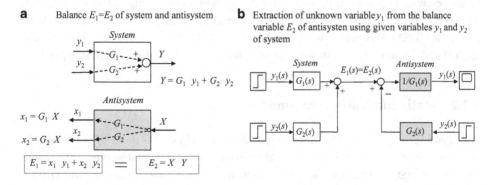

Fig. 9.2 ASA (Antisystem-Approach): **a** balance of system and antisystem; **b** extraction of one antisystem's variable from its balance

supposed to be given as well as the transfer functions $G_1(s)$ and $G_2(s)$. As it is shown in Fig. 9.2 the steps of known variables y_1 and y_2 of the system are applied to the system/antisystem connection, the variable y_1 is extracted and displayed by antisystem.

In the following are described two DSM developed in [5] – [11]:

- DSM *Terminator* for eliminating external disturbances,
- DSM *Plant Guard* for eliminating of internal disturbances caused by parameter's shift of linear time-varying plants (LTV).

9.2   DSM Terminator

9.2.1   Conception

It is assumed that during HOLD mode happens no change in the setpoint w, which corresponds to the term disturbance behavior of control theory. But contrary to conventional control the DSM Terminator and not a controller will take over the control, if a disturbance z takes effect.

If z acts, the controlled variable x is disturbed and become $w + \Delta x$. The "growth" of the controlled variable Δx will be determined by DSM Terminator (Fig. 9.3) comparing the real "disturbed" plant $G_S(s)$ with an "undisturbed" plant model (shadow plant).

Then the "growth" Δx of controlled variable is automatically converted into the "growth" of actuating variable Δy:

$$\Delta y(s) = \frac{1}{G_S(s)} \Delta x(s)$$

The "increase" Δy of actuating variable is added or subtracted by the Terminator to the "set point" of actuating variable y_w, so that "increase" Δx of the controlled variable x is eliminated. The DSM Terminator acts much faster than the classic controller by disturbance behavior because the inverse-controlled system $1/Gs$ has high D terms. The CLC with the disturbance z behaves exactly like the circle without disturbance. Significant is, that the DSM Terminator works independent of the place where the disturbance is applied and doesn't interfere in the setpoint behavior, i.e., if no disturbance is applied.

9.2.2   Mathematical Backgrounds

Let's assume that the transfer functions of the plant $G_S(s)$ and those of the controller $G_R(s)$ are given. The magnitude of the disturbance z and the point at which the disturbance is applied are unknown. In other words, the transfer function of the plant $G_S(s)$ is given by

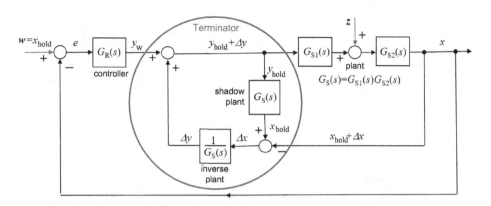

Fig. 9.3 Conception of DSM Terminator. *Source* [7, p. 8]

$$G_S(s) = G_{S1}(s)G_{S2}(s),$$

but the disassembly of the plant in terms $G_{S1}(s)$ und $G_{S2}(s)$ before and after application's point of disturbance z_0 is not known.

In Fig. 9.4 is explained schematically how the DSM Terminator is derived. The realization of DSM terminator in a CLC is shown in Fig. 9.5.

9.2.3 DSM Terminator's Application

DSM Terminator with MATLAB® /Simulink
In Fig. 9.6 is the MATLAB® /Simulink of the CLC with DSM Terminator presented, the step responses without DSM and with DSM are shown in Fig. 9.7. No comment is needed!

However, the elimination of disturbances is not always as perfect as theoretically expected and as shown in Fig. 9.7. It depends on ratios between plant's parameters and time delays of transfer functions $G_{z1}(s)$, $G_{z2}(s)$ and $G_{z2}(s)$, which are usually unknown. The smaller the time constants of $G_{z1}(s)$, $G_{z2}(s)$ and $G_{z2}(s)$ are compared to the plant's time constants of $G_{S1}(s)$ and $G_{S2}(s)$, the faster do effect the disturbances and the larger is the deviations from theoretical expected disturbance suppression. It should also be noted that the internal closed loop of the DSM Terminator (Fig. 9.6) has positive feedback between the inverse and shadow plants. This generates undamped oscillations of the actuating variable with small amplitude. To eliminate these oscillations in practical applications a low-pass filter should be installed after the DSM Terminator.

In summary, it can be concluded that the DSM Terminator has immense advantages over classic methods, because no information about disturbances and their application points is required.

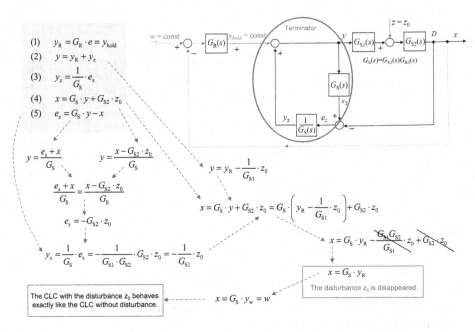

$$
\begin{aligned}
&(1) \quad y_R = G_R \cdot e = y_{hold} \\
&(2) \quad y = y_R + y_z \\
&(3) \quad y_z = \frac{1}{G_S} \cdot e_z \\
&(4) \quad x = G_S \cdot y + G_{S2} \cdot z_0 \\
&(5) \quad e_z = G_S \cdot y - x
\end{aligned}
$$

$$
y = \frac{e_z + x}{G_S} \qquad\qquad y = \frac{x - G_{S2} \cdot z_0}{G_S}
$$

$$
y = y_R - \frac{1}{G_{S1}} \cdot z_0
$$

$$
\frac{e_z + x}{G_S} = \frac{x - G_{S2} \cdot z_0}{G_S}
$$

$$
x = G_S \cdot y + G_{S2} \cdot z_0 = G_S \cdot \left(y_R - \frac{1}{G_{S1}} \cdot z_0 \right) + G_{S2} \cdot z_0
$$

$$
e_z = -G_{S2} \cdot z_0
$$

$$
x = G_S \cdot y_R - \frac{G_{S1} G_{S2}}{G_{S1}} \cdot z_0 + G_{S2} \cdot z_0
$$

$$
y_z = \frac{1}{G_S} \cdot e_z = -\frac{1}{G_{S1} \cdot G_{S2}} \cdot G_{S2} \cdot z_0 = -\frac{1}{G_{S1}} \cdot z_0
$$

$$
x = G_S \cdot y_R
$$

The CLC with the disturbance z_0 behaves exactly like the CLC without disturbance.

$$
x = G_S \cdot y_w = w
$$

The disturbance z_0 is disappeared.

Fig. 9.4 Derivation of DSM algorithm. *Source* [7, p. 10]

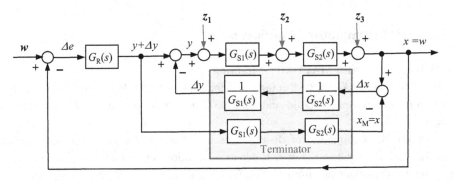

Fig. 9.5 Example of the DSM Terminator in the CLC with three disturbances z_1, z_2 and z_3. *Source* [11, p. 437]

DSM with MATLAB® script

The CLC with PT2 plant and PI controller is simulated in Fig. 9.8 and 9.9 with corresponding step responses. With this script is once more confirming the completely elimination of disturbance $z = 0.5$, which effects by 40 s after the controlled variable x has achieved its desired stead state of $x = w = 1$.

However, in the example above, an ideal inverse plant $G_{S\_inv}(s)$ was used (line 39 of the script, Fig. 9.9). In this case, G_{vzT} is equal to zero, $G_{vzT} = 0$ (line 43) and the distur-

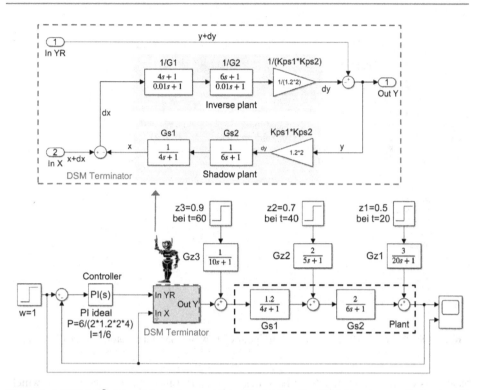

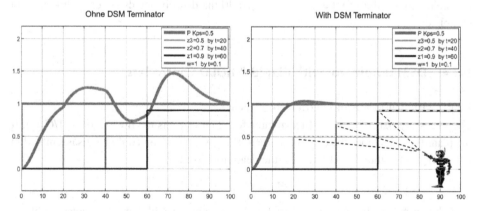

Fig. 9.6 MATLAB® /Simulink implementation of a CLC with DSM Terminator eliminating all three disturbances z_1, z_2 and z_3. *Source* [5, p. 188]

Fig. 9.7 Step responses without DSM Terminator (on the left) and with DSM Terminator. *Source* [11, p. 438]

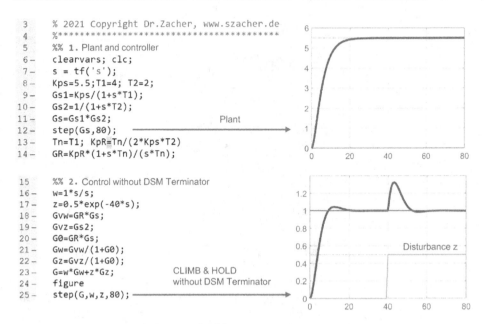

```
3     % 2021 Copyright Dr.Zacher, www.szacher.de
4     %********************************************
5     %% 1. Plant and controller
6 -   clearvars; clc;
7 -   s = tf('s');
8 -   Kps=5.5;T1=4; T2=2;
9 -   Gs1=Kps/(1+s*T1);
10 -  Gs2=1/(1+s*T2);
11 -  Gs=Gs1*Gs2;                    Plant
12 -  step(Gs,80);
13 -  Tn=T1; KpR=Tn/(2*Kps*T2)
14 -  GR=KpR*(1+s*Tn)/(s*Tn);

15    %% 2. Control without DSM Terminator
16 -  w=1*s/s;
17 -  z=0.5*exp(-40*s);
18 -  GVw=GR*Gs;
19 -  Gvz=Gs2;
20 -  G0=GR*Gs;
21 -  Gw=Gvw/(1+G0);
22 -  Gz=Gvz/(1+G0);
23 -  G=w*Gw+z*Gz;                 CLIMB & HOLD
24 -  figure                       without DSM Terminator
25 -  step(G,w,z,80);
```

Fig. 9.8 Sections 1 and 2 of MATLAB® script of CLC with DSM Terminator. *Source* [7, p. 11]

bance z is eliminated in line 46: it becomes $G_{zT}=0$. But the transfer functions, in which the order of the numerator polynomial is greater than the order of the denominator polynomial, like $G_{S\_inv}(s)$, are known to be unrealizable. In the practical implementations the real inverse plant according to line 40 is to be simulated. Nevertheless, also with the realization of the script according to line 40 the disturbance disappears although not complete.

9.2.4   DSM terminator with CLIMB & hOLD Controllers

Plants with dead time and D (deriving)-term
The necessary part of the DSM terminator for plant $G_S(s)$ is the inverse plant $1/G_S(s)$. If the plant $G_S(s)$ includes the dead time, then its transfer function could be disassembled in a term $G_{S\_0}(s)$ without dead time and the dead time itself:

$$G_S = G_{S\_0} \cdot e^{-sT_t} \tag{9.3}$$

It is shown below that the inverse of Eq. 9.3 become unstable and it is not possible to eliminate the disturbance with the DSM terminator:

$$\frac{1}{G_S} = \frac{1}{G_{S\_0} \cdot e^{-sT_t}} = \frac{1}{G_{S\_0}} \cdot e^{+sT_t}$$

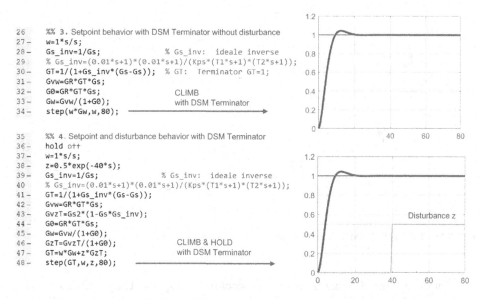

```
26    %% 3. Setpoint behavior with DSM Terminator without disturbance
27 -  w=1*s/s;
28 -  Gs_inv=1/Gs;                    % Gs_inv:  ideale inverse
29    % Gs_inv=(0.01*s+1)*(0.01*s+1)/(Kps*(T1*s+1)*(T2*s+1));
30 -  GT=1/(1+Gs_inv*(Gs-Gs));        % GT:  Terminator GT=1;
31 -  Gvw=GR*GT*Gs;
32 -  G0=GR*GT*Gs;
33 -  Gw=Gvw/(1+G0);
34 -  step(w*Gw,w,80);

35    %% 4. Setpoint and disturbance behavior with DSM Terminator
36 -  hold off
37 -  w=1*s/s;
38 -  z=0.5*exp(-40*s);
39 -  Gs_inv=1/Gs;                    % Gs_inv:  ideale inverse
40    % Gs_inv=(0.01*s+1)*(0.01*s+1)/(Kps*(T1*s+1)*(T2*s+1));
41 -  GT=1/(1+Gs_inv*(Gs-Gs));
42 -  Gvw=GR*GT*Gs;
43 -  GvzT=Gs2*(1-Gs*Gs_inv);
44 -  G0=GR*GT*Gs;
45 -  Gw=Gvw/(1+G0);
46 -  GzT=GvzT/(1+G0);
47 -  GT=w*Gw+z*GzT;
48 -  step(GT,w,z,80);
```

Fig. 9.9 Sections 3 and 4 of MATLAB® script of CLC with DSM Terminator. *Source* [7, p. 12]

The next problem arises if the plant has D-terms as shown below with derivative constant T_d:

$$G_S(s) = \frac{1 + sT_d}{1 + sT_1}$$

Such plants are not typical for industrial technological processes. But in the engineering of automotive industry especially by vehicle control are possible plants with PDT2-terms and dead time like below:

$$G_S(s) = \frac{K_{PS}}{1 + sT_1} \cdot \frac{1 + sT_d}{1 + sT_2} \cdot e^{-sT_t} \tag{9.4}$$

In [6] is suggested, mathematically derived, and verified the further development of DSM terminator for the plants like Eq. 9.4. Without going into details published in [6] the engineering solution for such plants is explained in Fig. 9.10.

CLIMB and HOLD controllers
The control system shown in Fig. 9.10 is applied for setpoint and disturbance behaviors. It consists of the following subsystems:

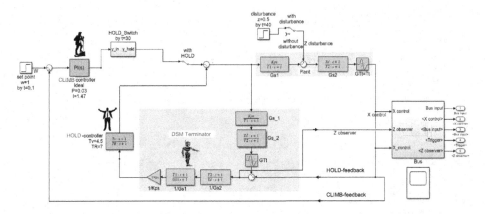

Fig. 9.10 CLC with DSM Terminator and CLIMB & HOLD controllers

- Plant to be controlled that is configured according to Eq. 9.4. The switch above plant allows to simulate the closed loop with or without disturbance.
- Main controller, that is called CLIMB-controller, because its task is to control the loop after step of set point w, i.e., the CLIMB-controller is used for setpoint behavior "climbing" to the set point w. As far as the controlled variable x achieved the steady state $x=w$ (by $t=30$ s in Fig. 9.10) the CLIMB-controller finished its function and is switched off with HOLD switch. The CLIMB-feedback is also switched off. The actuating value y is kept constant by its last value y_hold.
- HOLD-controller takes control of the entire loop as far as CLIMB-controller is switched off. The HOLD-controller is a part of the DSM Terminator. The function of HOLD-controller is to keep stable the internal loop with HOLD-feedback. That's why the fast PD controller is useful as HOLD- controller although it cannot eliminate the stead state error.
- DSM Terminator is built according to its conception mentioned above with parallel connected "disturbed" plant and "undisturbed" shadow plant. Then follows part $G_{S\_0}(s)$ of the inverse plant $1/G_S(s)$ without dead time according to Eq. 9.3. As far as the term $G_{S\_0}(s)$ is realized without dead time Tt, the algorithm of DSM terminator is "damaged". To improve it and to use the DSM Terminator correctly and stable, is the task of the HOLD-controller.

Finally, there is one blocks shown on the right side of Fig. 9.10. It is the bus with controller for supervising the data stream between control variable x and output of the entire control system. The bus controller consists of an observer, a trigger, and a saturation. With the observer is recognized the real disturbance z as variable $Z\_observer$.

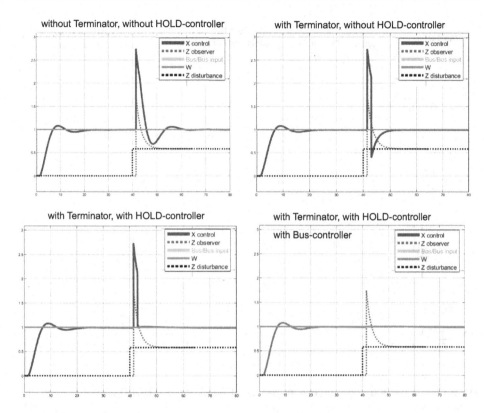

Fig. 9.11 Step responses of the control system for plants with D-terms and dead time. *Source* [7, p. 39]

It is the difference between outputs of plant and shadow plant. If met the condition $Z\_observer > 0.02$, then the saturation is triggered limiting the peak of the D-part of the plant. The bus controller is a kind of predictor, it reacts immediately after recognized disturbance without time delay.

The step responses of the experiments with the control system of Fig. 9.10 for the "worst case" of PDT2 plant with dead time are shown in Fig. 9.11, from which are seen the functions and the effectivity of the each part of the system: CLIMB-controller, DSM Terminator, HOLD-controller and Bus controller.

In the Fig. 9.12 are compared different options of control systems with DSM Terminator and CLIMB & HOLD-controllers for several types of plants. It is seen from this comparison that the main contribution to the quality of control is made from DSM Terminator, which is improved with the HOLD-controller.

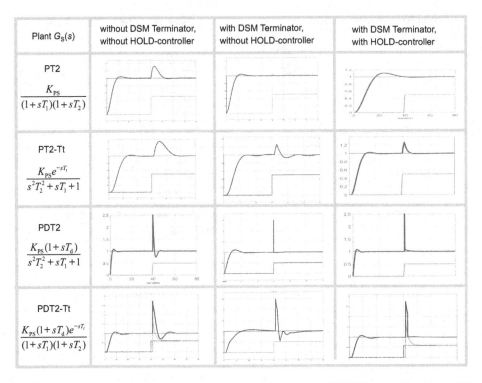

Plant $G_S(s)$	without DSM Terminator, without HOLD-controller	with DSM Terminator, without HOLD-controller	with DSM Terminator, with HOLD-controller
PT2 $\dfrac{K_{PS}}{(1+sT_1)(1+sT_2)}$			
PT2-Tt $\dfrac{K_{PS}e^{-sT_t}}{s^2T_2^2+sT_1+1}$			
PDT2 $\dfrac{K_{PS}(1+sT_d)}{s^2T_2^2+sT_1+1}$			
PDT2-Tt $\dfrac{K_{PS}(1+sT_d)e^{-sT_t}}{(1+sT_1)(1+sT_2)}$			

Fig. 9.12 Comparison of control options with DSM Terminator and HOLD-controller. *Source* [7, p. 40]

9.3   Management and Control of LTV Plants

9.3.1   LTI and LTV

Controlled plants described with transfer functions with constant coefficients accord-ing to [12] are called LTI (linear time-invariant) and those with time-dependent coef-ficients are called LTV (linear time-varying). The classic linear control theory is based exclusively on LTI plants. A simple example of LTI and LTV is shown in Fig. 9.13. The parameters K_p and T of the LTI in Fig. 9.13a are constants, the parameters $K_p(t)$ and $T(t)$ of the LTV in Fig. 9.13b are considered like input variables that are controlled externally or entered internally by a non-linear function.

The shift of K_{ps} and T_1 is much slower than the step response $x(t)$ of the plant, they are denoted as functions of the argument τ:

$$G_S(s,\tau) = \frac{x(s)}{y(s)} = \frac{K_{PS}(\tau)}{1+sT_1(\tau)} \tag{9.5}$$

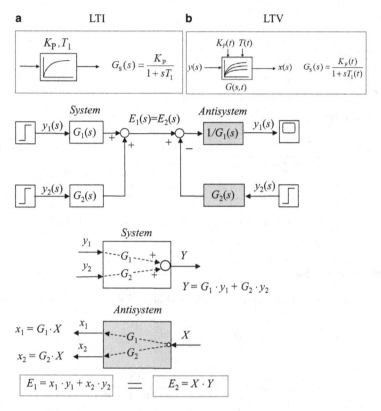

Fig. 9.13 Example of a PT1 plant: **a)** LTI with constant parameters; **b)** LTV with varying parameters. *Source* [11, p. 418]

A controller that is optimally set for time $\tau=0$ will no longer function optimally with the LTV plant at another time τ.

9.3.2   Gain Scheduling

If the functions $K_{PS}(\tau)$ and $T_1(\tau)$ of the LZV plant Eq. 9.5 are not given beforehand, the tuning of controller is done by methods of adaptive control, which are not discussed in this book.

If the functions $K_{PS}(\tau)$ and $T_1(\tau)$ are given, the parameters of a standard controller can be determined to each time τ. Such methods are known as *gain scheduling* [2].

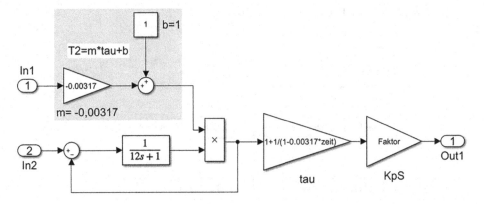

Fig. 9.14 Simulation of time delay upon Eq. 9.7 for LTV. *Source* [2, p. 124]

Example: Gain scheduling

The PT2 plant is the series connection of a LTI and LTV term: $T_2 = m \cdot \tau + b$

$$G_{S1}(s) = \frac{K_{PS1}}{1 + sT_1} = \frac{1}{1 + 3s}$$

$$G_{S2}(s) = \frac{K_{PS2}}{1 + sT_2} = \frac{0.8}{1 + (m \cdot \tau + b)s} \tag{9.6}$$

The time delay T_2 is given as linear function of time τ with coefficients m und b:

$$m = -0.00317$$
$$b = 1 \tag{9.7}$$

The LTV term of the plant is simulated in Fig. 9.14. The block DSM LZV, developed in [2] and shown in Fig. 9.15, generates the time delay T_2 to each value of processing time τ and the corresponding controller gain K_{PR}. The step responses to processing times $\tau 1 = 10$ s, $\tau 2 = 150$ s and $\tau 3 = 500$ s shown in Fig. 9.15 confirm that the control during entire time range is optimal, however with different damping. ◀

9.3.3 DSM Plant Guard

In [5] the shift of LTV parameters is considered as internal disturbance. It allows to applicate the ASA for control of LTV plants at the same way as it was done in the previous section for LTI plants with external disturbances. The corresponding DSM is in [5] referred to as "Plant Guard".

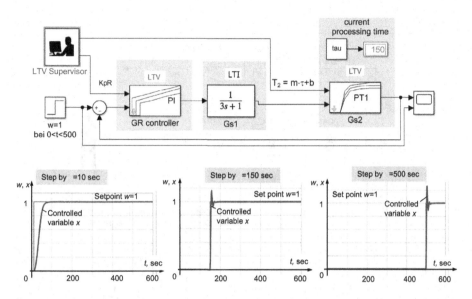

Fig. 9.15 CLC with the plant consisting of an LTI and an LTV term. The DSM LTV calculates to each value of the processing time *tau* the delay time T_2 of the plant upon Eq. 9.7 and the variable gain K_{PR} of the PI controller. The control is during entire time range of 600 s optimal. *Source* [5, p. 194]

The CLC with the DSM Plant-Guard is shown in Fig. 9.16 and the step responses in Figure Fig. 9.17.

The shift of all three parameters K_{PS}, T_1, T_2 of the LTV plant $G_S(s)$

$$G_S(s) = K_P(\tau) \cdot \frac{1}{1 + sT_1(\tau)} \cdot \frac{1}{1 + T_2(\tau)s}$$

with initial values by $\tau = 0$

$$K_{PS}(0) = 1$$

$$T_1(0) = 4 \text{ sec}$$

$$T_2(0) = 0.5 \text{ sec}$$

is implemented in Fig. 9.16 with MATLAB®/Simulink as input steps "Changing of parameters". The parameters of the PI controller are adjusted upon optimum magnitude

$$K_{PR} = 4$$

$$T_n = 4 \text{ sec}$$

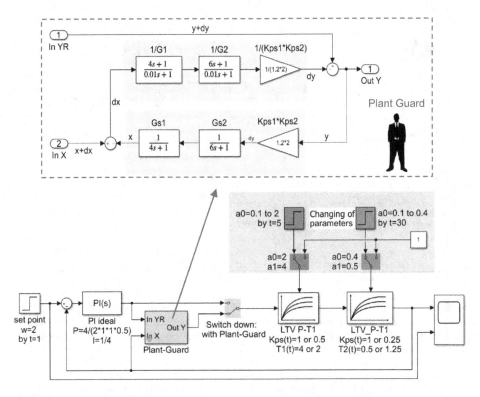

Fig. 9.16 CLC with the LTI plant and DSM Plant Guard. *Source* [11, p. 439]

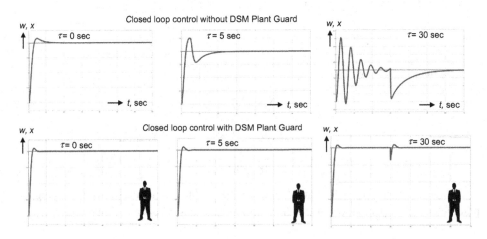

Fig. 9.17 Step responses of the CLC of Fig. 9.17 with LTI plant at different valuers of processing time $\tau = 0$; $\tau = 5$; $\tau = 30$ without DSM (above) and with DSM Plant Guard. *Source* [5, p. 439]

and are kept constant in opposite to the realization of LTV plant in Fig. 9.15. As it is seen from Fig. 9.17 the step responses of the CLC are "maintained" optimally with the DSM Plant Guard in almost entire time range between $\tau = 0$ s and $\tau = 30$ s based upon the same algorithm as DSM Terminator of Sect. 9.2.

9.3.4 Summary

All Data Stream Managers (DSM) and controllers for HOLD mode, described in this chapter, are shown in Fig. 9.18.

There are:

- CLIMB controller
  - It is the classic standard controller, which corresponds to the setpoint behavior, called CLIMB mode. The function of this controller is to bring the controlled value to its set point value. As far as the controlled value arrived the set point value the function of the CLIMB controller is finished.
- HOLD controller
  - The task of the HOLD controller is to supervise the HOLD mode, i.e. to keep constant the controlled values x keep constant or adapting the actuating value y. The HOLD devise is realized as well as controller, as the Data Stream Manager (DSM), or both.
- DSM Terminator
  - It is the unique DSM, proposed and developed in [1],[3], [6], [7], [8], [10], [11], to fully eliminate the external disturbances.
- DSM Plant Guard
  - It is the DSM based upon the same principle as terminator but used to compensate the internal disturbances of the controlled plant.
- LTV Supervisor
  - It is the DSM based upon the same principle as terminator but used to adapt the controlled loop for LTV (linear time varying plant.

All controllers and DSM, listed above, do act in the real world. The connection or the communication between real and virtual world is provided by human operator, as shown in Fig. 9.18.

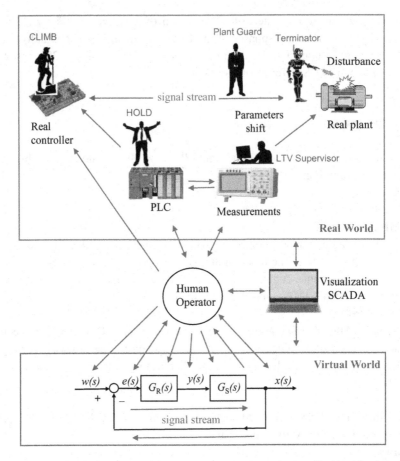

Fig. 9.18 Data Stream Managers and Controllers for HOLD-mode described in this chapter

Refereneces

1. Zacher, S. (2021). Antisystem-approach (ASA) for engineering of wide range of dynamic systems. *IJONEST, 1*(3), 52–66, ISSN: 2642–4088. https://ijonest.net/index.php/ijonest/issue/view/4. Accessed 21 Apr 2022.
2. Zacher, S. (2003). *Duale Regelungstechnik*. VDE Verlag.
3. Zacher, S. (2014). *Bus-approach for feedback MIMO-control*. Verlag Dr. S. Zacher.
4. Zacher, S. (2020). *Drei-Bode-Plots-Verfahren für Regelungstechnik*. Verlag Springer Vieweg.
5. Zacher, S. (2021). *Regelungstechnik mit Data Stream Management*. Verlag Springer Vieweg.
6. Zacher, S. (2021). Terminator im Regelkreis. *Automation-Letters, No. 43*. Verlag Dr. Zacher. https://www.zacher-international.com/Automation_Letters/43_Terminator.pdf. Accessed 21. Apr 2022.

7. Zacher, S. (2021). Schubert-terminator. *Automation-letters, No. 45*. Verlag Dr. Zacher. https://zacher-international.com/Automation_Letters/45_Schubert_Terminator.pdf. Accessed 21. Apr 2022.

8. Zacher, S. (2021). *ASA: Antisystem-approach for feedback control*. Verlag Dr. Zacher. https://youtu.be/UhUrWrx24Ag. Accessed 21. Apr 2022.

9. Zacher, S. (2021). *Closed loop control with data stream management*. Verlag Dr. Zacher. https://youtu.be/HyKQQOU1Low. Accessed 21. Apr 2022.

10. Zacher, S. (2021). *Terminator*. Verlag Dr. Zacher. https://youtu.be/EVHLY8RtSHQ. Accessed 21. Apr 2022.

11. Zacher, S., & Reuter, M. (2022). *Regelungstechnik für Ingenieure* (16th ed.). Verlag Springer Vieweg.

12. Föllinger, O., Dörrscheidt, F., & Klittich, M. (2008). *Regelungstechnik* (10th ed.). Verlag Hüthig.

Multivariable Control and Management

10

"Dimensionality reduction techniques address the "curse of dimensionality" by extracting new features from the data, rather than removing low-information features." (Quote: https://www. sciencedirect.com/topics/mathematics/curse-of-dimensionality *accessed April 24, 2022)*

10.1 Introduction: What is MIMO Control?

Let us first show two examples and explain what is the multivariable control, which is usually called MIMO control. MIMO means *Multi Input Multi Output*.

If the MIMO definition concerns plants, then inputs are actuating values $y_1, y_2, \ldots y_m$, and outputs are controlled variables $x_1, x_2, \ldots x_n$. In this chapter are considered MIMO plants with the number of variables $n = m$, while n is the order of the plant. If it is going on multivariable closed control loops, then the outputs are the same controlled variables $x_1, x_2, \ldots x_n$, but the inputs are either set points $w_1, w_2, \ldots w_n$ or disturbances z.

10.1.1 Examples of MIMO Control

Molecular filter ([3], p. 14)
The mixture of two products (A + B) is separated by a molecular filter Fig. 10.1 into two products A and B. The molecular filter consists of hundreds of hollow fiber membranes in a plastic cartridge. The mixture of products flows across the filter membrane and causes a pressure x_1, which determines the flow rate x_2 through the filter. The controlled variables x_1 and x_2 should exactly follow the values of set points w_1 and w_2 to achieve the qualitative separation.

The pressure x_1 is controlled by the controller R_1 via the valve V_1 as an actuator. The flow rate x_2 is controlled with the controller R_2 via the control valve V_2 as an actuator. The change of set point w_2 and the resulting change in flow affect the concentration of the mixture of substances, which in turn affects the filtrate rate x_2 and consequently affects the pressure x_1. The pressure controller R_1 recognizes the change of its controlled variable x_1 and after a short time will eliminate it as it happens usually by the disturbances. On the contrary, the change of set point w_1 changes the pressure x_1, which affects the changes in flow rate x_2.

If both controllers R_1 and R_2 work independent of each other, it is called *decentralized* MIMO control. In this case, the control of pressure and flow takes place but not immediately and not qualitatively.

To improve the quality of control in Fig. 10.1 is used so-called decoupled MIMO control. The mutual effect of x_1 and x_2 is compensated by controllers R_{12} and R_{21} of a decoupling block, which prevents the effect of actuating value y_2 on the controlled variable x_1 and contrarily the effect of actuating value y_1 on the controlled variable x_2. The decoupling achieves of course better control quality than two separate single control closed loops.

The identification of the filter as coupled MIMO plant and the engineering of the decoupled MIMO control of $n = 2$ controlled variables is discussed in this chapter.

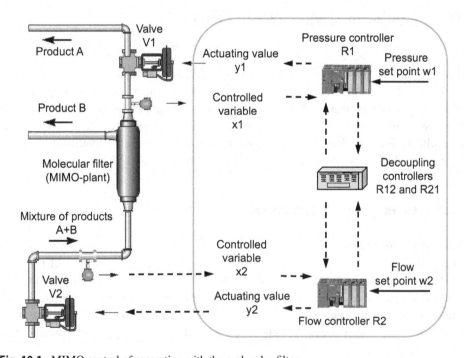

Fig. 10.1   MIMO control of separation with the molecular filter

Pumping stations ([2], p. 279)

A regional purification plant consists of seven pumping stations built in a 15 km long network (Fig. 10.2). Each pumping station has an upstream pump reservoir as storage space for the resulting wastewater and surface water.

The incoming wastewater is first treated with the corresponding pumping station, then it is temporarily stored before it is conveyed to the next station. The level in the storage room is determined by high resolution measurement. The capacity is recorded using a magnetic-inductive volume flow measurement and used to control the pumps with controllers R_1, R_2, ... R_7. This should ensure a constant delivery rate is guaranteed depending on the operating case.

Since each pumping station is equipped with a redundant pump system consisting of two identical pump units which are operated alternately, the utilization of both is almost the same pumps guaranteed. If a pump fails or is serviced, it will automatically replace the other pump, so that very high availability is reached.

Each pumping station affects all other stations in the network. All operating data are permanently sent to the central control room. The mathematical models of each station are previously defined and saved by the central control room, which made it possible to implement the MIMO control system for the entire network with automatic decoupling of working stations.

The engineering of the decoupled MIMO control for the network of n stations is described in this chapter.

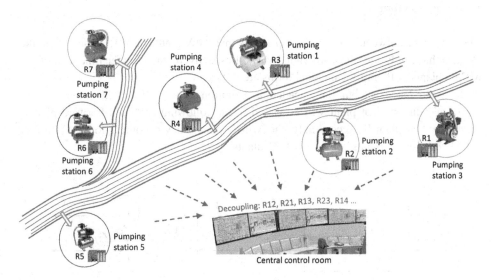

Fig. 10.2 MIMO control of the network with 7 pumping stations

10.1.2 Kinds of MIMO Plants

Generally, there are the following two kinds of MIMO plants (Fig. 10.3):

- Separated plants or MIMO plants with separated channels: each controlled variable is controlled with corresponding actuated value without influencing other controlled variables.
- Coupled plants or MIMO plants with coupled channels: each actuated value influenced many or all controlled variables.

Separated MIMO plants
The separated (not coupled) MIMO plants with n controlled variables $x_1, x_2, \ldots x_n$ and n actuating values $y_1, y_2, \ldots y_n$ are n separate plants. Each of them has respectively one controlled variable and one corresponding actuating value. The identification, mathematical description, stability analysis, and design of separated plants are well known in the classical control theory and are not discussed in this chapter.

Coupled plants
There are known two kinds of coupled MIMO plants, as shown in Fig. 10.3, whose academic exact definition is

- P-canonical form.
- V-canonical form.

The difference between both kinds of plants is that the V-plants have feedback while the P-plants have only feed-forward signal ways. Respectively, the transfer functions of the MIMO plants of P- and V-canonical forms are different.

To avoid the exact but long academic definition, like "MIMO plant in P-canonical form", this kind of plant will be called in the following as "Feed-Forward plants" or shortly "FF-plants". Accordingly, the MIMO plants in V-canonical form are further called "Feed-Back plants" or shortly "FB-plants".

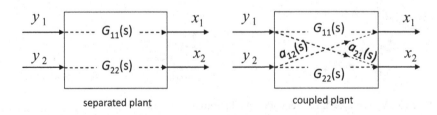

separated plant   coupled plant

Fig. 10.3 Kinds of MIMO plants with $n = 2$ controlled variables

10.1.3 Kinds of MIMO Closed Loop Control

The kinds of MIMO control are shown in Fig. 10.4.

All three kinds of MIMO control could be unified and mathematically described with input vectors $\mathbf{W}(s)$ and output vectors $\mathbf{X}(s)$, with transfer matrices of controllers $\mathbf{R}(s)$ and plants $\mathbf{G_S}(s)$, building the transfer matrices of the open $\mathbf{G_0}(s)$ and closed loop $\mathbf{G_w}(s)$ for reference behavior:

$$\mathbf{W} = \begin{pmatrix} w_1 \\ w_2 \end{pmatrix} \quad \mathbf{X}(s) = \begin{pmatrix} x_1(s) \\ x_2(s) \end{pmatrix} \tag{10.1}$$

$$\mathbf{G_0}(s) = \mathbf{R}(s)\mathbf{G_S}(s) \quad \mathbf{G_w}(s) = \frac{\mathbf{G_0}(s)}{1 + \mathbf{G_0}(s)} \tag{10.2}$$

$$\mathbf{X}(s) = \mathbf{G_w}(s)\mathbf{W}(s)$$

Separated MIMO control of separated plants (Fig. 10.4a)
Each separate plant will be controlled with a single controller. The engineering of such single control loops consists in the stability analysis and design of each single closed control loop, which is well known in the classical control theory and will not be discussed in this chapter. The mathematical description of the separated MIMO control is shown below

$$\mathbf{R}(s) = \begin{pmatrix} R_1(s) & 0 \\ 0 & R_2(s) \end{pmatrix} \quad \mathbf{G_S}(s) = \begin{pmatrix} G_{11}(s) & 0 \\ 0 & G_{22}(s) \end{pmatrix} \tag{10.3}$$

The plant and the controller are both diagonal matrices, that's why Eq. 10.1 results in two separate transfer functions, which should be separately treated.

Separated control of coupled MIMO plants (Fig. 10.4b)
A separated MIMO control of the coupled MIMO plant of the n-th order with n controlled variables is realized with n separate controllers. Each controller is tuned for its single control loop independent of other loops. But as far as single control loops of a coupled MIMO plant influence each other, the controlled variable of one loop acts as the disturbance for another loop.

The mathematical description of the separated MIMO control of the 2nd order is shown below

$$\mathbf{R}(s) = \begin{pmatrix} R_1(s) & 0 \\ 0 & R_2(s) \end{pmatrix} \quad \mathbf{G_S}(s) = \begin{pmatrix} G_{11}(s) & a_{12}(s) \\ a_{21}(s) & G_{22}(s) \end{pmatrix} \tag{10.4}$$

The controller is again a diagonal matrix as in Eq. 10.3, but the matrix $\mathbf{G_S}(s)$ of the MIMO plant is fulfilled with coupling blocks, which transfer signals from one loop to the other.

Fig. 10.4 Kinds of MIMO
control of $n = 2$ controlled
variables

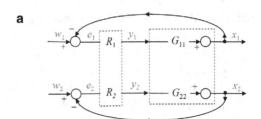

Separated MIMO control of separated plant

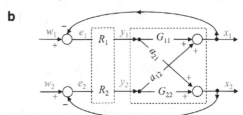

Separated MIMO control of coupled plant

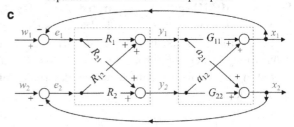

Decoupled MIMO control of coupled plant

Decoupled control of coupled MIMO plants (Fig. 10.4c)

The decoupling is clearly seen on the matrix $\mathbf{R}(s)$ of the MIMO controller below in Eq. 10.5:

$$\mathbf{R}(s) = \begin{pmatrix} R_1(s) & R_{12}(s) \\ R_{21}(s) & R_2(s) \end{pmatrix} \quad \mathbf{G}_S(s) = \begin{pmatrix} G_{11}(s) & a_{12}(s) \\ a_{21}(s) & G_{22}(s) \end{pmatrix} \tag{10.5}$$

The coupling blocks $a_{12}(s)$ and $a_{21}(s)$ of the plant are compensated with decoupling controllers $R_{12}(s)$ and $R_{21}(s)$. After such compensation, which is called decoupling, the matrix of controller $\mathbf{R}(s)$ in Eq. 10.5 is transformed to the diagonal matrix of Eq. 10.4. In this way, the entire decoupled MIMO control of Eq. 10.5 is converted to the single-loop MIMO control of Eq. 10.3, which has many advantages for control quality.

10.2 Coupled MIMO Plants

As it was already mentioned in Sect. 10.1.2 and shown in Fig. 10.3, there are the follow-ing two kinds of MIMO plants:

- MIMO plants with separated channels, which are not considered in this chapter.
- MIMO plants with coupled channels, which are divided into two types (Fig. 10.5) according to their structure:
  - Feed-Forward plants in P-canonical form, shortly FF-plants.
  - Feed-Back plants in V-canonical form, shortly FB-plants.

10.2.1 Feed-Forward Plants

The main plants are $G_{11}(s)$ and $G_{22}(s)$, and the coupling plants are $a_{12}(s)$ and $a_{21}(s)$:

$$
\begin{aligned}
G_{11}(s) &= \frac{x_1(s)}{y_1(s)} & a_{12}(s) &= \frac{x_1(s)}{y_2(s)} \\
a_{21}(s) &= \frac{x_2(s)}{y_1(s)} & G_{22}(s) &= \frac{x_2(s)}{y_2(s)}
\end{aligned}
\tag{10.6}
$$

Mathematical description of FF-plant as a system of equations:

$$
\begin{cases}
x_1(s) = G_{11}(s)y_1(s) + a_{12}(s)y_2(s) \\
x_2(s) = a_{21}(s)y_1(s) + G_{22}(s)y_2(s)
\end{cases}
\tag{10.7}
$$

A short description of the MIMO coupled systems by the use of vectors and matrices is possible. The single variables x_1, x_2 will be replaced through vectors **X** as was mentioned

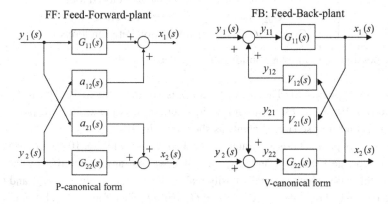

Fig. 10.5 FF-plants and FB-plants

in Sect. 10.1.3. In this case, the single transfer functions $G_{11}(s)$, $a_{12}(s)$, $a_{21}(s)$, and $G_{22}(s)$ of the plant build the plant matrix $\mathbf{G}_S(s)$:

$$\mathbf{G}_S(s) = \begin{pmatrix} G_{11}(s) & a_{12}(s) \\ a_{21}(s) & G_{22}(s) \end{pmatrix} \tag{10.8}$$

10.2.2 Interaction Transfer Function and Coupling Gain

The interaction between main plants $G_{11}(s)$, $G_{22}(s)$ of the feed-forward plants, given in Fig. 10.5, is dependent on the transfer functions $a_{12}(s)$, $a_{21}(s)$. To evaluate the intensity and dynamics between main plants is introduced so-called interaction transfer function $C(s)$:

$$C(s) = \frac{a_{12}(s)a_{21}(s)}{G_{11}(s)G_{22}(s)} \tag{10.9}$$

The interaction transfer function usually leads to a complicated expression, so by proportional PT1, PT2, ... PTn plants it is reasonable to simplify it to the steady state value by $t \to \infty$ or $s \to 0$, which is called static coupling factor $C(0)$ or simply coupling gain C_0:

$$C(0) = C_0 = \lim_{s \to 0} C(s) = \frac{a_{12}a_{21}}{K_{11}K_{22}} \tag{10.10}$$

K_{11} and K_{22} are static gains of main plants $G_{11}(s)$ and $G_{22}(s)$, and the a_{12} and a_{21} are static gains of coupling plants $a_{11}(s)$ and $a_{22}(s)$.

The value of the static interaction gain C_0 is a feature to recognize, and what kind of separated closed loop control of Fig. 10.4 happens as an actual control. Decisive for it are the stability conditions for systems of $n=2$ derived in Chap. 4:

$$\begin{cases} 1 + G_{01}(s) = 0 & 1^{st} \text{ separated closed loop} \\ 1 + G_{02}(s) = 0 & 2^{nd} \text{ separated closed loop} \\ 1 - C_0 G_{w1}(s)G_{w2}(s) = 0 & \text{interaction between loops} \end{cases} \tag{10.11}$$

There are possible the following options of separated control:

- $C_0=0$ and $a_{12}=a_{21}=0$. The MIMO plant has separate channels. The last stability condition in Eq. 10.11 escapes, and no interaction takes place. It corresponds to the separated control of separated plants as shown in Fig. 10.4a.
- $C_0<1$. The coupled plant is separately controlled, as given in Fig. 10.4b. Let us suppose that the separate loops G_{w1} and G_{w2} are stable without any static error by the steady state. In this case, which is usually easily realizable, the static values of G_{w1} and G_{w2} by $t \to \infty$ or $s \to 0$ should be $G_{w1}(0)=1$ and $G_{w2}(0)=1$. Therefore, the stability condition in Eq. 10.11 for interaction is by $C_0<1$ fulfilled, and the MIMO control is stable.
- $C_0>1$. The coupled plant is again as in the previous case separately controlled, but the interaction stability condition in Eq. 10.11 is not fulfilled and the system is unstable.

Example of stability check of separated control with coupled plant

According to MATLAB® script, given below, all poles of the system with $C_0 < 1$ have negative real parts, so the MIMO system is stable.

```
Kp1 = 2;  Kp2 = 12;
T1 = 1; T1a = 2; T2 = 3; T2b = 4;
a12 = 0.5; a21 = 0.2;
T12 = 0.5; T21 = 0.2;
Tn1 = T1a;
Tn2 = T2b;
C0 = (a12*a21)/(Kp1*Kp2);
K1 = Tn1/(2*Kp1*T1);
K2 = Tn2/(2*Kp2*T2);
b4 = T1*T1a*T2*T2b;
b3 = T1*T1a*T2 + T1a*T2*T2b + T2*T2b*T1 + T2b*T1*T1a;
b2 = T1*T1a + T1*T2 + T1*T2b + T1a*T2 + T1a*T2b + T2*T2b;
b1 = T1 + T1a + T2 +T2b;
b0 = 1;
a2 = a12*a21;  a1 = a12+a21; a0 = 1;
numC = [b4*C0   b3*C0   b2*C0   b1*C0   b0*C0];
denC = [a2 a1 a0];
C = tf(numC, denC);
Gw1= tf([K1*Kp1], [Tn1*T1   Tn1  K1*Kp1]); % first loop
Gw2= tf([K2*Kp2], [Tn2*T2   Tn2  K2*Kp2]); % second loop
G = 1 - C*Gw1*Gw2; % interaction
poles = tzero(G) % poles of entire MIMO-system
```

Let us run the same MATLAB® script, shown above, for the plant with the same parameters as above, but with

$$a12 = 3; a21 = 12;$$

which leads to

$$C_0 > 1.$$

It results in the following poles, which is a proof that the MIMO system with one positive pole is unstable.

$$0.0359 + 0.0000i$$

$$-0.2084 + 0.1308i$$

$$-0.2084 + 0.1308i$$

$$-0.3333 + 0.0000i$$

$$-0.5179 + 0.5138i \quad \blacktriangleleft$$

10.2.3 Feed-back Plants

A FB-plants are shown in Fig. 10.5. They have the following transfer functions:

$$G_{11}(s) = \frac{x_1(s)}{y_{11}(s)} \quad V_{12}(s) = \frac{y_{12}(s)}{x_2(s)} \quad V_{21}(s) = \frac{y_{21}(s)}{x_1(s)} \quad G_{22}(s) = \frac{x_2(s)}{y_{22}(s)} \quad (10.12)$$

Mathematical description of FB-plant as a system of equations with closed loops is

$$\begin{cases} x_1(s) = G_{11}(s)y_{11}(s) = G_{11}(s)\,[y_1(s) + V_{12}(s)\,x_2(s)] \\ x_2(s) = G_{22}(s)y_{22}(s) = G_{22}(s)[y_2(s) + V_{21}(s)\,x_1(s)] \end{cases} \quad (10.13)$$

10.2.4 Conversion of FB-Plants into FF-Plants

The closed loops of the FB-plants complicate its engineering. To simplify it, the mathematical description in Eq. 10.13 of a FB-plant could be converted into the transfer function of FF-plant, as given in Eq. 10.8. After tracking the signal way beginning from the input y_{11} of the FB-plant in Fig. 10.5 to the same variable y_{11}, we can define the transfer function of the open loop inside the FB-plant as a product of all transfer functions of the single blocks:

$$G_0(s) = G_{11}(s)V_{12}(s)G_{22}(s)V_{21}(s) \quad (10.14)$$

The same transfer function $G_0(s)$ as in Eq. 10.14 results by tracking the signal way from the input y_{22} to the same variable y_{22}.

The transfer function $G_{p11}(s)$ from input y_1 to the output x_1 of the FB-plant describes the disturbance behavior with the disturbance y_1, whose open loop transfer function $G_0(s)$ is given in Eq. 10.14.

The same relations are concerning all other transfer functions $G_{p12}(s)$, $G_{p21}(s)$, and $G_{p22}(s)$:

$$G_{p11}(s) = \frac{G_{11}(s)}{1 - G_0(s)} \quad G_{p22}(s) = \frac{G_{22}(s)}{1 - G_0(s)} \quad (10.15)$$

$$G_{p12}(s) = \frac{G_{11}(s)V_{12}(s)G_{22}(s)}{1 - G_0(s)} \quad G_{p21}(s) = \frac{G_{11}(s)V_{21}(s)G_{22}(s)}{1 - G_0(s)} \quad (10.16)$$

The transfer function of the FB-plant, converted in the FF-plant by Eq. 10.14, 10.15, 10.16, is given below

$$\mathbf{G_S}(s) = \begin{pmatrix} G_{p11}(s) & G_{p12}(s) \\ G_{p21}(s) & G_{p22}(s) \end{pmatrix} \quad (10.17)$$

Thus, the original mathematical description of FB-plant Eq. 10.13 is converted to the mathematical description of FF-plant without feedback:

$$\begin{cases} x_1(s) = G_{p11}(s)\, y_1(s) + G_{p12}(s)\, y_2(s) \\ x_2(s) = G_{p21}(s)y_1(s) + G_{p22}(s)y_2(s) \end{cases}$$

The functional block diagram of the FB-plant converted to the FF-plant is shown in Fig. 10.6.

10.2.5 Stability of Coupled MIMO Plants

Stability of FF-plants

An FF-plant of Eq. 10.7 is stable if all partial transfer functions $G_{11}(s)$, $a_{12}(s)$, $a_{21}(s)$, and $G_{22}(s)$, defined in Eq. 10.6, are stable.

Stability of FB-plants

An FB-plant of Eq. 10.13 is stable if all single transfer functions $G_{11}(s)$, $V_{12}(s)$, $V_{21}(s)$, and $G_{22}(s)$ are stable and besides this, the closed loops of Eqs. 10.15 and 10.16 are stable.

The closed loops of Eqs. 10.15 and 10.16 are stable, if the roots of the characteristic equation.

$$1 - G_0(s) = 0 \qquad\qquad (10.18)$$

have negative real parts.

Fig. 10.6 FF-plant converted from FB-plant. *Source* [1, p. 25]

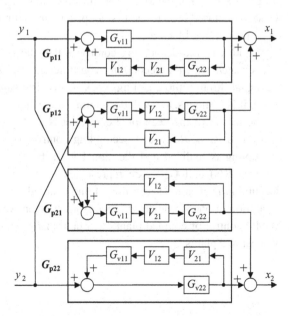

Example: stability of a Feed-back plant

The FB-plant

$$\mathbf{G_S}(s) = \begin{pmatrix} G_{11}(s) & V_{12}(s) \\ V_{21}(s) & G_{22}(s) \end{pmatrix} = \begin{pmatrix} \frac{1}{1+s} & 1 \\ 0,1 & \frac{2}{1+2s} \end{pmatrix}$$

is stable because

- single transfer functions $G_{11}(s)$, $V_{12}(s)$, $V_{21}(s)$, and $G_{22}(s)$ are stable.
- the characteristic equation of Eq. 10.18 is stable according to the Hurwitz stability criterion asshown below

$$1 - \frac{1}{1+s} \cdot 1 \cdot 0,1 \cdot \frac{2}{1+2s} = 0$$

$$2s^2 + 3s + 0,8 = 0 \quad \blacktriangleleft$$

10.3   Decoupled MIMO Control

10.3.1 Conception of Decoupling

The conception of decoupling control was mentioned in Sect. 10.1.3 and shown in Fig. 10.4 as the block diagram "decoupled MIMO control of coupled plant". The coupling block $a_{12}(s)$ of the MIMO plant, shown in Fig. 10.4, transfers the actuating value of the second closed loop to the first loop. The first loop in its turn affects the seconf closed loop via coupling block $a_{21}(s)$.

The idea of the decoupling control is that the influence of the second loop will be compensated with the decoupling controller $R_{12}(s)$, which transfers the same signal as the signal through the coupling block $a_{12}(s)$. The same is for the signal coupling block $a_{21}(s)$. It is compensated with decoupling controller $R_{21}(s)$.

The matrix of the decoupled controller $\mathbf{R}(s)$ is given in Eq. 10.5. It consists of main controllers $R_1(s)$, $R_2(s)$, and decoupling controllers $R_{12}(s)$ and $R_{21}(s)$. But as far as the decoupling is realized and the signal ways between both loops are compensated, the matrices $\mathbf{R}(s)$ and $\mathbf{G_S}(s)$ are represented as given in Eq. 10.3. The advantage of the decoupling is even the separation of one coupled plant from another which results in separation of one single-closed loop from another. According to Fig. 10.4, the "separated MIMO control of coupled plant" will be transformed into the "separated MIMO control of separated plants".

10.3.2 Decoupled MIMO Control Loops

Decoupled FF- and FB-controllers
The common representation of a decoupled controller is schematically shown in Fig. 10.4 and is described in Eq. 10.5. But the real implementation of decoupled controllers differs from this common representation. Each decoupled MIMO controller consists of two parts, as is shown in Fig. 10.7:

- Main controllers $R_1(s)$ and $R_2(s)$.
- Decoupling controllers $R_{12}(s)$ and $R_{21}(s)$.

There are possible the following kinds of MIMO controllers:

- Feed-Forward decoupled controller (FF-controller),
  - when $R_{12}(s)$ and $R_{21}(s)$ do feed-forward the actuating values of main controllers $R_1(s)$ and $R_2(s)$, as shown in Fig. (a) and
  - when $R_{12}(s)$ and $R_{21}(s)$ do feed-forward the controlled variables $x_1(s)$ and $x_2(s)$, as shown in Fig. B;
- Feedback decoupled controller (FB-controller), which is shown in Fig. c and which is constructed like the FB-plant.

Kinds of decoupled MIMO control loops
The decoupling MIMO control loops can be implemented as a combination of MIMO feed-forward (FF) and feed-back (FB) plants and controllers (Fig. 10.7):

- FF-plant
  - with FF-controller;
  - with FB-controller.
- FB-plant
  - with FF-controller.

FF-plant with FF-controller (Fig. 10.7a)
By the decoupling control, the transfer functions $R_{12}(s)$ and $R_{21}(s)$ should be chosen properly to eliminate the interaction between main loops. In other words, the decoupling controller $R_{12}(s)$ should compensate the signal of the second main loop through the coupling plant $a_{12}(s)$, and the decoupling controller $R_{21}(s)$ should compensate the signal of the first main loop transferred through the coupling plant $a_{21}(s)$.

Supposing

$$y_{1R} = y_1$$
$$y_{2R} = y_2$$

(10.19)

Fig. 10.7 Kinds of decoupled MIMO control

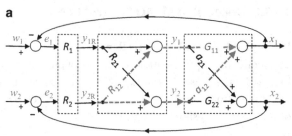

a

FF-plant with decoupled FF-controller

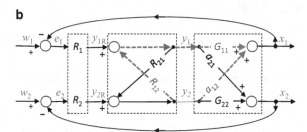

b

FF-plant with decoupled FB-controller

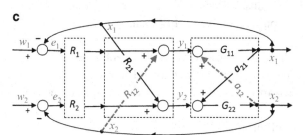

c

FB-plant with decoupled FF-controller

such compensation succeeds, if the signal from the lower loop to the upper loop via $a_{21}(s)$, picked up with red in Fig. 10.7a, is equal to the signal through $R_{12}(s)G_{11}(s)$, picked up with blue. The same is for a signal from the upper to the lower loop via $a_{12}(s)$:

$$a_{12}(s) = R_{12}(s)G_{11}(s)$$
$$a_{21}(s) = R_{21}(s)G_{22}(s)$$

(10.20)

From Eq. 10.20, follow the desired decoupling transfer functions:

$$R_{12}(s) = \frac{a_{12}(s)}{G_{11}(s)}$$
$$R_{21}(s) = \frac{a_{21}(s)}{G_{22}(s)}$$

(10.21)

The transfer functions of decoupling controllers of Eq. 10.21 are built with reciprocal transfer functions of the main plants $G_{11}(s)$ and $G_{22}(s)$. It leads by typical industrial to the big values of D-terms (derivative algorithms) by decoupling controllers. Besides this disadvantage, the transfer functions of Eq. 10.21 are not realized, because the order of the numerator of the transfer function is often greater than the order of the denominator.

To make the control realizable, we should complete the transfer functions of decoupling controllers with some small delay time constants. The order of the denominator increases and is equal to the order of the numerator, as it is done in the example below.

But because of these additional delay time constants, the decoupling conditions in Eq. 10.20 are disturbed and as a result the decoupling is not complete, as is seen in an example of Fig. 10.8.

Example: Decoupled control of FF-plant with FF-controller

The MATLAB®/Simulink model and step responses of controlled variable $x_1(t)$, $x_2(t)$ are shown in Fig. 10.8. But the decoupling is not perfect; the small negative reactions, picked up with red crosses, occur after each set point step. The matters are

- The assumption in Eq. 10.19 is not entirely correct.
- The big derivative term (short D-term) of decoupling controllers:

$$
\begin{aligned}
R_{12}(s) &= 0,05 \cdot \frac{(2s+1)(7s+1)}{(4s+1)} \\
R_{21}(s) &= 0,75 \cdot \frac{(3s+1)(8s+1)}{(5s+1)}
\end{aligned}
\tag{10.22}
$$

- The order of numerator by decoupling controllers in Eq. 10.22 is greater than the order of denominator, so the transfer function is practically not realizable. To realize $R_{12}(s)$ and $R_{21}(s)$ are arbitrary, the additional poles with possibly small values are applied, as is shown in Fig. 10.8. ◄

FF-Plant with FB-Controller (Fig. 10.7b)
The red and blue signal ways in Fig. 10.7b consist of the same transfer functions as in Fig. 10.7a. Therefore, the decoupling transfer functions are also the same as in Eq. 10.21. But instead of Eq. 10.19, the actuating values y_1 and y_2 are the same for red and blue signal ways. On the one hand, it is positive, because Eq. 10.19 escapes and the decoupling is perfect. On the other hand, it leads to the closed loop within decoupled FB-controller, so it could become instable. In the Sect. 10.3.3, is explained how to check the stability of such decoupling controllers.

FB-Plant with FF-Controller (Fig. 10.7c)
For a FB-plant, it is reasonable to feed the decoupling controllers $R_{12}(s)$ and $R_{21}(s)$ not from actuating values y_1 and y_2 as it was done in previous cases, but from the control-

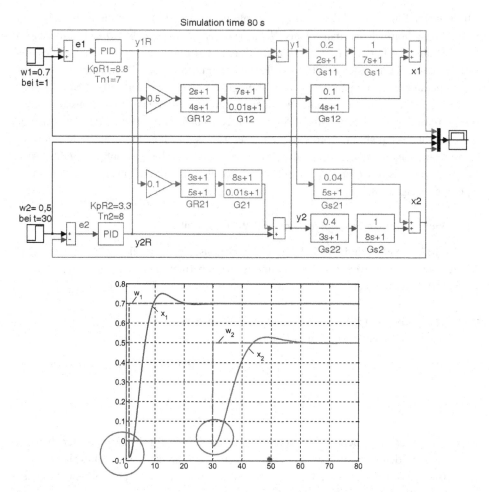

Fig. 10.8 Decoupling control of FF-plant with FF-controller. *Source* [1, p. 45]

ling variables x_1 and x_2 (Fig. 10.7c. It is easy to track the red and blue signal ways of Fig. 10.7c to determine that the following condition is needed for the fully decoupling:

$$R_{12}(s) = V_{12}(s)$$
$$R_{21}(s) = V_{21}(s)$$

(10.23)

However, a feedback system always runs the risk that it could be unstable.

10.3.3 Stability Proof of FF-Plant with FB-Controller

The control loops are shown in Fig. 10.7b. If the condition in Eq. 10.21 is fulfilled, the decoupling is prefect, and the upper and the lower loops are independent one of another. In this case, the entire MIMO control will be stable if each closed loop with the following characteristic equations is stable:

$$
\begin{cases}
1 + G_{01}(s) = 0 \rightarrow 1 + R_1(s)G_{11}(s) = 0 \ 1^{st} \text{ separated closed loop} \\
1 + G_{02}(s) = 0 \rightarrow 1 + R_2(s)G_{22}(s) = 0 \ 2^{nd} \text{ separated closed loop}
\end{cases}
\quad (10.24)
$$

In Fig. 10.7b is seen that also one closed loop between decoupling controllers $R_{12}(s)$ and $R_{21}(s)$ appears, therefore one more characteristic equation should be checked. Considering Eq. 10.21, this stability condition is given below

$$
1 - R_{12}(s)R_{21}(s) = 0 \rightarrow G_{11}(s)G_{22}(s) - a_{12}(s)a_{21}(s) = 0 \quad \text{interaction } R_1 \text{ and } R_2 \quad (10.25)
$$

▶ The entire MIMO control of an FF-plant with the FB-controller will be stable if poles of all characteristic equations Eq. 10.24 and 10.25 have negative real times.

Significant by the condition Eq. 10.25 is that the stability of the interaction between decoupling controllers $R_{12}(s)$ and $R_{21}(s)$ is dependent only on plant parameters, and not on decoupling controllers. Two cases are possible:

a) The closed loop of interactions has positive feedback, like Eq. 10.25, if the coupling plants $a_{12}(s)$ and $a_{21}(s)$ affect both with the same sign, e.g., positive as is given in Fig. 10.7b.
b) The feedback by Eq. 10.25 is negative if the coupling plants $a_{12}(s)$ and $a_{21}(s)$ affect by different signs.

The example below illustrates the stability check of both cases.

Example: Stability of FF-plant with FB-controller

The MATLAB®/Simulink model is shown in Fig. 10.9, in which the interaction closed loop is picked up gray. The decoupling controllers $R_{12}(s)$ and $R_{21}(s)$ are defined according to Eq. 10.21:

$$
\begin{aligned}
R_{12}(s) &= \frac{a_{12}(s)}{G_{11}(s)} = \frac{s+1}{1} \\
R_{21}(s) &= \frac{a_{21}(s)}{G_{22}(s)} = \frac{2s+1}{0,25}
\end{aligned}
\quad (10.26)
$$

Let us discuss two cases mentioned above:

a) *Unstable*: Both signs by coupling plants $a_{12}(s)$ and $a_{21}(s)$ are positive, as it is picked up with red color in Fig. 10.9. Therefore, both signs of decoupling controllers $R_{12}(s)$ and $R_{21}(s)$ are negative. The stability condition in Eq. 10.25 considering Eq. 10.26 is in this case:

$$1 - (s+1) \cdot 0,25(2s+1) = 0$$
$$0,5s^2 + 0,75s - 0,75 = 0 \tag{10.27}$$

   - As far as one coefficient in the characteristic equation Eq. 10.27 is negative, the system is according to the Hurwitz stability criterion unstable.

b) *Stable*: One sign by coupling plants $a_{12}(s)$ and $a_{21}(s)$ is positive, the other is negative, as it is picked up with blue color in Fig. 10.9. Accordingly, the sign of decoupling controller $R_{12}(s)$ is positive and the sign of $R_{21}(s)$ is negative. The system is stable because all coefficients in the characteristic equation Eq. 10.27 are positive:

$$1 + (s+1) \cdot 0,25 \cdot (2s+1) = 0$$
$$0,5s^2 + 0,75s + 1,25 = 0 \tag{10.28}$$

The step responses of controlled variable $x_1(t)$, $x_2(t)$ to cases (a) and (b) are given in Fig. 10.10. ◄

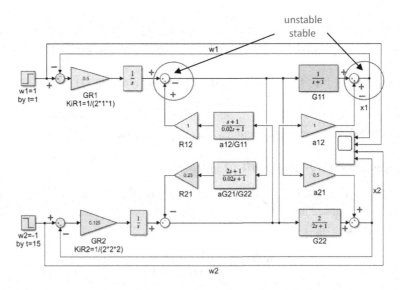

Fig. 10.9 Example of FF-plant with FB-controller. The system is unstable if both signs of the coupling plants are the same (red)

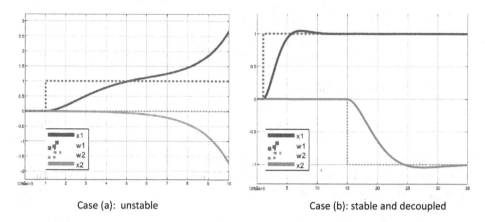

Case (a): unstable   Case (b): stable and decoupled

Fig. 10.10 Step responses of the MIMO control: **a** when coupling plants have the same sign; **b** when coupling plants have different signs

10.3.4 Bus-Approach

The functional block diagrams of the MIMO control systems for $n=2$ control variables x_1, x_2, shown in the figures, clearly illustrate the signal ways and let us track them. But the graphical diagrams of MIMO control systems with coupled plants of $n<2$ control variables are not understandable. To simplify the classical functional block diagrams of $n<2$ control variables, a virtual bus is introduced in [1], published in [2]-[6], and described in Chap. 4.

In Chap. 4 was shown that one of the advantages of the bus-approach is the possibility to track the signal ways by multivariable systems with coupled plants of high dimensionality n. By this means was proofed the stability of separated MIMO control of coupled plants of 2nd, 3rd, and 4th orders, which leads to generalizing the stability conditions for separated systems with FF-plants of order n with $j=1, 2, \ldots n$ and $k=1, 2, \ldots n$ as given below.

▶ **Decentralized MIMO control of a Feed-forward plant of order N**

- stability conditions for each separate loop and interactions between loops

$$\begin{cases} 1 + G_{01}(s) = 0 \\ 1 + G_{02}(s) = 0 \\ \dotfill \\ 1 + G_{0n}(s) = 0 \end{cases}$$

$$\begin{cases} 1 - C_{12}G_{w1}G_{w2} - C_{13}G_{w1}G_{w3} - \ldots - C_{1,n-1}G_{w1}G_{w,n-1} - C_{1n}G_{w1}G_{wn} = 0 \\ 1 - C_{21}G_{w2}G_{w1} - C_{23}G_{w2}G_{w3} - \ldots - C_{2,n-1}G_{w2}G_{w,n-1} - C_{2n}G_{w2}G_{wn} \\ \cdots \quad \cdots \quad \cdots \quad \cdots \quad \cdots \quad \cdots \quad \cdots \quad \cdots \quad \cdots \quad \cdots \quad = 0 \\ 1 - C_{n1}G_{wn}G_{w1} - C_{n2}G_{wn}G_{w2} - \ldots - C_{n,n-2}G_{wn}G_{w,n-2} - C_{n,n-1}G_{wn}G_{w,n-1} = 0 \end{cases}$$

- transfer functions of each separated loop

$$G_{wk}(s) = \frac{R_k(s)G_{kk}(s)}{1 + R_k(s)G_{kk}(s)}$$

- interaction transfer functions

$$C_{jk}(s) = C_{kj}(s) = \frac{a_{jk}(s)a_{kj}(s)}{G_{kk}(s)G_{jj}(s)}$$

The next advantage of the bus-approach is the possibility to use by the engineering of MIMO control the blocks "Bus Creator" and "Bus-Selector" of the well-known simulation-software MATLAB®/Simulink.

10.3.5 Bus-Approach for Decoupled MIMO Control

The conventional functional block diagrams of decoupling control of the order $n > 2$, based upon transfer functions, have no sense for practical use, because no signal way could be clearly tracked. It's totally different from applying the bus-approach for simplification for the MIMO system of the 3rd and higher orders.

MIMO system of the 3rd order
The FF-plant is controlled with the FB-controller, as shown in Fig. 10.11. The signal ways of the MIMO system with the bus are clearly seen and it is easy to track them. The signal way 1–2 through $a_{12}(s)$ will be compensated with the signal way 1-3-4 via $R_{12}(s)$, then the way 4–5 through main plant $G_{11}(s)$, which results in the first decoupling controller:

$$R_{12}(s)G_{11}(s) = a_{12}(s)$$

Fig. 10.11 Decoupled MIMO control of the 3rd order. *Source* [1, p. 84]

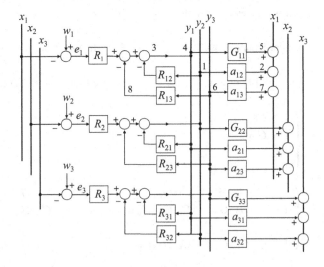

$$R_{12}(s) = \frac{a_{12}(s)}{G_{11}(s)} \tag{10.29}$$

Similarly, the signal way 6–7 through $a_{21}(s)$ will be compensated with the signal way 6–8 via $R_{13}(s)$, then the way 8-3-4-5 through main plant $G_{22}(s)$, which results in the next decoupling controller:

$$R_{13}(s)G_{11}(s) = a_{13}(s)$$

$$R_{13}(s) = \frac{a_{13}(s)}{G_{11}(s)} \tag{10.30}$$

Generalizing Eq. 10.29 and 10.30, we can derive the following expression to define each decoupling controller for $k = 1, 2, 3$ and $j = 1, 2, 3$:

$$R_{kj}(s) = \frac{a_{kj}(s)}{G_{kk}(s)} \tag{10.31}$$

For the MIMO system of the $n = 3$ order are needed $n = 3$ main controllers $R_1(s)$, $R_2(s)$, $R_3(s)$, and

$$n \cdot (n-1) = 3 \cdot 2 = 6 \tag{10.32}$$

decoupling controllers, i.e., a total of $n^2 = 3^2 = 9$ controllers.

MIMO system of the 4th order

Only a part of the MIMO system with FF-plant of the order $n = 3$ controlled with the FB-controller with the same order is shown in Fig. 10.12. Using Eq. 10.31, we can define all

$$n \cdot (n-1) = 4 \cdot 3 = 12 \tag{10.33}$$

decoupling transfer functions, i.e., together with the $n = 3$ main controllers, there are needed $n^{(2} = 4^2 = 16$ controllers.

Example: Decoupled MIMO control of the FF-plant with FB-controller of the 4th order

In Fig. 10.13 are given as matrices the FF-plant $\mathbf{G_S}(s)$ and the FF-controller $\mathbf{R}(s)$. All plants are of the type PT1 with $k = 1, 2, 3, 4$:

$$G_{PSk}(s) = \frac{K_{PSk}}{1 + sT_k}$$

The main controllers are configurated as I-controllers with $k = 1, 2, 3, 4$

$$R_k(s) = \frac{K_{IRk}}{s}$$

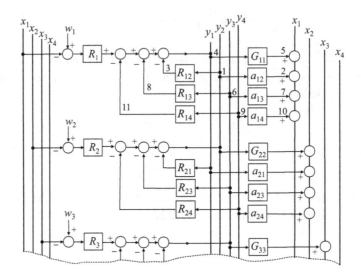

Fig. 10.12 Decoupled MIMO control of the 4th order. *Source* [1, p. 88]

$$R(s) = \begin{pmatrix} \dfrac{0{,}125}{s} & \dfrac{2{,}5(2s+1)}{4s+1} & \dfrac{1{,}5(2s+1)}{6s+1} & \dfrac{0{,}5(2s+1)}{s+1} \\[3mm] \dfrac{0{,}5(3s+1)}{5s+1} & \dfrac{0{,}042}{s} & \dfrac{1{,}75(3s+1)}{8s+1} & \dfrac{1{,}25(3s+1)}{2s+1} \\[3mm] \dfrac{4{,}5(s+1)}{2s+1} & \dfrac{2{,}5(s+1)}{5s+1} & \dfrac{0{,}5}{s} & \dfrac{2(s+1)}{6s+1} \\[3mm] \dfrac{0{,}6(4s+1)}{7s+1} & \dfrac{0{,}2(4s+1)}{3s+1} & \dfrac{0{,}8(4s+1)}{s+1} & \dfrac{0{,}025}{s} \end{pmatrix} \qquad G_S(s) = \begin{pmatrix} \dfrac{2}{2s+1} & \dfrac{5}{4s+1} & \dfrac{3}{6s+1} & \dfrac{1}{s+1} \\[3mm] \dfrac{2}{5s+1} & \dfrac{4}{3s+1} & \dfrac{7}{8s+1} & \dfrac{5}{2s+1} \\[3mm] \dfrac{4{,}5}{2s+1} & \dfrac{2{,}5}{5s+1} & \dfrac{1}{s+1} & \dfrac{2}{6s+1} \\[3mm] \dfrac{3}{7s+1} & \dfrac{1}{3s+1} & \dfrac{4}{s+1} & \dfrac{5}{4s+1} \end{pmatrix}$$

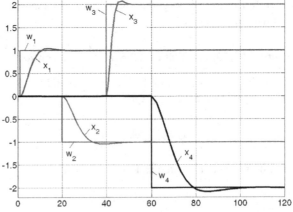

Fig. 10.13 Example of decoupled MIMO control of the $n=4$ order. *Source* [1, p. 89]

according to the optimum magnitude method, described in [2, 3]:

$$K_{IRk} = \frac{1}{2 \cdot K_{PSk} \cdot T_k}$$

The resulting parameters are given below

$$K_{IR1} = \frac{1}{2 \cdot 2 \cdot 1 \text{ sec}} = 0,125 \text{ sec}^{-1}$$

$$K_{IR2} = \frac{1}{2 \cdot 4 \cdot 3 \text{ sec}} = 0,042 \text{ sec}^{-1}$$

$$K_{IR3} = \frac{1}{2 \cdot 1 \cdot 1 \text{ sec}} = 0,5 \text{ sec}^{-1}$$

$$K_{IR4} = \frac{1}{2 \cdot 5 \cdot 4 \text{ sec}} = 0,025 \text{ sec}^{-1}$$

The step responses and simulation with MATLAB® shown in Fig. 10.13 confirm the optimum control and the best decoupling.

Note that the separated control of the same coupled MIMO plant is unstable, because according to Sect. 10.2.2 some of the static interactions C_0 are greater than one:

$$\mathbf{C}_0 = \begin{pmatrix} 0 & 2,5 & 1,5 & 0,5 \\ 0,5 & 0 & 1,75 & 1,25 \\ 4,5 & 2,5 & 0 & 2 \\ 0,6 & 0,2 & 0,8 & 0 \end{pmatrix} \qquad (10.34)$$
◀

10.3.6 MATLAB® Simulation of MIMO Control with Bus-Approach

In Fig. 10.14 and 10.15 are shown simulated with MATLAB® feed-forward MIMO plants with $n=2$ variables, which are controlled with a feed-forward controller, and step responses for two cases—with and without decoupling. The advantages of decoupling are clearly seen.

In Fig. 10.16 is simulated the MIMO control of the feed-back plant with $n=4$ controlled variables. The step responses are shown in Fig. 10.16 only for decoupled control because by the separated control without decoupling, the MIMO system is unstable and not able to be controlled.

In Fig. 10.16 is seen also the advantage of the bus-approach by the simulation of high dimensionality systems. Even if the simulation of such MIMO control without a bus-approach is possible, it is very difficult.

With these simulations, we finish the treatment of classic MIMO control and pass over in the next section to the description of the data stream management (DSM) of such multivariable systems.

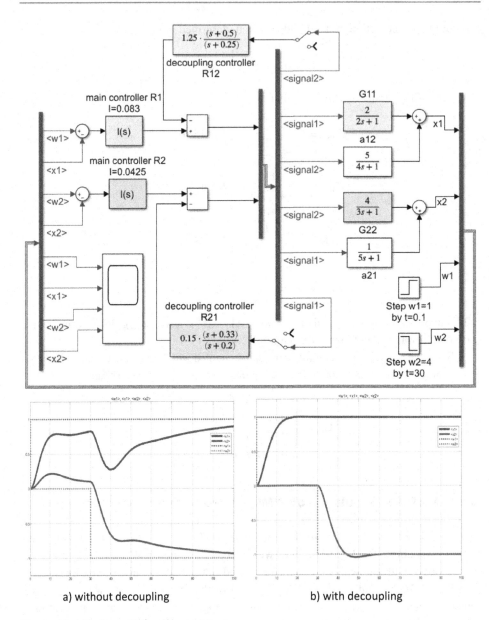

a) without decoupling   b) with decoupling

Fig. 10.14   FF-plant of 2nd order with FF-plant

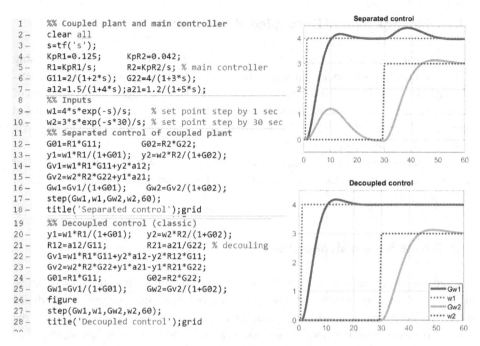

```
1     %% Coupled plant and main controller
2 -   clear all
3 -   s=tf('s');
4 -   KpR1=0.125;      KpR2=0.042;
5 -   R1=KpR1/s;       R2=KpR2/s; % main controller
6 -   G11=2/(1+2*s);   G22=4/(1+3*s);
7 -   a12=1.5/(1+4*s);a21=1.2/(1+5*s);
8     %% Inputs
9 -   w1=4*s*exp(-s)/s;     % set point step by 1 sec
10-   w2=3*s*exp(-s*30)/s; % set point step by 30 sec
11    %% Separated control of coupled plant
12-   G01=R1*G11;           G02=R2*G22;
13-   y1=w1*R1/(1+G01);  y2=w2*R2/(1+G02);
14-   Gv1=w1*R1*G11+y2*a12;
15-   Gv2=w2*R2*G22+y1*a21;
16-   Gw1=Gv1/(1+G01);      Gw2=Gv2/(1+G02);
17-   step(Gw1,w1,Gw2,w2,60);
18-   title('Separated control');grid
19    %% Decoupled control (classic)
20-   y1=w1*R1/(1+G01);  y2=w2*R2/(1+G02);
21-   R12=a12/G11;          R21=a21/G22; % decouling
22-   Gv1=w1*R1*G11+y2*a12-y2*R12*G11;
23-   Gv2=w2*R2*G22+y1*a21-y1*R21*G22;
24-   G01=R1*G11;           G02=R2*G22;
25-   Gw1=Gv1/(1+G01);      Gw2=Gv2/(1+G02);
26-   figure
27-   step(Gw1,w1,Gw2,w2,60);
28-   title('Decoupled control');grid
```

Fig. 10.15 MATLAB® script: bus-approach for FF-plant of 2nd order with FF-plant

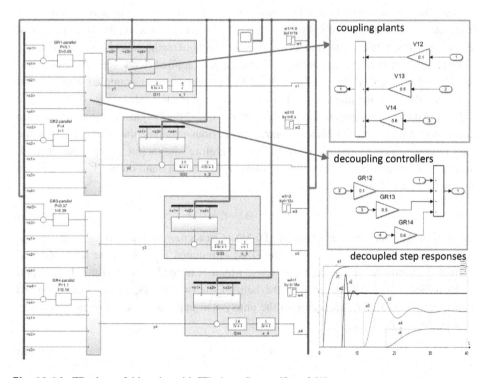

Fig. 10.16 FB-plant of 4th order with FF-plant. *Source* [2, p. 261]

10.4   Management of Decoupled MIMO Control

Bus-approach is a way to describe the MIMO systems from the new point of view using terms and definitions of industrial automation buses. In the previous sections, the bus-approach was used for the classic MIMO decoupling. The tracking of the signal ways of a bus made it possible to derive the stability conditions for separated control of coupled plants. The MIMO systems of the 3rd and higher orders are easy to simulate using MAT-LAB®, as shown in the previous section for $n = 7$.

Alone, these methods are a rather big contribution to the MIMO-control design. But except for these methods, the bus-approach opens new possibilities to build new kinds of MIMO systems.

10.4.1 Data Stream Management

The separation of channels of a MIMO closed loop control is based not on the classic method of decoupling, treated in previous sections, but on another conception introduced in [1] and described in [2]–[6]. A significant feature of this conception is that not the coupled plants of neighbor main control loops will be compensated by decoupling controllers, as is the case by classical MIMO control, but the data streams of own main controllers will be recognized and appropriately controlled.

This conception is based upon the antisystem approach, which is treated in Chap. 3 of this book and is mathematically illustrated in [2, 3, 7]. In the following, this conception will be explained upon data stream and its management without mathematical backgrounds.

To shortly explain how the antisystem approach is applied to the separating of channels of feed-forward MIMO plants, in Fig. 10.17 are shown as an example the main plant $G_{11}(s)$ and the coupling plant $a_{12}(s)$. Let us explain first the left side of Fig. 10.17.

Each plant has its separate input $y_1(s)$ and $y_2(s)$, and the outputs $x_{11}(s)$ and $x_{12}(s)$ are added. As known from the MIMO control of previous sections, the output $x_{11}(s)$ is the main output, which should not be disturbed. Therefore, the output $x_{12}(s)$ is not desired, but it acts as a disturbance via coupling plant $a_{12}(s)$ on the main output $x_{11}(s)$. The sum of both outputs is designated as $e_1(s)$:

$$e_1(s) = x_{11}(s) + x_{12}(s) \tag{10.35}$$

The aim of the Data Stream Manager (DSM) is to recognize $x_{12}(s)$ and to eliminate it from the sum $e_1(s)$. But instead of it, the DSM eliminated $x_{11}(s)$ from the sum $e_1(s)$ and calculated exactly the actuating value $y_1(s)$, which is needed to rebuild the desired and not disturbed $x_{11}(s)$. It happens with the help of the two blocks, shown shaded on the right side of Fig. 10.17. According to the antisystem approach, the following operation takes place:

$$x_{11} = e_2 - G_{12}y_{12} \tag{10.36}$$

Fig. 10.17 Elimination of the
variable y_1 from the sum e_1.
Source [2, p. 267]

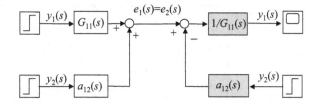

Under consideration

$$x_{11} = G_{11}y_1$$

the desired undisturbed actuating value will be calculated from Eq. 10.36:

$$y_1 = \frac{1}{G_{11}}(e_2 - G_{12}y_{12}) \tag{10.37}$$

On the scope of the right side of Fig. 10.17, this undisturbed actuating value $y_1(s)$ will be presented. Sending it further and applying it to the main plant $G_{11}(s)$, the desired undisturbed main controlling variable will be produced.

The block, which realized the data stream control according to Eq. 10.35, 10.36, 10.37, is called *DSM Router*.

10.4.2 DSM Router

"The MIMO-router is not a compensator of signal ways from one main loop to the other, but it is a distributor of two input signals:

– of the feedback x
– of the output of the own main controller.

The application of the MIMO-router simplifies the control system and reduces the number of control blocks. The router, once tuned for some main plant of the MIMO-system of low order, keeps its adjustment by identical plants of MIMO-system higher order. The MIMO-router promised to challenge the traditional developments of MIMO-control." (Citation, Source: [1, p. 152])

An excerpt from the functional block diagram of a MIMO closed loop with $n = 2$ controlling variable with the DSM Router consisting of transfer functions G_{11} and its inverse $1/G_{11}$ is shown in Fig. 10.18.

Let us first suppose that the actuating variable y_2 does not change, i.e., the lower control loop was already controlled and is in a steady state. The controlled variable x_2 has achieved its set point w_2 without error. The upper main controller R_1 is also in a steady state without error. The controlled variable $x_1 = w_2$. The upper main controller R_1 supplies a constant actuating variable y_1.

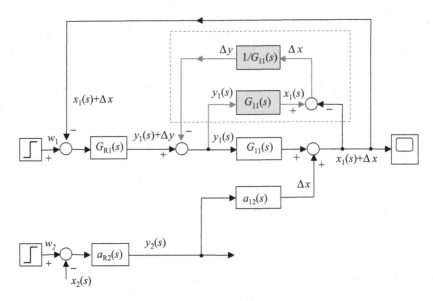

Fig. 10.18 DSM Router as connection of shaded block G_{11} and its inverse. *Source* [2, p. 267]

Further, let's suppose that the actuating variable y_2 changes. It happens if the set point w_2 of the lower loop changes. The controlled variable x_1 of the upper loops should be kept constant. But because of the coupling plant a_{12}, the controlled variable x_1 also changes and receives the value $(x_1 + \Delta x)$. The main controller R_1 reacts to this change and delivers a new actuating variable $(y_1 + \Delta y)$.

Figure 10.18 shows how the value Δy is recognized according to Eq. 10.35– 10.37 and subtracted from the "disturbed" actuating variable $(y_1 + \Delta y)$. The main controller R_1 will supply the same actuating variable y_1, so that the controlled variable x_1 remains undisturbed.

Example: DSM router for FF-plant of the 2nd order with FF-controller

For the MIMO control with $n=2$ controlled variables, shown in Fig. 10.19, are needed two main controllers and two DSM Routers. Each DSM Router has two inputs and one output. Significant is that each router takes inputs not from actuating signals of neighbor channels, but from the feedback x of the same main channel and from the actuating signal y of the main controller. Therefore, it is easy to tune the router, taking into account only the transfer function of the main plant, e.g., *Router 2* is adjusted corresponding only to the main plant G_{22}. The transfer functions of coupling plants a_{12} and a_{12} have no meaning for parameters of *Router 2*.

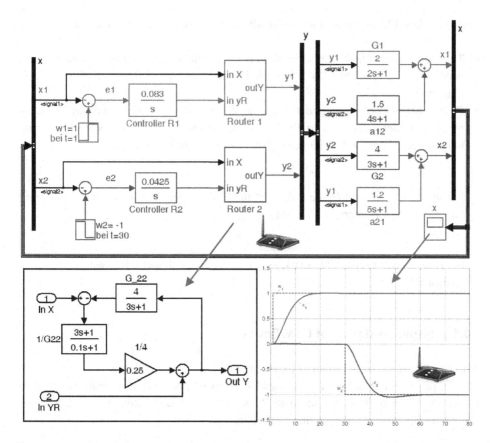

Fig. 10.19 DSM Router for MIMO control with $n = 2$ controlled variables. *Source* [1, p. 147]

In other words, the router is not a compensator of signal ways from one main loop to the other, as it was in the case of decoupling controllers in previous chapters, but is a distributor of two input signals:

- of the feedback x;
- of the output of the main controller yR.

As a result, the router predicts the input of its own main plant, e.g., of y_1 without receiving information about y_2 from another plant. ◀

The use of DSM Router simplifies the MIMO control and has many advantages:

- The router, once adjusted to some main plant, could be used by identical plants independent of the order of MIMO system, i.e., independent of the number of the coupling plants.
- For the MIMO decoupling control are needed n main controllers and $n(n$-$1)$ decoupling controller, according to Eq. 10.32 and 10.33. The total number N of control blocks, needed for the MIMO decoupling control of the order n, is

$$N = n + n \cdot (n - 1) = n^2$$

For the MIMO control of the order n with DSM Routers are needed n main controllers and n DSM Routers, i.e., equal to

$$N = 2n$$

In other words, the use of DSM Routers reduces significantly the total number N of control blocks.

10.4.3   Implementation of DSM Router

MIMO control of the 3rd order (Fig. 10.20)
The FF-plant with $n = 3$ controlled variables is perfectly controlled with the FF-controller consisting of 3 main controllers R_1, R_2, R_3, and 3 DSM Routers. For the classic decoupling are needed 9 decoupling controllers.

MIMO control of the 7th order (Fig. 10.21)
For the given MIMO feed-forward plant with $n = 7$ controlled variables are needed 7 DSM routers instead of 49 decoupling blocks. The step responses shown in Fig. 10.22 verify the full separation of channels with DSM Routers.

MIMO control of the 9th order (Fig. 10.23)
The convincing success of the use of the bus-approach and DSM Router, illustrated with MATLAB® script of Fig. 10.23, is achieved with the application of the coupled MIMO plant of 9th order.

In Figs. 10.24, 10.25, and 10.26 are portrayed parts of MATLAB® script of Fig. 10.23. The parameters of the main controllers are optimally calculated in the program part of Fig. 10.25 to deliver by all controlled variables the given settling time of $T_{\text{aus}} = 10$ s.

It is easy to understand the rules, and how scripts of Fig. 10.24–10.26 were programmed, and accordingly it is easy to extend these programs for more controlled variables, or is it possible to generalize these algorithms and fulfill these programs with "for"-loops and run half-automatic.

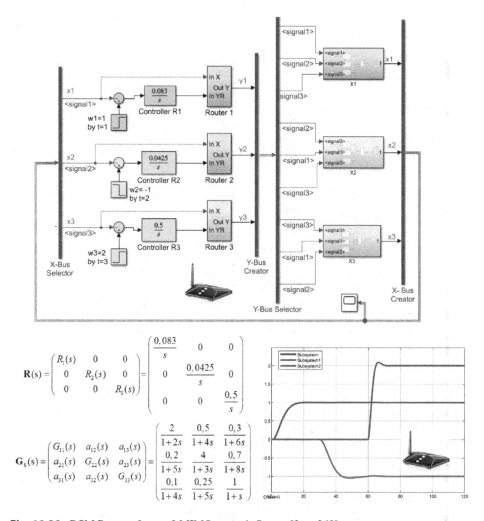

Fig. 10.20   DSM Routers for n = 3 MIMO control. *Source* [2, p. 268]

The separated control without DSM Router, shown in the upper diagram of Fig. 10.23, is inconceivable. To program of the classic control with 81 decoupled controllers is very expensive and practically impossible. Against it, the programming of only 9 DSM Routers of Fig. 10.26 causes no problems. The result is shown in the lower diagram of Fig. 10.23, so the use of DSM Router needs no comment.

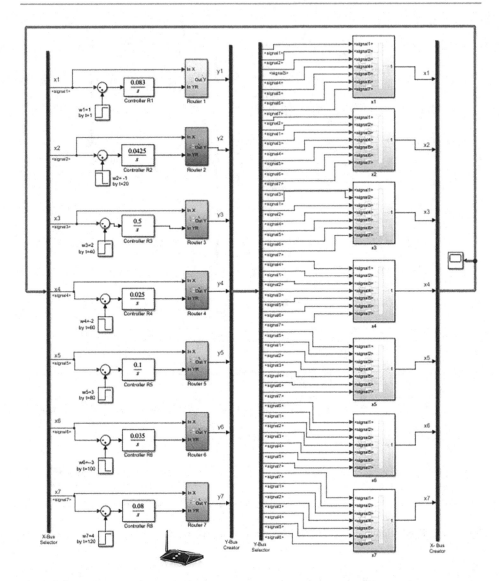

Fig. 10.21 MIMO control of the 7th order with 7 DSM Routers instead of 49 classical decoupling controllers

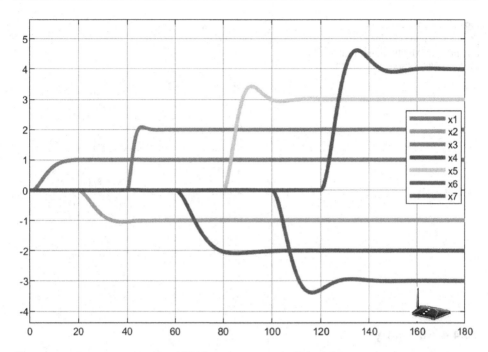

Fig. 10.22 Step responses of the MIMO control shown in Fig. 10.21

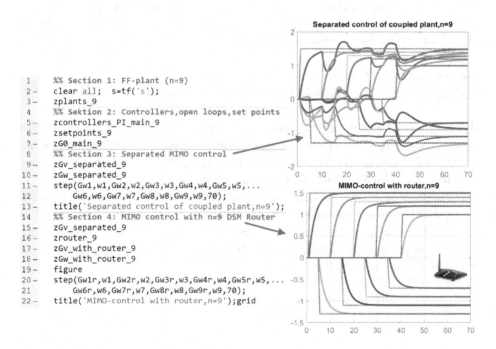

```
1    %% Section 1: FF-plant (n=9)
2    clear all;  s=tf('s');
3    zplants_9
4    %% Sektion 2: Controllers,open loops,set points
5    zcontrollers_PI_main_9
6    zsetpoints_9
7    zG0_main_9
8    %% Section 3: Separated MIMO control
9    zGv_separated_9
10   zGw_separated_9
11   step(Gw1,w1,Gw2,w2,Gw3,w3,Gw4,w4,Gw5,w5,...
12       Gw6,w6,Gw7,w7,Gw8,w8,Gw9,w9,70);
13   title('Separated control of coupled plant,n=9');
14   %% Section 4: MIMO control with n=9 DSM Router
15   zGv_separated_9
16   zrouter_9
17   zGv_with_router_9
18   zGw_with_router_9
19   figure
20   step(Gw1r,w1,Gw2r,w2,Gw3r,w3,Gw4r,w4,Gw5r,w5,...
21       Gw6r,w6,Gw7r,w7,Gw8r,w8,Gw9r,w9,70);
22   title('MIMO-control with router,n=9');grid
```

Fig. 10.23 MATLAB script of MIMO control with DSM routers of the 9th order

Fig. 10.24 Entering the parameters and transfer functions of MIMO plant into MATLAB® script of Fig. 10.23

```
        FF-plants
 1 –    s=tf('s');     % zplants_9
 2 –    Kps1=2;        Kps2=4;     Kps3=1.28;
 3 –    Kps4=2.25;     Kps5=1.9;  Kps6=2.3;
 4 –    Kps7=1.15;     Kps8=1.6;  Kps9=0.95;
 5 –    T1=2;          T2=3;       T3=16;
 6 –    T4=21.35;      T5=4.5;     T6=6.25;
 7 –    T7=6;          T8=5.8;     T9=4.3;
 8 –    G11=Kps1/(1+T1*s); G22=Kps2/(1+T2*s);
 9 –    G33=Kps3/(1+T3*s); G44=Kps4/(1+T4*s);
10 –    G55=Kps5/(1+T5*s); G66=Kps6/(1+T6*s);
11 –    G77=Kps7/(1+T7*s); G88=Kps8/(1+T8*s);
12 –    G99=Kps9/(1+T9*s);
13 –    a12=0.5/(1+4*s);      a13=-0.9/(1+8.6*s);
14 –    a14=-0.87/(1+8.7*s);  a15=-0.8/(1+8.9*s);
15 –    a16=-0.7/(1+9.3*s);   a17=-0.6/(1+9.9*s);
16 –    a18=-1.66/(1+5.9*s);  a19=-1.43/(1+7*s);
17 –    a21=0.4/(1+5*s);      a23=-0.11/(1+16*s);
18 –    a24=-0.9/(1+15*s);    a25=-0.08/(1+14.5*s`

43 –    a85=-0.35/(1+4.7*s);  a86=-0.9/(1+5.5*s);
44 –    a87=-0.87/(1+5*s);    a89=-1.64/(1+4.9*s);
45 –    a91=-3.1/(1+2.5*s);   a92=-0.24/(1+2.8*s);
46 –    a93=-0.5/(1+3.1*s);   a94=-0.31/(1+1.6*s);
47 –    a95=-0.73/(1+1.8*s);  a96=-2.72/(1+7.4*s);
48 –    a97=-1.78/(1+8.6*s);  a98=-1.95/(1+9.1*s);
```

10.4.4 Summary

The Data Stream Manager, named *Router* in [1], is a new unique development for MIMO control, resulting from bus-approach (Chap. 4) and antisystem approach (Chap. 5). The DSM Router is not a compensator of signal ways from one main loop to the other, but it is a distributor of two data streams:

- of the feedback x of the whole loop;
- of the actuating value y, produced by the "own" main controller.

Therefore, a DSM Router gets its inputs only from its "own" main loop and controlled its "own" actuating value without getting information from all other control loops. As a result, each closed loop is separated from all other loops with the help of one single DSM Router, i.e., for n main loops are needed n DSM Routers.

The application of DSM Routers reduces the number of control blocks. For comparison, by the classical MIMO control there are needed $n(n-1)$ decoupling controllers. It is convinced by the example, given in the previous section, that namely for $n = 9$ main control loops are needed only 9 DSM Routers, while for the classic MIMO control there are needed 81 decoupling controllers.

```
        FF-main-PI-controllers, tuned for controling time Taus
 1 -    s=tf('s');       Taus=10;  % zcontrollers_PI_main_9
 2 -    Tn1=T1; Tn2=T2; Tn3=T3; Tn4=T4; Tn5=T6; Tn6=T6;
 3 -    Tn7=T7; Tn8=T8; Tn9=T9;
 4 -    KpR1=3.9*Tn1/(Kps1*Taus); KpR2=3.9*Tn2/(Kps2*Taus);
 5 -    KpR3=3.9*Tn3/(Kps3*Taus); KpR4=3.9*Tn4/(Kps4*Taus);
 6 -    KpR5=3.9*Tn5/(Kps5*Taus); KpR6=3.9*Tn6/(Kps6*Taus);
 7 -    KpR7=3.9*Tn7/(Kps7*Taus); KpR8=3.9*Tn8/(Kps8*Taus);
 8 -    KpR9=3.9*Tn9/(Kps9*Taus);
 9
10 -    R1=KpR1*(1+s*Tn1)/(s*Tn1); R2=KpR2*(1+s*Tn2)/(s*Tn2);
11 -    R3=KpR3*(1+s*Tn3)/(s*Tn3); R4=KpR4*(1+s*Tn4)/(s*Tn4);
12 -    R5=KpR5*(1+s*Tn5)/(s*Tn5); R6=KpR6*(1+s*Tn6)/(s*Tn6);
13 -    R7=KpR7*(1+s*Tn7)/(s*Tn7); R8=KpR8*(1+s*Tn8)/(s*Tn8);
14 -    R9=KpR9*(1+s*Tn9)/(s*Tn9);
```

```
 1 -    G01=R1*G11; %zG0_main   Main open loops
 2 -    G02=R2*G22;
 3 -    G03=R3*G33;
 4 -    G04=R4*G44;
-----------------------------------------
10 -    y1=w1*R1/(1+G01);       Actuating values
11 -    y2=w2*R2/(1+G02);
12 -    y3=w3*R3/(1+G03);
-----------------------------------------
17 -    y8=w8*R8/(1+G08);
18 -    y9=w9*R9/(1+G09);
```

```
 1      % zrouter_9            DSM Routers
 2 -    Rou1=Gv1-w1*R1*G11;
 3 -    Rou2=Gv2-w2*R2*G22;
 4 -    Rou3=Gv3-w3*R3*G33;
 5 -    Rou4=Gv4-w4*R4*G44;
 6 -    Rou5=Gv5-w5*R5*G55;
 7 -    Rou6=Gv6-w6*R6*G66;
 8 -    Rou7=Gv7-w7*R7*G77;
 9 -    Rou8=Gv8-w8*R8*G88;
10 -    Rou9=Gv9-w9*R9*G99;
```

Fig. 10.25 Entering the transfer functions into MATLAB® script of Fig. 10.23 and calculating the parameters of main controller and DSM Routers

The use of DSM Router simplifies the MIMO and lets to easily extend the programs o 3rd or 4th order into high dimensionality applications of the DSM Router, as is illustrated in examples of the 7th and 9th orders. The DSM Router, once tuned for some main plant of the MIMO system of low order, keeps its adjustment by identical plants of MIMO system of higher order.

"The MIMO-router promised to challenge the traditional developments of MIMO-control." (Quote: Source: [1, p. 152])

```
       Forward data stream by separated control without routers
1      % zGv_separated_9
2 -    Gv1=w1*R1*G11+y2*a12+y3*a13+y4*a14+y5*a15+y6*a16+y7*a17+y8*a18+y9*a19;
3 -    Gv2=w2*R2*G22+y1*a21+y3*a23+y4*a24+y5*a25+y6*a26+y7*a27+y8*a28+y9*a29;
4 -    Gv3=w3*R3*G33+y1*a31+y2*a32+y4*a34+y5*a35+y6*a36+y7*a37+y8*a38+y9*a39;
5 -    Gv4=w4*R4*G44+y1*a41+y2*a42+y3*a43+y5*a45+y6*a46+y7*a47+y8*a48+y9*a49;
6 -    Gv5=w5*R5*G55+y1*a51+y2*a52+y3*a53+y4*a54+y6*a56+y7*a57+y8*a58+y9*a59;
7 -    Gv6=w6*R6*G66+y1*a61+y2*a62+y3*a63+y4*a64+y5*a65+y7*a67+y8*a68+y9*a69;
8 -    Gv7=w7*R7*G77+y1*a71+y2*a72+y3*a73+y4*a74+y5*a75+y6*a76+y8*a78+y9*a79;
9 -    Gv8=w8*R8*G88+y1*a81+y2*a82+y3*a83+y4*a84+y5*a85+y6*a86+y7*a87+y9*a89;
10 -   Gv9=w9*R9*G99+y1*a91+y2*a92+y3*a93+y4*a94+y5*a95+y6*a96+y7*a97+y8*a98;
```

```
       Forward data stream by control with DSM Router
1      % zGv_with_router_9
2 -    Gv1=w1*R1*G11-Rou1+y2*a12+y3*a13+y4*a14+y5*a15+y6*a16+y7*a17+y8*a18+y9*a19;
3 -    Gv2=w2*R2*G22-Rou2+y1*a21+y3*a23+y4*a24+y5*a25+y6*a26+y7*a27+y8*a28+y9*a29;
4 -    Gv3=w3*R3*G33-Rou3+y1*a31+y2*a32+y4*a34+y5*a35+y6*a36+y7*a37+y8*a38+y9*a39;
5 -    Gv4=w4*R4*G44-Rou4+y1*a41+y2*a42+y3*a43+y5*a45+y6*a46+y7*a47+y8*a48+y9*a49;
6 -    Gv5=w5*R5*G55-Rou5+y1*a51+y2*a52+y3*a53+y4*a54+y6*a56+y7*a57+y8*a58+y9*a59;
7 -    Gv6=w6*R6*G66-Rou6+y1*a61+y2*a62+y3*a63+y4*a64+y5*a65+y7*a67+y8*a68+y9*a69;
8 -    Gv7=w7*R7*G77-Rou7+y1*a71+y2*a72+y3*a73+y4*a74+y5*a75+y6*a76+y8*a78+y9*a79;
9 -    Gv8=w8*R8*G88-Rou8+y1*a81+y2*a82+y3*a83+y4*a84+y5*a85+y6*a86+y7*a87+y9*a89;
10 -   Gv9=w9*R9*G99-Rou9+y1*a91+y2*a92+y3*a93+y4*a94+y5*a95+y6*a96+y7*a97+y8*a98;
```

	Separated control		Control with Router		Set points
1	% zGW_separated_9	1	% zGW_with_router_9	1	% setpoints_9
2 -	GW1=Gv1/(1+G01);	2 -	GW1r=Gv1/(1+G01);	2 -	w1= 1.5*s*exp(-s*1)/s;
3 -	GW2=Gv2/(1+G02);	3 -	GW2r=Gv2/(1+G02);	3 -	w2=-1.3*s*exp(-s*5)/s;
4 -	GW3=Gv3/(1+G03);	4 -	GW3r=Gv3/(1+G03);	4 -	w3= 1.4*s*exp(-s*10)/s;
5 -	GW4=Gv4/(1+G04);	5 -	GW4r=Gv4/(1+G04);	5 -	w4=-1.1*s*exp(-s*15)/s;
6 -	GW5=Gv5/(1+G05);	6 -	GW5r=Gv5/(1+G05);	6 -	w5= 1.3*s*exp(-s*20)/s;
7 -	GW6=Gv6/(1+G06);	7 -	GW6r=Gv6/(1+G06);	7 -	w6=-0.9*s*exp(-s*25)/s;
8 -	GW7=Gv7/(1+G07);	8 -	GW7r=Gv7/(1+G07);	8 -	w7= 1.2*s*exp(-s*30)/s;
9 -	GW8=Gv8/(1+G08);	9 -	GW8r=Gv8/(1+G08);	9 -	w8=-0.7*s*exp(-s*35)/s;
10 -	GW9=Gv9/(1+G09);	10 -	GW9r=Gv9/(1+G09);	10 -	w9= 1.0*s*exp(-s*40)/s;

Fig. 10.26 Calculation of forward data streams Gv and closed loops Gw with and without DSM Routers, used in MATLAB® script of Fig. 10.23

References

1. Zacher, S. (2014). *Bus-approach for feedback MIMO-control*. Verlag Dr. S. Zacher.
2. Zacher, S. (2021). *Regelungstechnik mit Data Stream Management*. Verlag Springer Vieweg. https://link.springer.com/book/10.1007/978-3-658-30860-5. Accessed 20 Jan 2022
3. Zacher, S., & Reuter, M (2022). *Regelungstechnik für Ingenieure*. Verlag Springer Vieweg. https://link.springer.com/book/10.1007/978-3-658-36407-6. Accessed 10 Feb 2022
4. Zacher, S. (2019). Bus-Approach for Engineering and Design of Feedback Control. Denver, CO, USA: *Proceedings of ICONEST*, October 7–10, 2019, published by ISTES Publishing, pp. 26–27
5. Zacher, S. (2020). Bus-Approach for Engineering and Design of Feedback Control. *International Journal of Engineering, Science and Technology*, **1**(2), 16–24. https://www.ijonest.net/index.php/ijonest/article/view/9/pdf. Accessed Jan 2022

6. Zacher, S., Saeed, W. (2010). Design of Multivariable Control Systems using Antisystem-Approach. *AALE 2010*, 11th–.12th February. ISBN 9-3-902759-00-9
7. Zacher, S. (2003). *Duale Regelungstechnik*. VDE Verlag.

Model-Based Control and Management

11

"Where science comes to an end... art comes in."

(Quote: Richard Wagner. http://www.kunstzitate.de/bildendekunst/
kunstimblickpunkt/wagner_richard.htm *accessed April 24, 2022)*

A controller whose transfer function $G_R(s)$ contains the model of the controlled plant is known as a model-based controller. An exact model of the plant in the form of a transfer function $G_S(s)$ is required for the design and tuning of a model-based controller.

There are known many types of model-based controllers, among them are

- *Compensating control,* when the controller consists of the inverse transfer function of the plant $G_S(s)$ and the transfer function of the desired behavior of closed loop. One option of compensating controllers is called the *Smith predictor* and is developed for the control of plants with dead time.
- *Control with a reference model,* when the plant is adapting step by step to its desired behavior given by a pattern.
- *Predictive Functional Control* (PFC) adapting step by step the actuating value of the real controller to the predicted value of the simulated control loop.
- *Dead-Beat control,* when the control is realized with steps of a two-point controller bringing the controlled variable from the initial point to the given end point in the definitive settling time.
- *Methods of artificial intelligence* like Fuzzy- and Neuro control with control rules derived from human behavior.

This chapter describes the concept of a conventional compensating controller and shows its advantages over standard PID controllers. After they are discussed, the model-based controllers are based upon the antisystem approach (ASA), which was described in . Two kinds of ASA control are presented in this chapter:

- balance ASA control, and
- compensating ASA control.

Finally, the data stream management is developed according to the ASA concept.

11.1   Classic Compensating Control

11.1.1   "Invalid" Plant

Let us repeat the very well-known goal of control, namely to keep the controlled variable $x(t)$ constantly equal to the desired set point w (reference value). The error $e(t)$ of control is the difference between the set point w and the actual value $x(t)$. The desired control should be stable, as fast as possible, without or only with small overshoots and without error. An example of the closed loop control (CLC) with the PID controller is shown above in Fig. 1.1. In Fig. 11.1 below is given the step response with the settling time $T_{aus} = 17$ s without retained error $e(\infty) = 0$, with the maximum overshoot of 20% and with $N = 2$ half-waves, i.e., with the damping $\vartheta = 1/N = 0,5$. The quality parameters

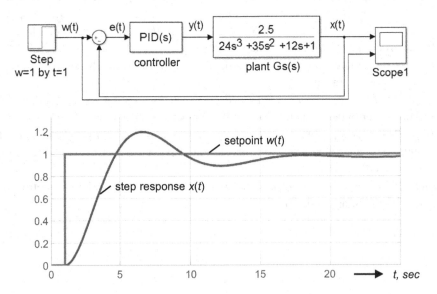

Fig. 11.1 CLC with standard PID controller and step response. *Source* [10, p. 207]

of control are not bad but let us suppose that we want to improve control by reducing settling time T_{aus} and increasing damping ϑ. It is impossible to do with the tuning of the PID controller because the quality of control depends on the dynamic parameters of the plant. The decreasing of the settling time will reduce damping and vice versa.

To achieve the desired step response, we eliminate the standard PID controller from the closed loop and put instead some microcontroller or CPU programmed as the inversed transfer function of the plant (Fig. 11.2):

$$G_R(s) = \frac{1}{G_S(s)}$$

As a result, the transfer function of the plant is compensated:

$$G_0(s) = G_{\text{Srez}}(s)G_S(s) = \frac{1}{G_S(s)}G_S(s) = 1$$

In other words, the plant is invalid. The time delay disappeared; the dynamics of the closed loop is equivalent to the P behavior as is shown by the step response of Fig. 11.2.

Unfortunately, the very fast control of the invalid plant has no use because the error $e(s)$ achieved 50% of the set point which is of course not allowed:

$$G_w(s) = \frac{x(s)}{w(s)} = \frac{G_R(s)G_S(s)}{1 + G_R(s)G_S(s)} = \frac{1}{1+1} = 0,5$$

$$x(s) = G_w(s)\hat{w} = 0,5 \cdot \hat{w} \rightarrow e(s) = \hat{w} - x(s) = 0,5 \cdot \hat{w}$$

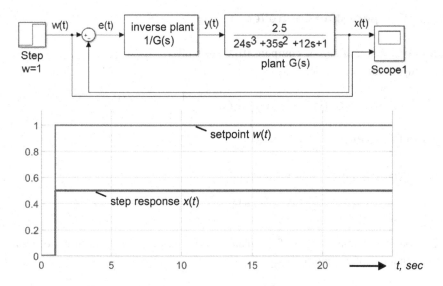

Fig. 11.2 CLC with "invalid" plant. *Source* [10, p. 208]

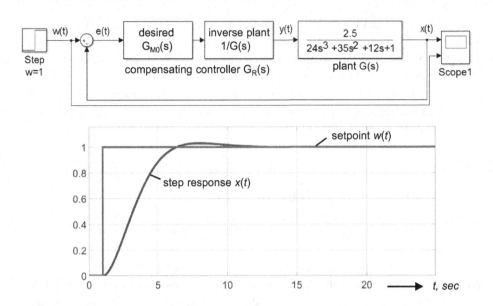

Fig. 11.3 Compensating controller consisting of the "invalid" plant and desired transfer function of the open loop $G_M(s)$. *Source* [10, p. 209]

The solution to the problem of how to achieve the desired step response using an "invalid" plant is shown in Fig. 11.3 and is called a model-based compensating controller.

11.1.2 Compensating Controller

Transfer function of classic compensating controller

The compensating controller

$$G_R(s) = G_{Srec}(s)G_{M0}(s) = \frac{1}{G_S(s)} \cdot \frac{G_M(s)}{1 - G_M(s)} \tag{11.1}$$

consists of two parts:

- the inverse (reciprocal) transfer function

$$G_{Srec}(s) = \frac{1}{G_S(s)}$$

for complete compensation of the plant $G_S(s)$, i.e., to make it "invalid" to influence the dynamics of the control loop.

- the desired transfer function $G_M(s)$ of the closed loop control $G_W(s)$, which is to be previously defined by the user. $G_{M0}(s)$ denotes the transfer function of the desired open loop behavior, e.g., if the desired behavior of the closed loop is PT1

$$G_M(s) = \frac{1}{1 + sT_M}, \tag{11.2}$$

then the transfer function of the desired open loop is

$$G_{M0}(s) = \frac{G_M(s)}{1 - G_M(s)} = \frac{1}{sT_M} \tag{11.3}$$

Comparison standard PID and compensating controller

The PID controller tuned according to the optimum magnitude and compensation controller were simulated in [10] with MATLAB ®/Simulink by controlling the same plant, and their step responses were compared to each other (Fig. 11.4) The actuating value was limited by saturation $y_{min} < y < y_{max}$. The results of the comparison are summarized below

- Fig. 11.4a: The CLC with the compensation controller has a shorter settling time T_{aus} than with the PID controller, but only by reference behavior. The settling time by disturbance behavior is the same for both controllers because the input signal is no longer a step.
- Fig. 11.4b: The PID controller produces a very strong pulse of actuating value y arriving at $y = 400$ that exceeds the maximum limit $y_{max} = 100$
- Fig. 11.4c: As a result of the limitation, the linearity of the control loop is lost, and the behavior of the non-linear closed loop depends on the size of the set point. The behavior of the CLC with compensating controller after the step of set point $w = 1$ differs from other w values, e.g., of w = 3. The compensating controller has no significant benefits against the PID controller.

Disadvantages of compensating controllers

The industrial plants are typically P-terms with time delay. They are usually stable and phase-minimal, so that no unstable poles/zeros occur even after inversion $1/G_S(s)$. But since the numerator polynomial of such plants has a lower order than the denominator polynomial, the inverse transfer function is not realizable. As an example, a PT3 plant and its inversion are shown below

$$G_S(s) = \frac{K_{PS}}{(1 + sT_1)(1 + sT_2)(1 + sT_3)}$$

$$\frac{1}{G_S(s)} = \frac{(1 + sT_1)(1 + sT_2)(1 + sT_3)}{K_{PS}}$$

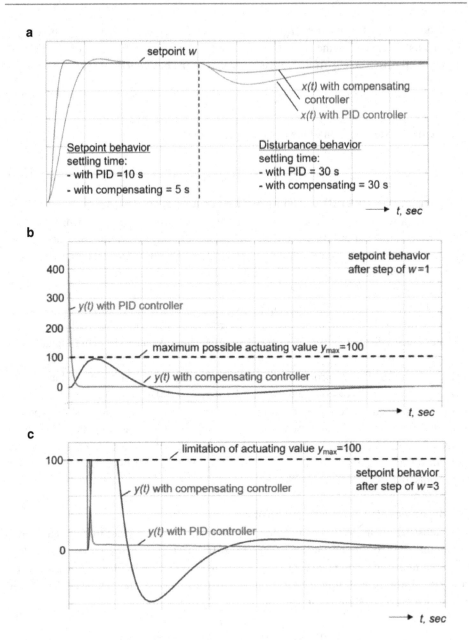

Fig. 11.4 Comparison of CLC with PID controller and compensating controller: **a** step responses of controlled variable $x(t)$ by setpoint behavior and disturbance behavior; **b** step responses of actuating values $y(t)$ by setpoint behavior without actuator limitation; **c** the same as (b) with actuator limitation

To make the inverse transfer functions realizable, the denominator polynomial should be supplemented with small time constants T_{R1}, T_{R2}, and T_{R3}, which usually amount to approximately 1% of the original time constants T_1, T_2, and T_3 of the plant:

$$\frac{1}{G_S(s)} = \frac{(1 + sT_1)(1 + sT_2)(1 + sT_3)}{K_{PS}(1 + sT_{R1})(1 + sT_{R2})(1 + sT_{R3})}$$

Because of these additionally settled time constants T_{R1}, T_{R2}, and T_{R3}, the plant is not more "invalid".

The expected desired behavior $G_M(s)$ is realized with errors, if at all.

Another serious disadvantage of classic conventional controllers with reciprocal transfer functions are the large D-terms due to time delays T_1, T_2, and T_3 which causes disturbances by control.

11.2 ASA Compensating Control

To avoid the above-mentioned disadvantages of the classic compensating controller, another method to get the "invalid" plant is proposed in [9]–[11]. This method is based upon the antisystem approach ([1]–[8]) described in .

11.2.1 Antisystem Approach (ASA)

According to ASA, for any dynamic system an antisystem can be formed in such a way that a balance between the two systems applies to any antisystem input. For every dynamic system, which transfers its inputs into outputs with an operator A in one direction, there is an equal system with the same operator A, which transfers other inputs into outputs in opposite direction.

As an example, a simple system consisting of one dynamic block with transfer function $G_S(s)$ is shown in Fig. 11.5. The system has one input $y(s)$, which is usually given, and one output $x(s)$. The antisystem is the transfer function $G_A(s)$ with input w_x, which could be arbitrarily chosen, and output w_y. If $G_S(s) = G_A(s)$ then there is a balance between inputs and outputs of the system and antisystem as was shown in Chap. 5:

$$y(s)w_y(s) = x(s)w_x(s)$$

The variables

$$e_x = x(s)w_x(s)$$
$$e_y = y(s)w_y(s)$$

are called "energy" or "intensity".

In other words, between the system and the antisystem arises the balance of "energies" by every arbitrary input function $w_y(s)$:

$$e_x = e_y \qquad (11.4)$$

The "energies" of the system and antisystem compensate each other:

$$e_x - e_y = 0$$

It corresponds with the famous third physical Newton's law, "for every action there is an equal and opposite reaction".

Regarding control loops, there are two options for implementation of ASA:

- balance control, based on the balance of variables $e_y = e_x$ as shown in Fig. 11.6, developed in [6, 16] and described in [9, 11].
- compensating control ([6, 9]–[15, 17]), which is further described in detail.

Balance control (Fig. 11.6)

The transfer function $G_M(s)$ of the desired behavior in Fig. 11.16 is considered as an antisystem. It is the mathematical model or hardware model of the plant $G_S(s)$ that is to be controlled. Or, if it is possible, the antisystem is implemented as the second physical plant identical to the "system". As far as both transfer functions are equal

$$G_S(s) = G_M(s)$$

then there is a balance between "energies" $e_y(s)$ and $e_x(s)$

If the plant $G_S(s)$ is disturbed, then the balance of Eq. 11.4 is also disturbed. The difference between "energy" of a system and antisystem

$$e(t) = e_y(t) - e_x(t) \qquad (11.5)$$

will be controlled through the actuating value $y(t)$ and eliminated bringing the "energies" $e_y(s)$ and $e_x(s)$ again to balance Eq. 11.4.

An example of the closed loop control based on the conception of Fig. 11.6 eliminating the error of Eq. 11.5 was developed in [6] and is given in Fig. 11.7.

In this case, the "system" is not the plant $G_S(s)$ but the closed loop built by the plant and a standard controller $G_R(s)$. The antisystem is the desired behavior $G_M(s)$ of the closed loop. The error $e(t)$ arises if the actual behavior of the closed loop differs from its desired behavior. Significant is that according to ASA conception of Fig. 11.6, the value of input w_x has no influence on the balance of Eq. 11.4 and error of Eq. 11.5.

Another implementation of ASA conception was developed in [16] and applied in [18] for the control of two identical motors, one of which was a plant, and the other was

Fig. 11.5 Example of simple system and antisystem. *Source* [9, p. 57]

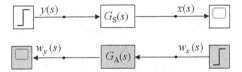

Fig. 11.6 Conception of balance control

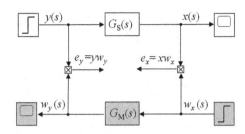

Fig. 11.7 Balance control with system $G_R(s)G_S(s)$ and antisystem $G_M(s)$

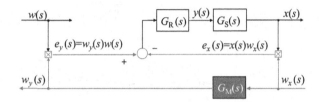

used as a shadow plant. Figure 11.8 illustrates this application; more details are given in [16] and [18]. The testing of Fig. 11.7 and 11.8 confirms the quality of the balance control. However, more research for a detailed study of this new conception is needed.

Multivariable ASA balance control

The advantage of the balance control is especially large in multivariable MIMO systems (Multi Input Multi Output), because the "energies" $e_y(s)$ and $e_x(s)$ are scalar values, while set point w and process value x are vectors.

Many industrial processes can be divided into subprocesses, and variables can be grouped into several sets corresponding to each subsystem, building a MIMO plant. This distribution corresponds to engineered subsystems and in many cases, it is just a conceptual framework for control design. The MIMO control for such processes is very well studied; the design methods are developed and described, for example by [19]–[21], but the practical use is complicated because of the dimension of the system. Most of the known industrial applications do the control of separate subsystems ignoring interaction among subsystems inside the MIMO plant. The commonly used methods for MIMO closed loop control are state space feedback and observer design.

Based upon the pole placing concept, these methods are universal and efficient, but the plant dimension more as $n = 3$ makes difficult the description of loops. The decoupling of MIMO subsystems brings the best results, but the realization is complicated because of derivative D-terms by decoupling. The industrial MIMO plants consist usually of proportional blocks with delay or with dead time. The reciprocal of such blocks, which are needed for decoupling, are not realizable. For this reason, the decoupling blocks will be simplified or completed with additional time delays, which reduces the quality of control.

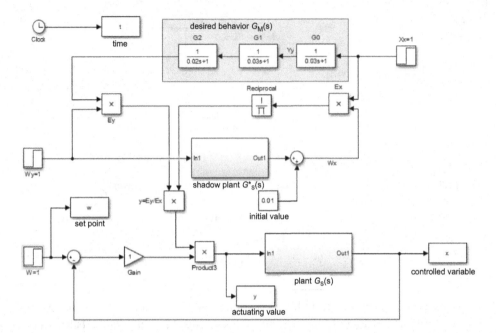

Fig. 11.8 Balance control of two identical motors, one of which is a plant, and the other is a shadow plant. *Source* [16, p. 9]

The application of balance control for MIMO control free of some problems mentioned above is introduced in [6] and briefly explained in the following. As an example, the MIMO plant in P-canonical form is given in Fig. 11.9 above. The "system" consists of "subsystems" with transfer functions $G_{11}(s)$, $G_{12}(s)$, $G_{21}(s)$, and $G_{22}(s)$

$$G_{11}(s) = \frac{x_1(s)}{y_1(s)} = \frac{K_{PS11}}{1 + sT_{11}} \quad G_{12}(s) = \frac{x_1(s)}{y_2(s)} = \frac{K_{PS12}}{1 + sT_{12}} \tag{11.6}$$

$$G_{21}(s) = \frac{x_2(s)}{y_1(s)} = \frac{K_{P21}}{1 + sT_{21}} \quad G_{22}(s) = \frac{x_2(s)}{y_2(s)} = \frac{K_{P22}}{1 + sT_{22}} \tag{11.7}$$

The "antisystem" is shown in Fig. 11.9 below. It consists of the same subsystems in V-canonical form, i.e., with feedback from inputs W_{x1}, W_{x2} to the outputs W_{y1}, W_{y2}. According to ASA and analog in Eq. 11.4, the following balance between "energies" $e_y(s)$ and $e_x(s)$ of the system and antisystem takes place for all arbitrarily chosen inputs W_{x1}, W_{x2}, except $W_{x1} = W_{x2} = 0$:

$$e_y(s) = e_x(s) \tag{11.8}$$

$$e_x(s) = W_{x1}(s)x_1(s) + W_{x2}(s)x_2(s)$$
$$e_y(s) = W_{y1}(s)y_1(s) + W_{y2}(s)y_2(s)$$

Fig. 11.9 MIMO plant as "system" (above) and shadow plant as "antisystem"

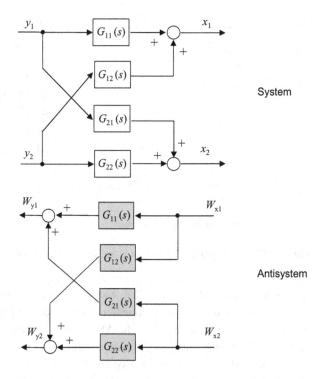

The step responses of "energies" e_x and e_y by the arbitrarily chosen input steps y_1, y_2 and W_{x1}, W_{x2} are shown in Fig. 11.10. The difference between e_x and e_y is less than 10^{-14} and is practically equal to zero.

In Fig. 11.11 is shown how the ASA is applied by the MIMO system with two controlled variables.

The balance control for the real MIMO plant is applied in [6], the step responses are seen in Fig. 11.11 for the I controller on the left and for the PI controller on the right. The control quality by balance control with PI controller is comparable with classical decoupling control.

Compensating ASA control

The term ASA refers to a loop made up of two blocks (real controlled system and its model, called the shadow system). The transfer function of the parallel connection

$$1 + G_S(s)$$

applies as a "system". The series connected feedback with the transfer function

$$\frac{1}{1 + G_S(s)}$$

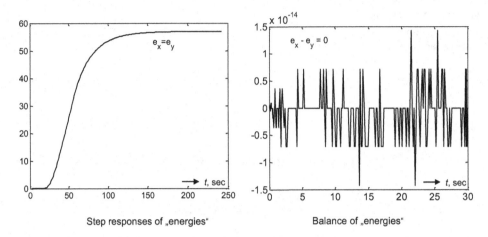

Fig. 11.10 MIMO plant and shadow plant: step responses of "energies" ex and ey of balance of "energies"

is viewed as an "antisystem". The system and the antisystem compensate each other without forming the reciprocal transfer function of the plant as was in the case of the classic compensating controller. The ASA controllers are therefore free from the disadvantages of conventional compensation control.

The "invalid" plant formed as a system of parallel and feedback connections is shown in Fig. 11.12:

$$G_0(s) = [1 + G_S(s)] \cdot \frac{1}{1 + G_S(s)} = 1 \tag{11.9}$$

The full compensation of the plant is achieved according to Eq. 11.4 and has the following advantages against the classic compensating controller:

- no reciprocal transfer function of the plant,
- the compensation is realizable,
- no additional delay time constants T_{R1}, T_{R2}, ... are needed, as a result the expected desired behavior $G_M(s)$ is realized without error,
- no D-term are generated, so the actuating value is well below the maximum limit.

In the next section is described the model-based controller, based on Eq. 11.4, and called the ASA controller. It is shown how to use the advantages of ASA control including the possibility of designing new options for model-based controllers (prefilter, compensator) as well as creating new control structures such as turbo control, control with bypass, and ASA predictor. The last option is a modified Smith Predictor for the model-based control of the plants with dead time.

However, the connection of Fig. 11.12 cannot be used directly in a compensating controller because the output of the loop is not the controlled variable $x(t)$ as usual, but the sum $x(t) + y(t)$.

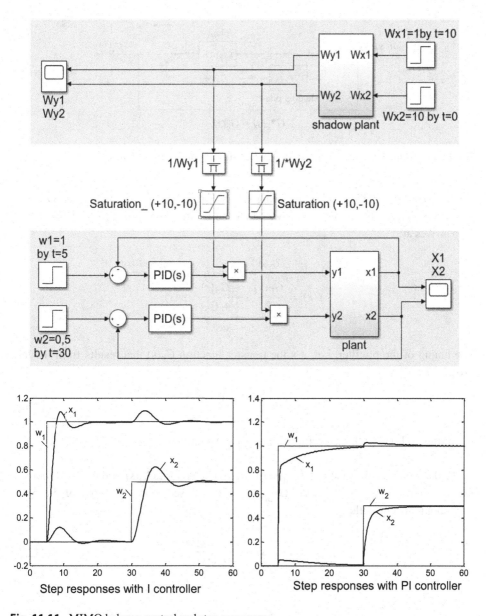

Fig. 11.11 MIMO balance control and step responses

11.2.2 ASA Prefilter

The CLC with ASA prefilter results from Fig. 11.12 of the "invalid" plant. The aid of the ASA prefilter is to avoid the immense large error of about 50% which is shown in Fig. 11.2. The solution is given in Fig. 11.13.

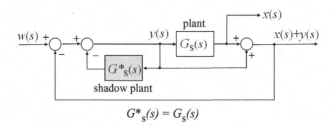

$$G^*_S(s) = G_S(s)$$

Fig. 11.12 Plant and shadow plant as system and antisystem. *Source* [12, p. 4]

Going out from Eq. 11.9 and considering $G^*_S(s) = G_S(s)$, the transfer function $G_F(s)$ of prefilter is defined below in such a way that the desired behavior $G_M(s)$ of the closed loop arrives at:

$$x(s) = G_F(s) \cdot \frac{\frac{1}{1+G_S(s)} \cdot G_S(s)}{1+G_0(s)} w(s) = G_F(s) \cdot \frac{\frac{1}{1+G_S(s)} \cdot G_S(s)}{1+1} w(s)$$

$$x(s) = \underbrace{\frac{G_F(s)}{2} \cdot \frac{G_S(s)}{1+G_S(s)} w(s)}_{G_M(s)}$$

The tuning of the prefilter, i.e., it's the transfer function $G_F(s)$ that results from the last expression:

$$G_F(s) = 2G_M(s) \frac{1+G_S(s)}{G_S(s)} \tag{11.10}$$

According to Eq. 11.10, the prefilter consists of two terms:

- the doubled transfer function $2G_M(s)$ of desired behavior $2G_M(s)$ of the closed loop;
- the reciprocal transfer function $G_{wSrec}(s)$ of the closed loop consisting of the plant without any controller:

$$G_{wSrec}(s) = \frac{1+G_S(s)}{G_S(s)} \tag{11.11}$$

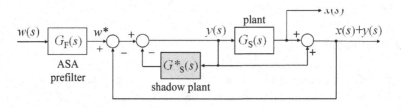

Fig. 11.13 ASA prefilter. *Source* [12, p. 6]

Finally, considering Eq. 11.10 and 11.11 the transfer function $G_F(s)$ of ASA prefilter is

$$G_F(s) = 2G_M(s)G_{wSrec}(s) \qquad (11.12)$$

The functional block diagram of the prefilter is shown in Fig. 11.14. It looks a bit complicated but has no reciprocal transfer functions as is the case of a classic compensating controller. Therefore, the ASA prefilter is free of D-term and can be practically implemented.

The disadvantage of the ASA prefilter is that the disturbance cannot be eliminated because the prefilter is outside the control loop and gets no information about a controlled variable. The shadow plant, which is included in the control loop, applies only to the compensation of the plant, and cannot prevent a disturbance.

Example: ASA prefilter

For comparison, the PT3 plant

$$G_S(s) = \frac{2}{1+s} \cdot \frac{1}{1+8s} \cdot \frac{1}{1+3s}$$

is simulated with the ASA prefilter for desired behavior with $T_M = 0{,}5$ s

$$G_M(s) = \frac{1}{(1+sT_M)^3}$$

and with a PID standard controller, which is optimally tuned according to the optimum magnitude. It is supposed that the actuating value must not exceed the $y_{max} = \pm 100$.

The simulation model is given in Fig. 11.15. The step response of ASA control is a little bit better than of the PID controller (Fig. 11.16). But the main advantage of the ASA prefilter is that it needs the maximum actuating value of $y = 30$ while the actuating value of the PID controller arrived at $y = 400$ and exceeds the allowed maximum $y_{max} = \pm 100$. ◄

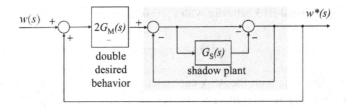

Fig. 11.14 Functional block diagram of prefilter (closed loop with the plant is not shown)

11.2.3 ASA Controller

The ASA controller $G_K(s)$ given in Fig. 11.17 is placed inside a closed loop, unlike the ASA prefilter, which is located outside the loop (Fig. 11.13). That's why the ASA controller is free of the disadvantage of the ASA prefilter mentioned above regarding the disturbance. Namely, the ASA controller eliminates the retained error $e(\infty)$.

The transfer function $G_K(s)$ of the ASA controller is defined below going out from Eq. 11.9 and considering $G^*_S(s) = G_S(s)$. Analog ASA prefilter of the ASA controller is tuned to arrive at the desired behavior $G_M(s)$ of the closed loop:

$$x(s) = \frac{G_K(s) \cdot \frac{1}{1+G_S(s)} \cdot G_S(s)}{1+G_0(s)} w(s) = G_K(s) \cdot \frac{\frac{1}{1+G_S(s)} \cdot G_S(s)}{1+G_K(s)} w(s)$$

$$x(s) = \underbrace{\frac{G_K(s)}{1+G_K(s)} \cdot \frac{G_S(s)}{1+G_S(s)} w(s)}_{G_M(s)}$$

From the last expression results the transfer function $G_K(s)$:

$$G_K(s) = \frac{G_M(s) \frac{1+G_S(s)}{G_S(s)}}{1 - G_M(s) \frac{1+G_S(s)}{G_S(s)}}$$

Finally, using Eq. 11.11 of reciprocal transfer function $G_{Srec}(s)$ the transfer function $G_K(s)$ is simplified to

$$G_K(s) = \frac{G_M(s) G_{wSrec}(s)}{1 - G_M(s) G_{wSrec}(s)} \tag{11.13}$$

The functional block diagram of the ASA controller, shown in Fig. 11.18, is like Fig. 11.14 but differs from it as far as regarding input and output as of desired behavior: $2G_M(s)$ in Fig. 11.14 and only one $G_M(s)$ in Fig. 11.18.

Example: ASA controller

The same PT3 plant, which was used in the previous example with the ASA prefilter,

$$G_S(s) = \frac{2}{1+s} \cdot \frac{1}{1+8s} \cdot \frac{1}{1+3s}$$

with the same desired PT3 behavior with $T_M = 0{,}5$ s

$$G_M(s) = \frac{1}{(1+sT_M)^3}$$

is simulated this time with the ASA controller and compared with the PID standard controller optimally set according to the optimum magnitude (Fig. 11.19). The corresponding step responses are given in Fig. 11.20. The step responses of ASA con-

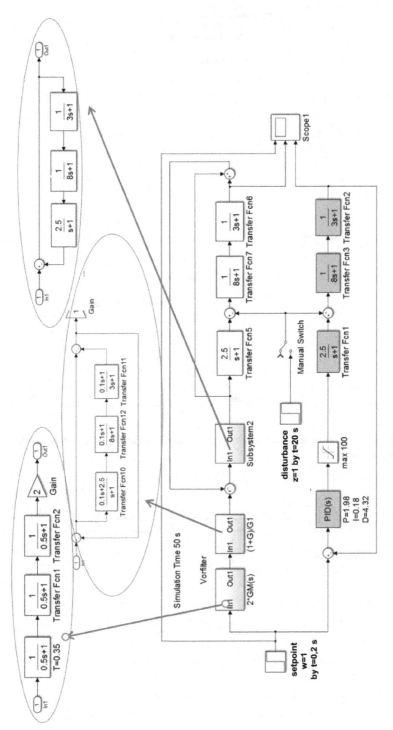

Fig. 11.15 ASA prefilter compared with PID controller. *Source* [12, p. 9]

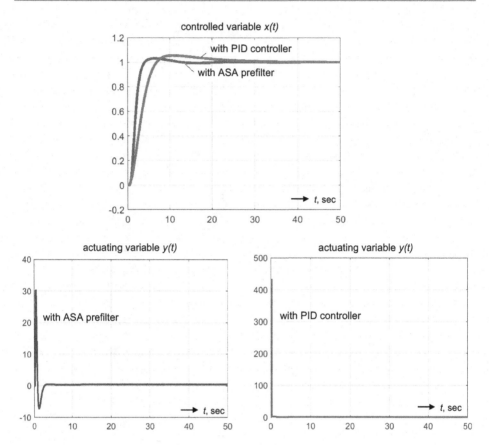

Fig. 11.16 Step responses and actuating values confirm the advantage of the ASA prefilter by set-point behavior. *Source* [12, p. 10]

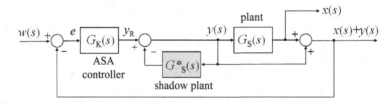

Fig. 11.17 ASA controller $G_K(s)$

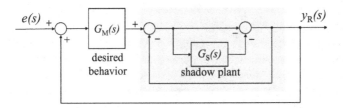

Fig. 11.18 Functional block diagram of ASA controller

trol regarding settling time and overshoot are superior both in setpoint behavior and in disturbance behavior compared with the PID control. The maximum actuating value of the ASA controller is $y = 30$ and does not exceed the allowed maximum $y_{max} = \pm 100$. ◄

11.2.4 Shadow Plant

The success of ASA control primarily depends on how precisely the real controlled plant $G_S(s)$ is identified and used as a shadow plant $G^*_S(s)$ in the CLC.

Three options for the shadow plant implementation by control are possible:

a) as software model $G^*_S(s)$ of the plant $G_S(s)$, inserted in the algorithm of ASA controller $G_R(s)$ (Fig. 11.21, above), i.e., $G^*_S(s) = G_S(s)$. This option is commonly used by all model-based controllers.

b) as hardware model, when the software model of the plant $G_S(s)$ is implemented as a program, e.g., as the C code, with hardware like a microprocessor (Fig. 11.21, middle). Usually, this option is not used for control, but it is typical for hardware-in-the-loop (HWL) simulation of a real controller with the software mode of the physical plant and by rapid control prototyping (RCP), when the software model of the newly developed controller is tested upon the real physical plant.

c) the shadow plant $G^*_S(s)$ is applied in the real loop as the second real physical plant $G_S(s)$ (Fig. 11.21, below). The advantages of the last solution, namely the double-plant implementation, are:

- If both identical plants are non-linear and both are in the same working point, both systems behave identically, which is hardly expected from a software model of the plant.
- If both identical plants act under the same conditions, both are affected identically by any disturbances and consequently are identically controlled according to the ASA concept.

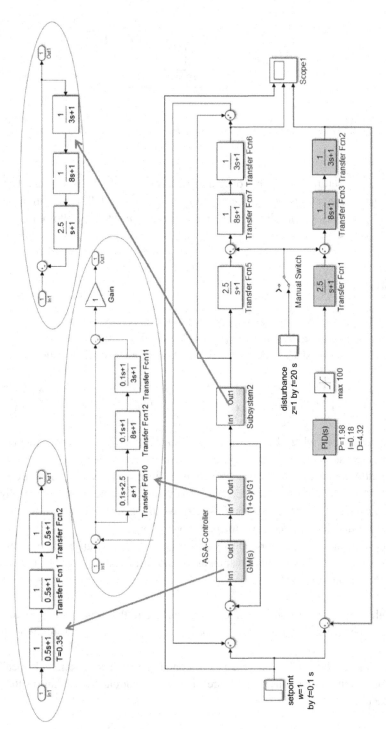

Fig. 11.19 ASA controller compared with PID controller. *Source* [12, p. 16]

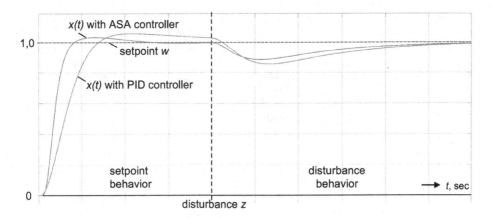

Fig. 11.20 Step responses of closed loop control with ASA and PID controller. *Source* [12, p. 17]

However, for technological and economic reasons, the second plant can be used as a shadow plant only in special cases, for example, with the speed control of a motor proposed in [12] and [18]. The options (a) and (c) were implemented in [18] for revolution (speed) control of a motor and compared with the control of the same motor with a standard PID controller. Two identical DC Mclennan drives were used as plant and shadow plant. Both drives were identified very precisely at one or the same for both working points:

$$G_S(s) = \frac{1,03}{(1+0,01s)^2(1+0,008s)} e^{-0,012s}$$

$$G_S^*(s) = \frac{0,93}{(1+0,01s)^2(1,008s)} e^{-0,012s}$$

Since the gains of both transfer functions according to the ASA concept must be the same, a gain correction factor k was taken into account by the controller algorithm:

$$k = \frac{1,03}{0,93} = 1,08$$

The results are given in Fig. 11.22. From these step responses is seen that the realization of the "two-engine" ASA control is practically possible.

The further test results of [6, 12]-[18] confirm that the control quality of all three options of shadow plant with the ASA controller shown above are better than the classical model-based compensating controller. However, further investigations are needed to evaluate the effectiveness of such control and to define the possible areas of application.

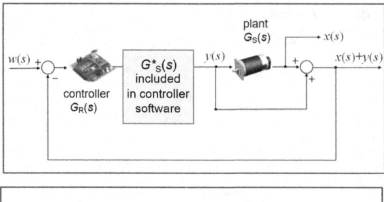

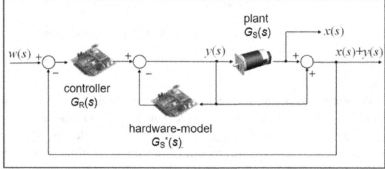

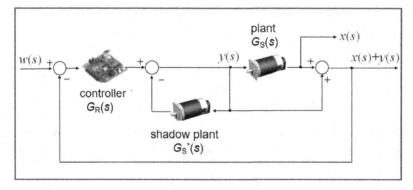

Fig. 11.21 Shadow plant as software model included in the controller algorithm (above), shadow plant as hardware model (middle), and a second plant used as shadow plant (below)

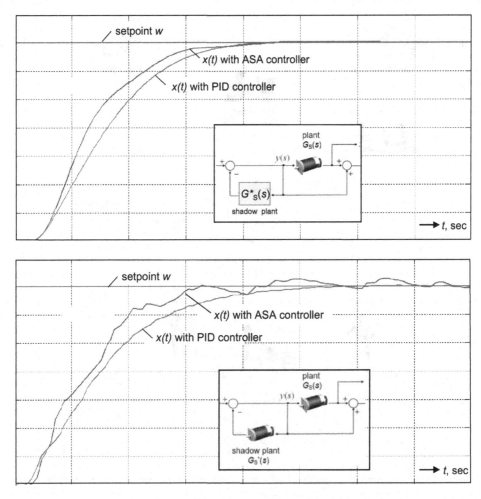

Fig. 11.22 Step responses of the engine speed control with the shadow plant as software model (above) and with the second engine. *Source* [12, p. 21]

11.3   Date Stream Management of ASA Control

11.3.1 Data Stream by ASA Control

The application of the ASA controller opens new signal streams and as a result opens the possibility to use more data stream managers (DSM). The DSM applied in this chapter are divided into two groups asshown in Fig. 11.23:

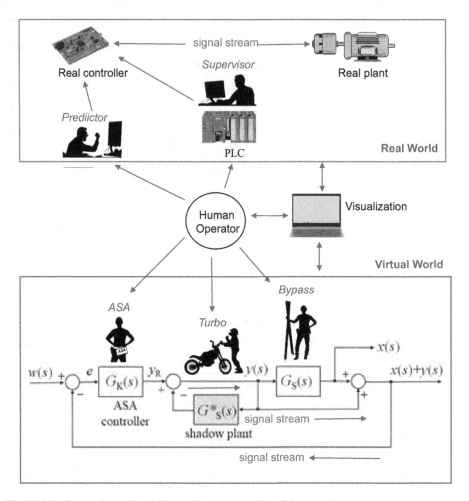

Fig. 11.23 Conception of Data Stream Management for ASA control

- Design manager for the virtual world:
  - Supervisor and
  - Predictor.
- Control manager for the real world
  - Bypass and
  - Turbo.

The brief overview of goals and features of DSM Manger is given below after that in the
next sections follows a detailed description.

Supervisor
In Sect. 11.2.2 was shown that the ASA prefilter is located outside of the control loop and gets no information about the controlled variable. Therefore, it cannot eliminate the disturbances. The aim of DSM Supervisor is to improve the disturbance behavior by combining ASA prefilter with PLC (programmable logical control).

Bypass
The aim of DSM Bypass is the simplification of the ASA controller, which was described in Sect. 11.2.3. As a result of this description, the universal block is programmed in [17] for MATLAB®/Simulink Library "MyLib".

Predictor
The DSM ASA predictor is simplified and improved with means of ASA option of the well-known Smith predictor, which is used by control of PTn plants with dead time. The Smith predictor [11] is applied with the classic compensating controller, the ASA predictor based upon DSN Bypass.

Turbo
The aim of DSM Turbo is to accelerate a standard controller like P, I, PI, PD, or PID, using a closed loop with bypass and settling the additional feedback to the input of the controller. The settling time by the use of DSM Turbo is decreased comparing to the classic control of about 50%.

11.3.2 DSM Supervisor

The DSM supervisor is illustrated in Fig. 11.24. It consists of the standard PID controller, which is switched on in the loop if the controlled $x(t)$ variable changes, but the set point w has no change or the abbreviation of it is zero, i.e., $w = 0$. In other words, the PID controller takes over the control by disturbance behavior. The setpoint behavior is controlled with an ASA prefilter, whose control quality is better than that of the PID controller. Furthermore, the DSM Supervisor improved the safety of control using the redundancy of the PID controller and ASA prefilter.

The test results of Fig. 11.25 completely confirm the advantages of DSM Supervisor.

11.3.3 DSM Bypass

The aim of the DSM Bypass is to improve the ASA controller of Sect. 11.2.3 although ASA controller has no significant disadvantages to be improved. As mentioned above,

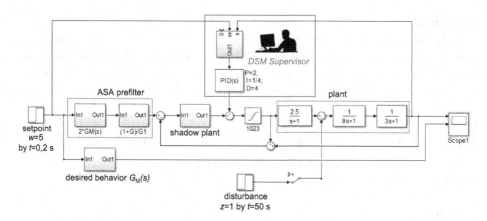

Fig. 11.24 DSM Supervisor with ASA prefilter by disturbance behavior

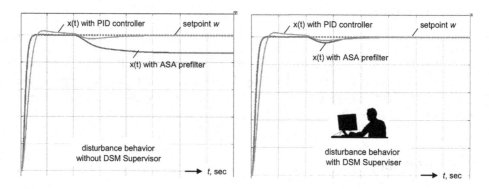

Fig. 11.25 Improved step response of disturbance behavior by using DSM Supervisor (on the right)

the ASA controller has advantages over the ASA prefilter. However, the complicated structure of the ASA controller, given in (Fig. 11.17 and 11.19), and the complexity of its transfer functions in Eq. 11.13 are disadvantageous.

To simplify the ASA controller, the shadow plant is settled in one unit with the controller as shown in Fig. 11.26a. The shadow plant is not more in the loop, the parallel connection of the plant $1 + G_S(s)$ is not more compensated, and the plant is not more "invalid". Therefore, the signs of signal streams are changed as given in Fig. 11.26b. Also is changed the ASA controller transfer function in Eq. 11.13, which is not more applicable.

The transfer function of the DSM Bypass ASA is determined as follows:

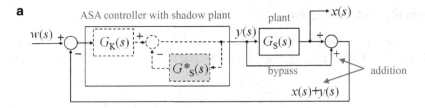

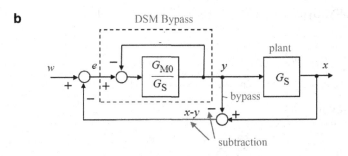

Fig. 11.26 Two steps to convert an ASA controller to DSM Bypass: **a** ASA controller and shadow plant are settled in one unit; **b** the bypass is creating changing signal addition to subtraction

$$G_R(s) = \frac{G_{R0}(s)}{1 + G_{R0}(s)} \qquad (11.14)$$

The transfer function $G_{R0}(s)$ of Eq. 11.14 is denoted in [10] and [11] as an "opened loop with controller $G_R(s)$":

$$G_{R0}(s) = \frac{G_{M0}(s)}{G_S(s)} \qquad (11.15)$$

The transfer function $G_{M0}(s)$ of Eq. 11.15 is called in [10] and [11] "opened loop of desired behavior $G_M(s)$":

$$G_{M0}(s) = \frac{G_M(s)}{1 - G_M(s)} \qquad (11.16)$$

The transfer function $G_R(s)$ of the DSM Bypass ASA is defined upon given transfer functions of the plant $G_S(s)$ and of the desired behavior $G_M(s)$ by calculating Eqs. 11.16, 11.15, and 11.14 one after another.

Given are the IT1-plant and the desired PT2 behavior $G_M(s)$ with the same time delays $T_M = 0,05$ s, i.e., the desired settling time is expected about $T_{aus} = 5T_M = 0,25$ s:

$$G_S(s) = \frac{0,0328}{s(0,4s+1)} \qquad G_M(s) = \frac{1}{(0,05s+1)^2} \tag{11.17}$$

The tuning of DSM Bypass upon MATLAB® script is given below

```
s=tf('s');                          % Laplace-Operator
Gs=0.0328/(s*(1+0.4*s));            % plant
GM=1/(1+0.05*s);                    % desired behavior
GM0=GM/(1-GM);                      % opened desired behavior Gl. 11.16
GR0=GM0/Gs;                         % opened loop with controller Gl. 11.15
GR=GR0/(1+GR0)                      % DSM Bypass Gl. 11.14
% Step response of closed loop with DSM Bypass
Gv=GR*Gs;                           % forward signal stream by setpoint behavior
G0=GR*(Gs-1);                       % open loop
Gx=Gv/(1+G0);                       % transfer function of closed loop
step(Gx, GM)                        % step response of closed loop and of desired behavior
grid
```

The simulated step response follows exactly the given desired behavior. ◄

In Fig. 11.27 is shown the MATLAB®/Simulink block *ASAController_eigen_lib.sltx*, which was developed in [17] as the element of users' Simulink Library "MyLib".

After MATLAB "Create" command, the GR block is transferred from the Simulink library to a Simulink file as "DSM Bypass" and can then be used in every closed control loop with bypass. In Fig. 11.27 is explained an example of the plant

$$G_S(s) = \frac{0,7}{(1+0,5s)(1+0,5s)} = \frac{0,7}{0,25s^2 + s + 1}$$

how DSM Bypass is configured for the example of the desired behavior of closed loop:

$$G_M(s) = \frac{1}{(1+0,3s)(1+0,1s)} = \frac{1}{0,03s^2 + 0,4s + 1}$$

The simulated step response, shown in Fig. 11.27, has no difference from the desired behavior.

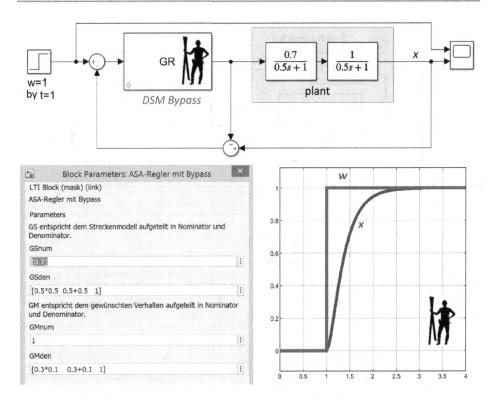

Fig. 11.27 DSM Bypass as LTI-Block of the user Simulink Library "MyLib", developed in [17]. *Source* [10, p. 227]

11.3.4 ASA Predictor

The ASA Predictor is a data stream manager which, like a well-known Smith Predictor, is used as a compensating controller for plants with dead time. Below is given the definition for Smith predictor:

"If the controlled system has a dead time T_t, the algorithm of the compensating controller should be modified. The algorithm was named Smith predictor after the name of the developer (Berkeley-University, 1957) and consists in considering a part of the path $G_S(s)$ without dead time." (Quote: Source: [11], page 389)

The functional block diagram of the Smith predictor is given in Fig. 11.28 upon [11]. The plant consists of two parts, without dead time $G_S(s)$ and dead time T_t.

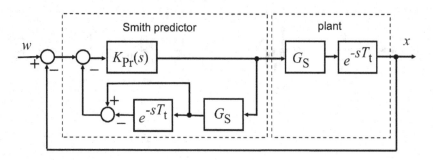

Fig. 11.28 Smith predictor. *Source* [11, p. 390]

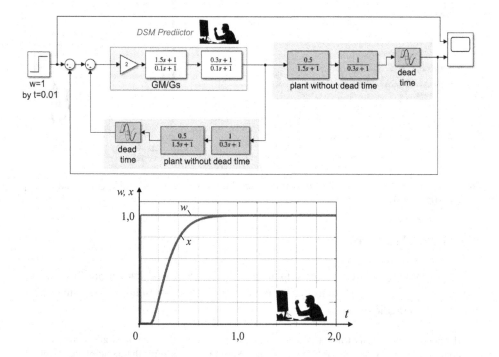

Fig. 11.29 ASA Predictor. *Source* [10, p. 229]

The transfer function of the classic compensating controller $K_{Pr}(s)$ for the part of plant $G_S(s)$ without dead time is defined in Eq. 11.1 depending on desired behavior $G_M(s)$:

$$K_{Pr}(s) = \frac{1}{G_S(s)} \cdot \frac{G_M(s)}{1 - G_M(s)} \qquad (11.18)$$

The transfer function of the whole Smith predictor includes Eq. 11.18:

$$G_R(s) = \frac{K_{Pr}(s)}{1 + K_{Pr}(s)\, G_S(s)\,(1 - e^{-sT_t})} \qquad (11.19)$$

Smith Predictor was analyzed in [10] and converted into a simpler structure using ASA, which is referred to as ASA predictor. The transfer function of the data stream manager ASA predictor is described with shadow plant as illustrated in Fig. 11.29.

Against Eq. 11.19, the transfer function $G_{\text{Pred}}(s)$ of the ASA predictor is very simple:

$$G_{\text{Predictor}}(s) = \frac{G_M(s)}{G_S(s)} \tag{11.20}$$

Example: ASA predictor (Source: [10, pp 227, 228])

The given plant

$$G(s) = \frac{0,5}{(1 + 1,5s)(1 + 0,3s)} \cdot e^{-sT_t}$$

consists of two parts:
- $G_S(s)$ without dead time

$$G_S(s) = \frac{0,5}{(1 + 1,5s)(1 + 0,3s)}$$

- dead time $T_t = 0,1$ s

According to Eq. 11.20 follows the transfer function $G_{\text{Pred}}(s)$ of the ASA predictor:

$$G_{\text{Predictor}}(s) = 2 \cdot \frac{1 + 1,5s}{1 + 0,1s} \cdot \frac{1 + 0,3s}{1 + 0,1s}$$

The step response of the control loop with the ASA predictor is shown in Fig. 11.29. It follows the desired step response of $G_M(s)$ as it is by Smith predictor, but the ASA Predictor has an advantage, namely the transfer function $G_{\text{rPredictor}}(s)$ is simpler and can be easily implemented in practice; it does not contain any D-terms, as is it by Smith predictor. ◀

11.3.5 DSM Turbo

The DSM Turbo has the same goal as the DSM Bypass and is realized in the closed loop with bypass like Fig. 11.26b. But there are two differences illustrated in Fig. 11.30a:

- instead of ASA controller is used the standard PID controller;
- the bypass of Fig. 11.26b is with the feedback "Turbo" completed.

The turbo module $G_{\text{Turbo}}(s)$ of Fig. 11.30b is nothing more than the reciprocal transfer function $G_{\text{Srec}}(s)$ of the closed loop consisting of the plant without any controller according to Eq. 11.11:

$$G_{\text{Turbo}}(s) = G_{\text{wSrec}}(s) = \frac{G_S(s) + 1}{G_S(s)} \qquad (11.21)$$

Since the reciprocal of the given plant

$$G_S(s) = G_{S1}(s)G_{S2}(s)G_{S3}(s) = \frac{2,5}{s+1} \cdot \frac{1}{3s+1} \cdot \frac{1}{8s+1} \qquad (11.22)$$

practically cannot be realized because the order of the numerator polynomial is greater than the order of the denominator polynomial, additional time delays as small as possible are introduced for the shadow plant:

$$G_S(s) = \underbrace{\frac{0,01s + 2,5}{s+1}}_{G_{S1}(s)} \cdot \underbrace{\frac{0,01s + 1}{3s+1}}_{G_{S2}(s)} \cdot \underbrace{\frac{0,01s + 1}{8s+1}}_{G_{S3}(s)}$$

The actuating value $y(s)$ of Fig. 11.30a consists of two terms:

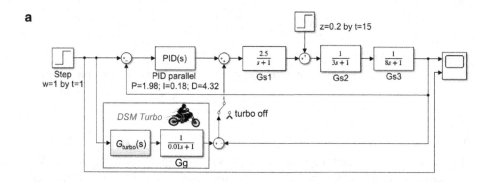

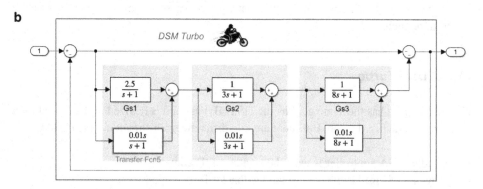

Fig. 11.30 a) Closed loop with PID controller and DSM Turbo; b) DSM Turbo. *Source* [10, p. 230]

- y_R from PID controller

$$y_R(s) = G_R(s)\hat{e} = G_R(s)(\hat{w} - x(s)),$$

- y_T from Turbo module

$$y_T(s) = G_T(s)G_g(s)\hat{w} - x(s)$$

The entire actuating value $y(s)$ and the controlled variable $x(s)$ are defined below

$$y(s) = y_R(s) + y_T(s) = G_R\hat{w} - G_R x(s) + \frac{1 + G_S}{G_S}G_g\hat{w} - x(s)$$

$$x(s) = G_S y(s) = G_R G_S \hat{w} - G_R G_S x(s) + (1 + G_S)G_g\hat{w} - G_S x(s)$$

It results in the transfer function of the whole closed loop:

$$G_w(s) = \frac{x(s)}{\hat{w}} = \frac{G_g + G_g G_S + G_R G_S}{1 + G_S + G_R G_S}$$

The $G_g(s)$ is defined as going out from the desired behavior $G_M(s)$ of the closed loop:

$$G_M(s) = \frac{G_g + G_g G_S + G_R G_S}{1 + G_S + G_R G_S} \rightarrow G_g(s) = G_M + G_R \cdot \frac{G_S}{1 + G_S}(G_M - 1)$$

Supposing that a very fast behavior is desired

$$G_M(s) = \frac{1}{0,01s + 1} \approx 1$$

follows the simplified transfer function $G_g(s)$

$$G_g(s) \approx G_M(s)$$

The standard PID controller of Fig. 11.30 is set optimally without considering the turbo circuit, e.g., according to the optimum magnitude, discussed in . The step responses by the setpoint and disturbance behavior shown in Fig. 11.31 have the optimal damping $\vartheta = 0.707$ as it is expected by optimum magnitude. But control with DSM Turbo is twice as fast as without turbo.

Example: DSM turbo

Given is the plant $G_S(s)$ that should be controlled with a PI controller $G_R(s)$ upon desired PT2 behavior $G_M(s)$, i.e., without oscillations:

$$G_S(s) = \frac{0,5}{(12s + 1)(10s + 1)} \qquad G_M(s) = \frac{1}{(5s + 1)^2}$$

Fig. 11.31 Step responses to
Fig. 11.30 of control with and
without DSM Turbo. *Source*
[10, p. 231]

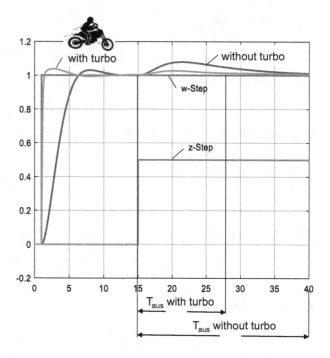

Since the desired damping is $\vartheta = 1$, the PI controller is tuned upon the best-known rule, namely by the optimum magnitude, Type A, as shown in , without considering the desired settling time:

$$G_0(s) = G_R(s)G_S(s) = \frac{K_{PR}(1 + sT_n)K_{PS}}{sT_n(1 + sT_1)(1 + sT_2)}$$

$$T_n = T_1 = 12 \text{ sec} \quad K_{PR} = \frac{T_n}{4\vartheta^2 K_{PS}T_2} = 0,6$$

The expected settling time T_{aus} of the closed loop with this tuning is

$$T_{aus} = 11 \cdot T_1 = 11 \cdot 10 = 110 \text{ sec}$$

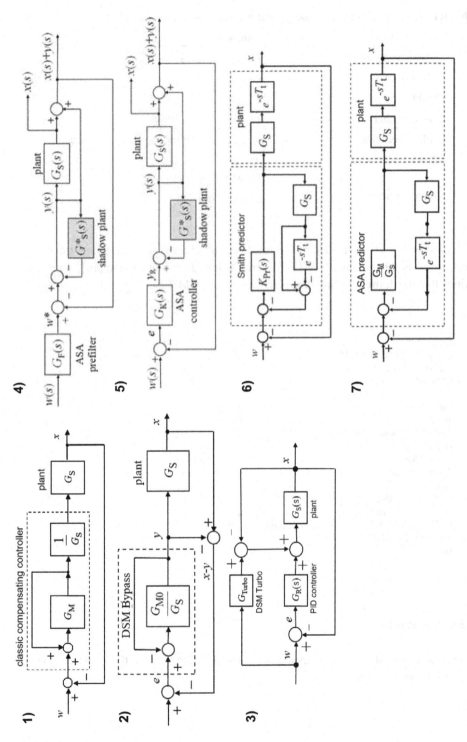

Fig. 11.32 Overview of model-based compensating controllers: 1) Classic controller; 2) DSM Bypass; 3) DSM Turbo; 4) ASA Prefilter; 5) ASA Controller; 6) Smith Predictor; 7) ASA Predictor

Table 11.1 #L Overview of tuning rules, model-based compensating controllers

No.	Controller type	Tuning rule
1	Classic compensating controller	$G_R(s) = \frac{1}{G_S(s)} \cdot \frac{G_M(s)}{1-G_M(s)}$
2	DSM Bypass	$G_{R0}(s) = \frac{G_{M0}(s)}{G_S(s)} G_{M0}(s) = \frac{G_M(s)}{1-G_M(s)} G_R(s) = \frac{G_{R0}(s)}{1+G_{R0}(s)}$
3	DSM Turbo	$G_{Turbo}(s) = \frac{G_S(s)+1}{G_S(s)}$
4	ASA Prefilter	$G_F(s) = 2G_M(s)\frac{1+G_S(s)}{G_S(s)}$
5	ASA Controller	$G_{wSrec}(s) = \frac{1+G_S(s)}{G_S(s)} G_K(s) = \frac{G_M(s)G_{wSrec}(s)}{1-G_M(s)G_{wSrec}(s)}$
6	Smith Predictor	$K_{Pr}(s) = \frac{1}{G_S(s)} \cdot \frac{G_M(s)}{1-G_M(s)}$ $G_R(s) = \frac{K_{Pr}(s)}{1+K_{Pr}(s)\,G_S(s)\,(1-e^{-sT_t})}$
7	ASA Predictor	$G_{Predictor}(s) = \frac{G_M(s)}{G_S(s)}$

But desired is T_{aus} of 30 or 40 s, as follows from the given $G_M(s)$. Therefore, the DSM Turbo is used as shown in the MATLAB® script below. The simulation in [10], page 233, confirms that the desired settling time is reached with DSM Tuner: $T_{aus} = 30$ s.

```
%% ---------- Plant and desired behavior ----------------------------------
s=tf('s');                             % Laplace-Operators
Kps=0.5; T1=12;T2=10;TM=5;
Gs=Kps/((1+T1*s)*(1+T2*s));            % plant
GM=1/(1+s*TM)^2;                       % desired behavior
%% ---------- PI controller -----------------------------------------------
theta=1;                               % damping
Tn=T1;                                 % compensation
KpR=Tn/(4*theta^2*Kps*T2);             % tuning upon optimum magnitude, type A
GR=KpR*(1+s*Tn)/(s*Tn);                % PI controller
G0=GR*Gs;                              % open loop
Gw=G0/(1+G0);                          % closed loop without DSM Turbo
%% -------DSM Turbo --------------------------------------------------------
GwS=(1+Gs)/Gs;                         % reciprocal loop
Gg=GM+GR*GwS*(GM-1);                   % Gg
GwT=(Gg+Gg*Gs+GR*Gs)/(1+Gs+GR*Gs);     % closed loop with DSM Turbo
step(Gw, GwT, GM)                      % step responses
grid
```

◄

11.3.6 Summary

The overview of model-based compensating controllers described in this chapter is given in Fig. 11.32 and in Table 11.1.

References

1. Zacher, S. (2000). *SPS-Programmierung mit Funktionsbausteinen*. VDE Verlag.
2. Zacher, S. (2003). *Duale Regelungstechnik*. VDE-Verlag.
3. Zacher, S. (2008). *Mobile Mathematik*. Verlag Dr. S. Zacher.
4. Zacher, S. (2008). *Existentielle Mathematik*. Verlag Dr. S. Zacher.
5. Zacher, S. (2008). *Verbotene Mathematik*. Verlag Dr. S. Zacher.
6. Zacher, S., & Saeed, W. (2010). Design of Multivariable Control Systems using Antisystem-Approach, pp. 201–208. *7th AALE Angewandte Automatisierung in der Lehre und Forschung*, FH Technikum Wien, 10./11 Feb. 2010.
7. Zacher, S. (2014). *Bus-approach for feedback MIMO-control*. Verlag Dr. S. Zacher.
8. Zacher, S. (2020). *Drei Bode-Plots Verfahren für Regelungstechnik*. Verlag Springer Vieweg.
9. Zacher, S. (2021). Antisystem-Approach (ASA) for engineering of wide range of dynamic systems. *Ames, Iowa, USA: International Journal on Engineering, Science and Technology IJONEST, 3*(1), 52–66.
10. Zacher, S. (2021). *Regelungstechnik mit Data Stream Management*. Verlag Springer Vieweg.
11. Zacher, S., & Reuter, M. (2022). *Regelungstechnik für Ingenieure* (16th ed.). Verlag Springer Vieweg.
12. Zacher, S. (2016). *ASA-Implementierung*. Verlag Dr. S. Zacher. https://www.zacher-international.com/Automation_Letters/08_ASA-Implementierung.pdf. Accessed 20 Feb 2022.
13. Zacher, S. (2016). *ASA-Regler für I-Strecke*. Verlag Dr. S. Zacher. https://www.zacher-international.com/Automation_Letters/25_ASA_Regler_OSLO.pdf. Accessed 20 Feb 2022.
14. Zacher, S. (2016). *ASA-Regler mit Bypass*. Verlag Dr. S. Zacher. https://www.zacher-international.com/Automation_Letters/29_ASA_Regler_mit_Bypass.pdf Accessed 20 Feb 2022.
15. Zacher, S. (2017). *ASA-Regler. Test und Nachbesserung*. Verlag Dr. S. Zacher. https://zacher-international.com/Automation_Letters/32_ASA_Regler_Test.pdf. Accessed 20 Feb 2022.
16. Zacher, S. (2017). *ASA-Bilanzregelung*. Verlag Dr. S. Zacher. https://zacher-international.com/Automation_Letters/33_ASA_Bilanzregelung.pdf. Accessed 20 Feb 2022.
17. Mille, R. (2017). *Rapid Control Prototyping eines ASA controllers mit MATLAB PLC Coder*. Verlag Dr. S. Zacher.
18. Groß, D. (2015). Entwurf und Untersuchung einer modellbasierten prädiktiven Regelung mit ASA-Konzept. *Master-Thesis*, Hochschule Darmstadt, FB EIT, Fernstudium. https://www.zacher-international.com/C22_Team_Projekt/Zwei_Motoren_gross.pdf. Accessed 20 Feb 2022.
19. Korn, U. (2000). Mehrgrößenregelung. *«Automatisierungstechnik kompakt»*, Hrsg. S. Zacher. Vieweg Verlag.
20. Lunze, J. (2005). *Regelungstechnik 2. Mehrgrößensysteme, Digitale Regelung*. Springer-Verlag.
21. Schulz, G. (2002). *Regelungstechnik: Mehrgrößenregelung-Digitale Regelungstechnik-Fuzzy-Regelung* (Vol. 2). Oldenbourg-Verlag.

Printed in the United States
by Baker & Taylor Publisher Services